DESTRUCTIVE CLIMATE CHANGE, ABERRANT THINKING AND IMPENDING ECONOMIC DISRUPTIONS

DESTRUCTIVE CLIMATE CHANGE, ABERRANT THINKING AND IMPENDING ECONOMIC DISRUPTIONS

by

Andre E. Laferriere

ISBN 1-58721-631-0

1st books – rev. 5/12/00

About the Book

An eye-opening, straight from the horse's mouth 400 pages, 160,000 words the way it should be told, no holds barred, critical of credentialled but biased experts pumping out a variety of truths mixed with misinformation, the subject of Global Warming and it's effects to worsen is presented in plain English. No titillating apocalyptic theme, however. Fundamentally provides rules-based explanations of increasing calamitous weather compiled from actual NOAA recorded data, thirty five easy to evaluate illustrations proving what earth warming looks like, covering episodic atmospheric derangements from Russia to Vietnam, from the western Pacific, across America to Africa. Challenges anti-environmentalists spurious concepts, determined to smoke-screen the truth. Establishes the reactive nature of our confined atmosphere to pollution, its increasing periods of death-dealing floods and associating extremes in torrential downpours with a God ordained programmed physics-based cleansing of its air. Describes the pressure balance between poles and equator, consequently the role hurricanes play in speeding up restoration of that balance. Includes out experiencing twice as many tornadoes this past decade compared to 30 years ago, deadly examples graphically illustrated in weather map form, distinguished the usual twister from high speed monsters, the latter's cause clearly described. In yearly frequency, 1424 tornadoes in 1998, a record smasher that too many of our citizens seem oblivious of. Because of political ramifications, party philosophies and my own and why, takes the reader on a ride from the Great Depression through interesting accounts of naval aviation, GI's murdered in Casablanca, Black markets, interesting personalities such as Bear Bryant, wonderful pilots, death trap fighter planes poorly engineered, the making of a weatherman to current weather phenomena and effects on economics, unwary private plane pilots and so forth. In short, such a heavy dose of academia needed variety. Includes peculiarities of Morocco and experiences, some sinister.

Destructive Climate Change, Aberrant Thinking and
IMPENDING ECONOMIC DISRUPTIONS

Insidiously, at least not disturbing vulnerable types, preponderous in numbers, the damaging effects of a Nature, reactive as it is and usually cyclical between long periods of what might be termed respites characterized by reasonable seasonal weather, have taken on an unprecedented alarming trend – there is now the obvious change in pattern; we now are being treated to repeated bouts of harmful, often very dangerous weather at an increasing rate of frequency just as we are also, gas pedal to the floor, inflicting harm along with other industrial nations on a power we should do well to respect. Wrathful it is not but pummel the human species it will.

In my anxiety to be heard, like all human beings wanting to express myself by sharing what I have learned I offer facts, recorded facts, pushed aside by the clever whose motives are clear to me, albeit nothing that requires the depth of Einstein. I tried reaching out to the powers that be, despite the issue of environmental problems to include ecological damage, never quite thought of as warranting crisis prioritizing, there prevailing but contemplations of something distant and remote, rather than now, expectations of rising sea levels 50 years in the future bothering but a few. Although I read John W. Kingdon's book, "Agendas, Alternatives and Public Policies" in which he decries in respectful terms, i.e., Congressman do not consider a crisis unless 50 thousand die and furthermore have to be hit on the side of the head before they react, I did give it a try – replies contained herein. However I do respect many in Washington, DC and if anyone needs to learn it is the public in general, after which democracy becomes more efficient at election time.

PREFACE

"Everybody talks about the weather but nobody does anything about it", Mark Twain. Yes we can!

"You may delay but time will not", Ben Franklin.

One need not be a devout patriot always on the lookout for travesties against our nation not to recognize the environmental, political and organized interference melee, the White House capitulating to Big Business, as it just did (explained later) which has taken up a lot of newspaper space this past decade, ample excerpts from this source distributed in the ensuing material, thus it cannot be said my writing is contentious. Week by week, article after article, and one TV show after another convinced me there cannot be a consensus in view of a public subjected to a variety of sincere renditions mixed with false premises. Therefore, the impression this "fricassee" made on me warranted I give it the old "college try" and dissect what threatens to take us for a long walk on a short pier in terms of a reactive Nature – it's mysteries exaggerated by "Puzzle Peddlers". I have endeavored, successfully I believe, to break them down in easy to understand components. Supplementing this topic, adequately illustrated, I also would like the young (under the age of 65 – which is young for me) to understand "Pop" and "Grandpa" and perhaps feel a thoughtfulness, a needed perspective as to what molded our character, political inspirations and "bummers", surviving in grim times, patriotism and also the humorous side of various episodes. I especially want to highlight the wonderment and grandeur of naval aviation, the nobility of such selected people and their faces still clearly remembered. I shan't forget either, "truck loads of soldiers' bodies stacked like cordwood", to be air freighted back home. And Morocco, the strange land that it was.

Destructive Climate Change, Aberrant Thinking and Impending Economic Disruptions

Feeling more like Don Quixote, a study in futility, at least his act has been heralded; I pursue the truth and have tried to pass it on. Frivolous of me to insert xeroxed copies of letters from several dignitaries in the White House, who gratifying to me in my anxiety to at least be taken, I think, a little more seriously. Are they but perfunctory "Atta Boys"? I like to feel the Vice President and two senators generally do not answer letters from the demented and just maybe, I had something interesting to pass on – whatever. Tired from all the conflicting views on Earth Warming, I sent them some pertinent weather data and my views, gems of reality if I may say so and based on some fifty five years of experience, ten of which I had the opportunity in naval aviation to be a weatherman, seven of those years forecasting. It was particularly gracious of Mr. Al Gore to respond as he did and while two senators also wrote me, allow me to leave their names out simply because those missives I sent them were not open letters. When one senator added a hint of recognition of whatever talent I earned and to this day still learning what I can, by adding the word insight, I must confess I feel a little better.

Washington D.C. does have a problem relative to pursuing cleaner air while avoiding upsetting both the economy and our citizenry. It is quite easy to blame the government, if as any environmentalist may be inclined to do, for this or that form of pollution and expressing frustration over what they deem as insufficient action. I also am well aware from reading the news and listening to radio talk shows that just perhaps the public is very much to blame also. And yet, let me mollify this a little, it's quite understandable if clever opponents of what really has to be done, are recognized for their inaccurate and specious material and expansive denunciations by the likes of Russ Limbaugh aimed at Americans who are for now receptive to anti-government verbiage. While Mr. Limbaugh, who has quite a following is extremely busy with his microphone although I sense he is foaming at the mouth in his unsuppressed desire, frenetically expressed at times, to not only see President Clinton impeached but to suffer eternal damnation, he has in the past made it clear environmentalists are "Psycho-babblers" and "Earth Warming is a myth!" Quote, unquote. Of course, progress in mandating rules to reduce pollution does not mean drastic measures that could send the economy into a tailspin, that we all realize but there are moderate means to be pursued if only the public could be conditioned to accept what I know to be inevitable. Basically I want to offer as much material as I can and let you be the judge rather than my deluding myself into believing whatever this book has to say will be accepted at face value.

Nurtured as a very young man in military directness I'm not into finesse and pedantic trappings when I write. I am a technician and not a poet. Some writers, whose anti-environmentalists themes are suspect having collected as I have their articles for some six years, and a few trim-bearded professorial-looking hosts on violent weather TV shows are impressive. It has become very obvious to me they are rather misinformed, inexperienced or darn liars when it comes to meteorological functions. I chafe at blunderous statements. One TV narrator, although not representing any scientific establishment said, "America averages six hundred tornadoes a year." In the ensuing pages pleas note the real count of tornado frequencies since 1961 published in graph form by NOAA (National Oceanic and Atmospheric Administration). Actually we have, these past five years, experienced close to an average exceeding 1100 twisters per year. I have, in my letters to both the vice president and one prominent senator, insisted without being too assertive that tornado frequencies are the best and most precise index of where our

climate is headed. That wavering upward climb corresponds also with other destructive weather which I attribute, and again, I'll let you be judge and jury, to the gradual **Africanization*** of our climate while Canada will enjoy weather more suitable to increased agricultural output. One propagandist, as usual, doctorated and I'm not and I really don't care, stretching the influence of El Nino to the nauseating point, declaring it, last May of this year, 1998, as a boon to our wallets, and of course the theme, embellished with derision tossed at Al Gore, designed to smokescreen Earth Warming out of your mind, as unproven, has not since May and it is now early December, showed any willingness to explain the fact that warm winters really portend deranged seasonal weather the rest of the year. The blessing he expounded upon with such delicious literacy ado turned out to be a major deadly and very expensive spring, summer and Fall of the 1998 year. The Red Cross supposedly (per the radio) saying they had spent more money this year than ever in their past years as a total. As far as financial benefits from El Nino, there were, of course, but the cost of catastrophic weather far exceeded any monetary gain – soon, if it continues this climate change, and I'm sure it will, given there are yearly respites as I'm of the opinion, not wishing to come across as having some heavenly purview, that Nature, which I sometimes think has some Godly reason, per Einstein's definition of God *(God is a power that controls all)*. Nature does not want to fry or drown us into extinction, but having digressed too much, soon, changing weather patterns will become a great financial burden for this country. And so, hopefully you will not see me afflicted with some form of the Teacher Syndrome.
**I've coined a new word; remember it.*

All of us, especially literary pundits should pry themselves away from their Dot Com World, take a ride somewhere, look over the sky, keep count of disasters or perhaps talk to an old man in Texas who said in sixty one years he had never seen so much flooding. Let us start with one very basic lesson: Cold* winters are much more desired than balmy ones, in general, simply because the rest of the seasons will perform their duties within a normal time limit and not, as it being my pet theme, warm winters cause an earlier start to the tornado season and so on. The remaining upset seasons and what they offer as wacky weather as some magazine titled a depiction of American weather these past few years, the details of which I am both happy and saddened to elaborate on in this book. However, I am not going to bore you with the age-old humdrum of cold and warm fronts, blah, blah. Instead, I will illustrate what is happening with accompanying data so that you can be in the position of analyzing reality without the help of disingenuous experts.

Now, let me continue some more with my obvious contempt for people who display a lack of common sense or in all probability have never acquired skill at perceiving what is obvious. One expert claimed that El Nino had lasted for one hundred years off the West Coast of South America and the resulting drought had brought about the extinction of the Mayan Indian population. I can't quite conceive such a large populace lying around and plucking sustenance from dried bushes of whatever when history has shown us that people have enough sense to pack up and head for greener pastures. I'm sure the Mayans knew where north was but I also know as you also should know what Indian features look like and for heaven's sake where do all these folks with visages you are familiar with come from? Millions of Indians in those countries starting with Columbia, across all of Central America, Mexico and thousands of them in America? Right here in little old Rhode Island we have quite a few- I know South American type Indians when I see them! Enough with sarcasm or I'll be emulating several writers subscribing in their really interesting articles to tossing derision at Al Gore. If I continuously use the vice president's name it does not mean that I'm politically oriented, rather I view him as symbolizing the environmentalist. I really believe this man knows a lot about the environment and the growing problem in a technical sense but finds himself bucking recalcitrant members of Congress.
**Includes the Southland and the Arctic.*

        Thank you for your thoughtful letter. I appreciate your kind words and you
forwarding me materials detailing your recent work.

        I am pleased to hear that the information you received from the National Climatic
Data Center was interesting. In addition, as you requested, I have forwarded your letter
about meteorological research to Vice President Gore.

        Again, thank you for your continued correspondence. Please do not hesitate to call
or write me if you have further questions or if I may be of additional assistance on this or
any other matter.

                                Sincerely,

                                United States Senator

Dear Mr. Laferriere:

        Recently, Senator   sent me a copy of your letter regarding
natural disasters. I welcome this opportunity to respond to you.

        I am overwhelmed by the many letters and cards I have received from
people, like you, who are concerned about the crucial challenges we are
facing as a nation. I am grateful for your genuine concern for the
effectiveness and success of this Administration. Please be assured that
the President and I will be working hard to accomplish the goals of
particular importance to the American people.

        Again, thank you for letting me hear from you.

                                Sincerely,

                                Al Gore

Politicians are very special people. I draw no derogatory analogy if I compare them to the Alpha lead dog and we are the yipping pack behind them, such is their ambition. Most are lawyers and they know where the power is. Their egos, if you have had experience with them on a personal level, next to doctors, shines through with just a look as you speak. Without uttering a word they are saying, "Hurry up, little one, my time is valuable". Off the subject, I've always wanted to knock doctors. I knew a New York City head waiter whose income on a daily basis was fattened by generous tipping. "Worse tippers in the world, these doctors" that's what he told me. I feel better now. But politicians need us and gracious they are, always, their letters ends with "Feel free to contact me". Two things I remind myself of: They will not anoint me and while I definitely know more about climate changes than they do, they will never admit it. You should note in Al Gore's letter I must have been brave enough to hint at the government's feeble scare tactic, e.g. rising ocean levels 30 years from now and that as best as I can tell no one is abandoning their abodes near the sea. While you may not agree I will, in later pages, tell you why this concept is a moot question. After some 55 years I know good from the bad in the world of scientists and to an extent that's what this book is about.

The next five pages are xeroxed copies of information sent to the two senators. Whether it is important or not to these gentlemen is irrelevant for now. Frankly, I have concluded from reading newspapers and listening to invited politicians on radio talk shows the subject of tornadoes arouses no one.

𝔘nited 𝔖tates 𝔖enate

WASHINGTON, DC 20510–2101

October 13, 1998

Mr. Andre Laferriere
30 Washington St. Apt. 514
Central Falls, RI 02863

Dear Mr. Laferriere:

Thank you very much for your letter regarding the dangers of air pollution and their effect on the weather. I appreciate your concerns, as it is through correspondence such as yours that I am able to keep abreast of the worries and insights of my constituency. Only by staying informed can I continue to be an effective and responsible Senator.

During my time in the Senate, I have worked hard to enact legislation to combat pollution, preserve our environment and natural resources, and provide protection to endangered species. I feel strongly that we must work harder to decrease the amount of pollutants going into our air and water, and must also do more to eliminate other factors contributing to serious environmental problems in our society.

Thank you again for your concern about the environment, and rest assured, I will continue to work to ensure that our environmental laws are adhered to. Please feel free to contact my office in the future if you have any other questions or concerns. It is nice to hear that people like yourself care about our world.

With best wishes and warm regards,

Sincerely,

October 26, 1998.

During the 70s and 80s as a result of the oil embargo the impetus to conserve energy prompted by rising cost was set in motion through governmental urging. Everyone was asked to lower thermostats, for example, to 68 degrees. Small economy car sales boomed, capturing near 30% of the market as compared to today's 10%. A campaign to insulate homes caught on nationwide thanks to economic incentives and liberal loans.

Interestingly, tornado frequencies declined from 1985 to 1989 inclusive. Diagram A is Xeroxed from NOAA's brochure on tornadoes. Diagram B is Xeroxed from the very comprehensive "The Wiley Encyclopedia of Energy and the Environment" containing the works of some 50+ researchers in various universities.

In short, less energy was used. Many weather observation sites, their recorded data in the archives of

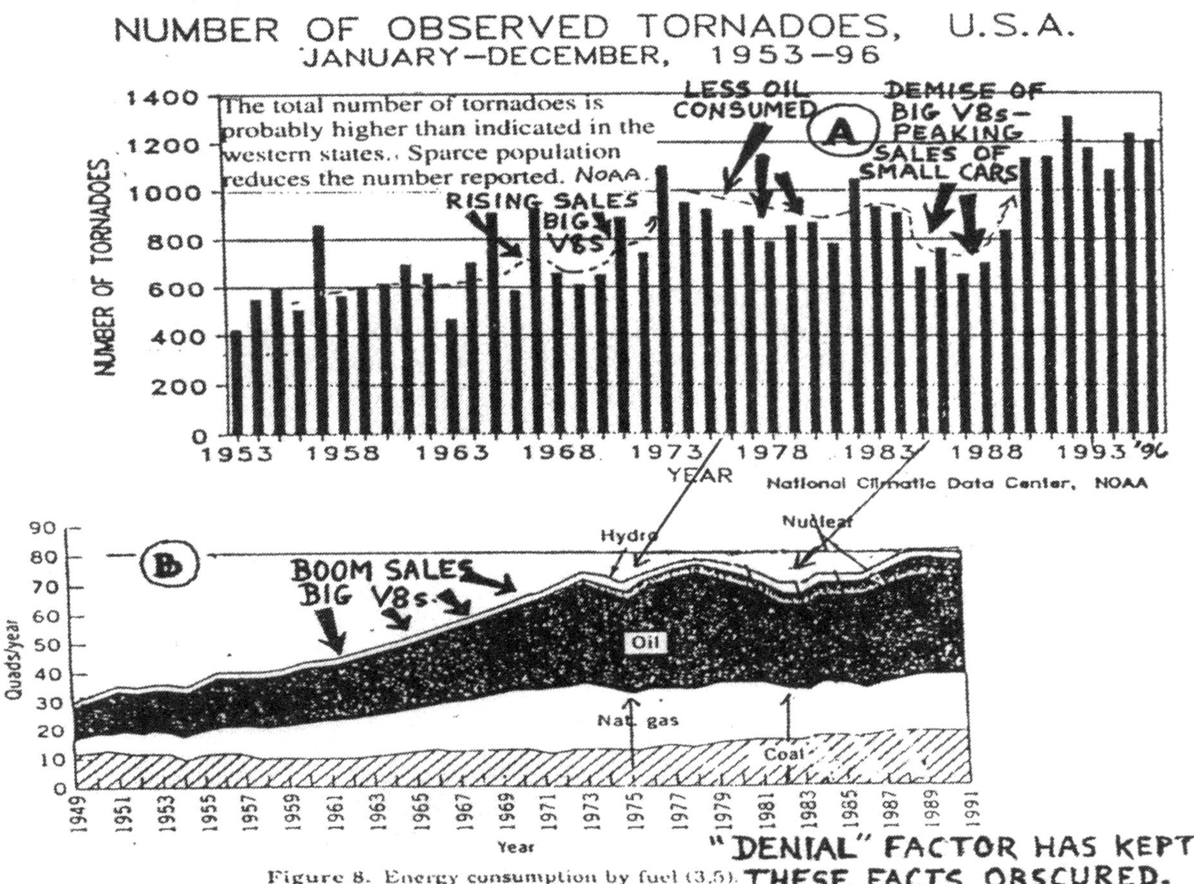

Figure 8. Energy consumption by fuel (3.5).

NOOA show yearly temperature averages with a decided drop in the mid 80s. Temperature trend graphs on following page of a 29-year period of yearly averages for cities named from 1966 to 1995 were transposed into a more recognizable illustration from a multitude of NOAA data. There are over 2000 observers in the US, 276 of them in New England alone.

The relationship between higher tornado frequencies and a rising temperature trend over the southeast quadrant of the US indicates one of the <u>more</u> <u>serious</u> aspects of earth-warming. Is the public, not at all enthusiastic over government intrusion, as some spokesmen refer to it, actually aware of the unusual nature of weather trends? For example, do they know it is very abnormal for hurricanes to be occurring in late October and nine tornadoes in Oklahoma, just two weeks ago? Given that the fuel supply for tornadoes, being semi-tropical moist air, has it's source in the south Atlantic, moving from east to west as it does, over Florida and the Gulf of Mexico, then turning northward where the land mass from Texas to Florida sharply elevates temperatures over the south and southeast (from 70+ d. to 90+ d.) <u>no better indicators exist</u> to show what is transpiring than yearly temperature averages for the past three decades in Miami, Tampa, Daytona Beach, Atlanta, Birmingham and New Orleans.

THERMOMETERS DO NOT LIE !!
ACTIVE EL NINO YEARS
LA NINA YEARS

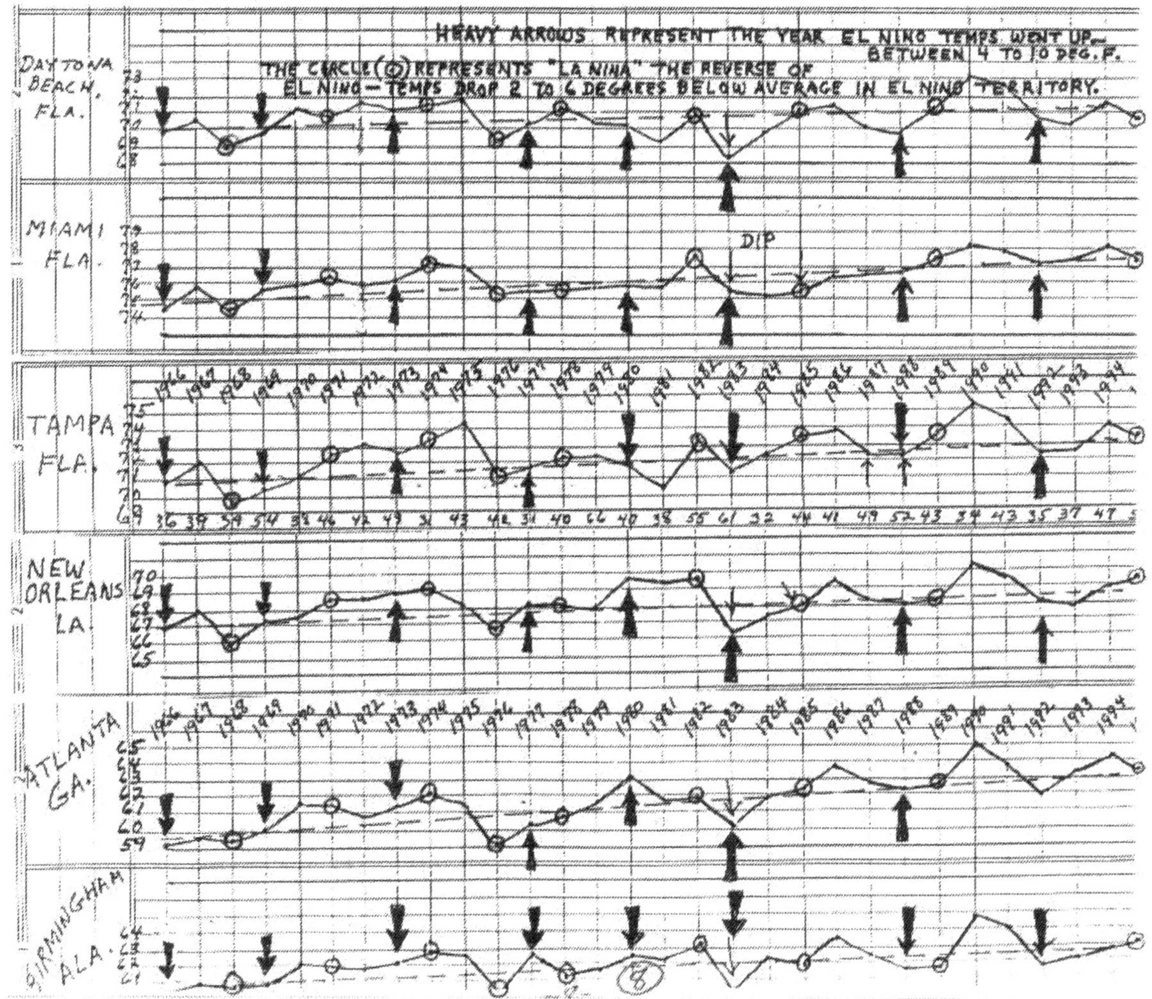

HEAVY ARROWS REPRESENT THE YEAR EL NINO TEMPS WENT UP BETWEEN 4 TO 10 DEG. F.
THE CIRCLE (⊙) REPRESENTS "LA NINA" THE REVERSE OF EL NINO — TEMPS DROP 2 TO 6 DEGREES BELOW AVERAGE IN EL NINO TERRITORY.
DAYTONA BEACH, FLA.
MIAMI FLA.
DIP
TAMPA FLA.
NEW ORLEANS LA
ATLANTA GA.
BIRMINGHAM ALA.

If there is any question as to solar radiation having increased temperatures? Records over arid regions such as Tucson and Las Vegas are worthy of serious thought. Increased ultra-violet intensity is world-wide. Australian authorities, for example, expressed fear of increasing skin cancer.

Two paths that polar outbreaks of cold air follow thrusting down across Canada to the US. One originates in the Arctic Ocean near the North Pole, sweeps down across Alaska and into our northwest. The other, blessing New England with crisp dry air in beautiful October travels from the North Pole across Central Canada and thus the Great Lakes, New England, etc…

Some idiot called Russ Limbaugh today on his frenetic radio talk show and said, "You know, earth-warming Al Gore…har, har, har." Twice I have heard the Big One say "Earth warming is a myth." But I'm sure all of Washington knows this.

Juneau's temperature records clearly show that there is earth warming which is not an original thought on my part. That city is certainly close to the Arctic Circle. Then of course, further south, usually getting brushed by Fall and Winter cold air invasions also shows what is happening; Fresno is a fair indicator of warming.

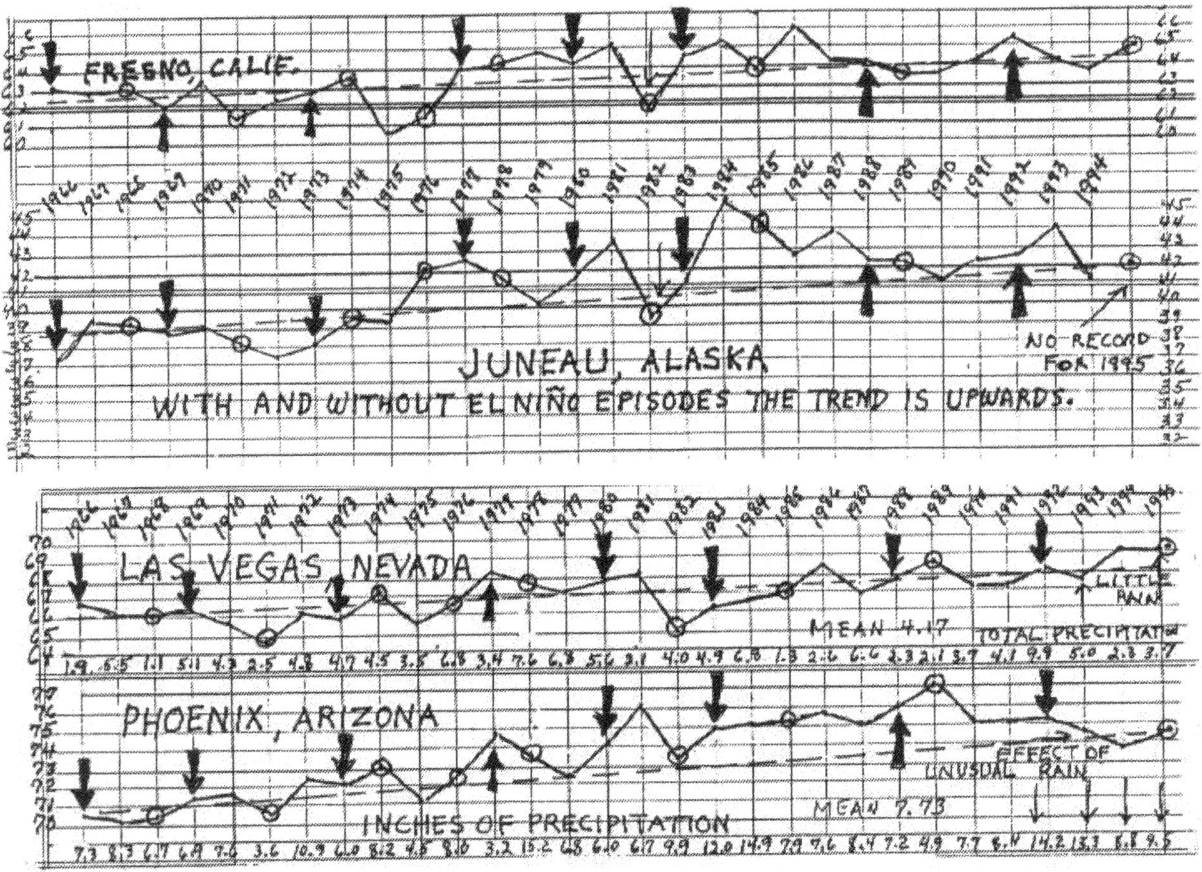

Given that nature mercifully, or so it seems, gives periods of cooling us off by virtue of volcanic dust at the upper levels of our troposphere or resumes interludes of El Nino's warming effect, the anti-environmentalists leap upon these two factors and, skillfully composing articles, stretch the afore-mentioned phenomena from one or two years to four or five. Apparently nature's system of rapidly transporting air via an expressway route at high levels back to the poles, with the obvious purpose of recycling and cleansing, so to speak, is conveniently overlooked.

In short the upward climb of temperatures on balance goes on and with it more tornadoes, thunderstorms, lightning, hailstones, droughts, hurricanes and flooding.

When the cost hits home, as it surely will, recalcitrant anti-environmentalists will retreat from their campaigns it may be too late, not that I expect catastrophic changes in weather.

The problem of carbon dioxide, unique, is quite different than that of particulate pollutants and toxics. It will be with us for a long time.

I've presented this in a non-scientific manner and think of it as more suitable for public consumption rather than coming across as one doling out guidance to esteemed learned men. If only someone would consider some day the need to initiate an education program for our citizenry in the form of videos, movies, TV shows sponsored by the government, brochures, whatever. It will be sorely needed if politicians want support for what surely will be in the very near future.

Andre Laferriere

Ten Years Naval Weatherman.
20 years combustion engine technician

Politicians are not a bad bunch* and it annoys me to hear derogatory comments, as if they were not worthy of our trust. I urge you to write and not necessarily over the subject I am most interested in. Some years ago I wrote Senator Chafee a letter protesting a planned scooping up all Social Security money you working people send in and dumping it into the General Fund. This pending bill called the Balanced Budget Act put up for a senatorial vote by Senator Paul Simon of Illinois had designs on a sacred fund, already dipped into for a stupendous amount, a loan they called it. Unlike my present attitude over impending pollution problems which is basically take it or leave it while I spend my remaining years getting my sleep, this really outraged me! Senator Chafee sent a one and one half page letter how the government was hard pressed, etc., but he reminded me he had voted yes but with stipulations. The bill failed to pass by a mere four votes. It was then a matter of Senator Paul Simon to try again and the chances of passing alarmed me. I then drew the illustration that follows and sent it to the AARP, American Legion, Veterans and Ladies Auxiliary plus seventeen newspapers and magazines, particularly those located in Illinois. Had I been some prominent figure that campaign would have been a recipe for disaster but I didn't like the smell of this. To call your Social Security dues a tax and therefore treat it like standard taxation was to me, a criminal idea. Instinctively from my experience and studying as I always have, the crash of 1929 was a sheer show of what Laisser-Faire economics is all about which should not to be confused with Supply and Demand, the latter being untethered with no diligent sheriffs to watch for weaknesses in character. Senator Chafee is a Republican as we all know and I realized after reading his thoughtful letter not all Republicans are tough Vikings. Let me just type word for word from this elderly statesman please, just two paragraphs:
    "Therefore, on March 1, 1994, I supported passage of Senate Joint Resolution 41, which would have provided for a balanced budget Constitutional amendment. As you may know, however, the proposal received only 63 votes, short 4 of the total required for the Senate to approve a Constitutional amendment."

End of one paragraph, imagine, a Constitutional amendment no less. But, the Senator had a better idea and wrote:

"You mentioned your specific concerns about the effect this legislation would have had on the
 Social Security program. I hope you will be pleased to know that prior to the vote on S.J.
 Res. 41, I voted for a substitute measure offered by Senator Reid of Nevada that would have
 exempted the Social Security Trust Funds from consideration under a balanced budget amendment."
However, campaign funding has with it's increases a need to be reformed.

Did I influence anyone? I rather doubt it. Nevertheless, for some unknown reason Mr. Simon went into civilian endeavors. Dear reader, do write. You may be surprised as I was. Mr. Simon, a Democrat, disappointed me.

Going back to the letter from another senator pertaining to his affirming a record of backing environmental issues I may have touched a sore point.   I did gently hop on a confused and improperly informed public, which is in need of some cohesive and factual familiarization, at least something more then what is in circulation now. I know the senator need not defend his record and there are many legislators who have joined him, or otherwise the EPA would not be flexing a little more muscle. Not being William Buckley, Jr., I must have failed to clearly define the objective I have in mind. I did not wish to be an annoyance and told him so and made it clear it would be unreasonable for me to expect reply after reply, since what I had to say required a lot of detail and numerous letters. The subject is both national and international but predicting rising ocean levels thirty years from now doesn't quite hit home with anyone – it doesn't compare to what has occurred and will occur again. Knowing full well there has to be a guarded approach to mandate clean air acts I went so far as to suggest ways and means of reducing the heavy pollution produced by diesel tractor trailer trucks without putting the trucking industry's backs to the wall. None of my ideas are unique but I just believe there is a lack of inertia and reluctant voters. Furthermore, I'm convinced the evolution of environmental purposes promises great things to come but it is the nature of the human species to fear change. I suppose the oil industry is justified in resisting.

The Kyoto Treaty is not only a noble endeavor but urgent. Nature is not going to stand still for ten, fifteen or twenty years. Besides America, the major polluters are China, India and a few of lesser responsibility for loading our Pacific Ocean air with every conceivable kind of particulate pollutants and of course, carbon dioxide. The two nations above are reluctant to spend on corrective measures, inured I suppose to seeing trucks loaded with mud-spattered corpses but if war was called for I'm sure they would find the money to wage it. Belaboring what I wrote to the senator or not, telling him the subject must be wearing thin and he and others so busy. But I did claim the subject of causes and effects was not complicated at all but it has and is being rendered complicated by assassins of truth, those were my very words, who embroider causes that are elusive and mysterious. All kinds of garbage I have read, seen on TV and occasionally over the radio. I do not want any more mail from Washington but I suspect some inkling of troubled waters ahead has affected both Al Gore and members of Congress. Ostensibly, if a radioman was not exaggerating, Mr. Gore lost his cool and told Congress they had their heads in the sand. Despite his attempts of the past to stir up more approval of his obvious pro-environmental philosophies he is bucking an indolent public, some confused, some adamantly opposed to anything the government wants to do, some on his side of course. The public, fear not, have been busy passing on to their congressional representatives their views, misled, surely to be taken for a long walk on a short pier, by the numerous ulterior motivated writers and the articulate, such as this clipping reveals. In attempting to gather a consensus with you I will try to itemize the good and the bad by maintaining, I hope, some semblance to flowing continuity while worrying over the possibility of being pegged as a nagging bore. Making weather, but from a different approach, as you will see, glowing with optimism as I am, interesting is not at all easy. "We've got to let the people know!" A voice in my head just shouted that. Now and then, some humor is needed.

> Between 1990 and 1995 alone, Brazil's emissions grew by 20 percent, India's by 28 percent, Indonesia's by 40 percent and China's by 27 percent. Over the next two decades, developing nations will be responsible as much as 60 percent of all greenhouse emissions.
>
> Perhaps the greatest contradiction, however, is that global warming theory advocates' dire warnings about the dangers of global warming have simply not matched reality. Newly calculated satellite data indicate that the planet has not been warming at all, but cooling slightly since 1979. ← ??

Spin-doctors excel in half-truths – this clipping as an example. No hint volunteered as to the oil embargo of the late 70's, $2.00 a gallon for gasoline. No kids peeling rubber till midnight.

It would be dishonorable of me to cut out an excerpt from a lengthy article whose title connotes there was a lot of hot air at a Kyoto Treaty meeting in South America just because it backs up my conviction there are spin doctors out there doing a number on public opinion. This excerpt reveals the obvious. The man has to belong to the Carbon Club as depicted by a pro-environmentalist writer, who also, was brave enough to suggest in an article that the deadly "Mitch" storm should be named the Exxon Hurricane. Now, I greatly respect NOAA's satellite up there, named Nimbus 7, in it's endeavor to measure the Ozone layer's state of health, carbon dioxide percentages of high level atmospheric air and finally to me, at least, the rather nebulous suspicion of it all inasmuch as I don't know how earth's surface temperatures can be sensed by infra-red beaming, punching through a million different varying conditions of temperature fluctuations in clouds, storms, rain, snow and so on. However, I'll give them the benefit of the doubt but I'm sure there is an immense amount of computer extrapolation. All I know, if you refer back again to pages 7 to 11, we here on earth, there probably being 100,000 thermometers all over the world being looked at, <u>see</u> some <u>rather</u> <u>different</u> <u>values</u>. Still, nature is not out to kill us all, therefore a worldly average temperature rise of 1 or 2 degrees is a reasonable rise over a long time. But, it is foolish to believe we have nothing to worry about as the anti-environmentalist pundits, and believe me, they work for well-backed institutes, want you to accept simply because there are pollution problems which manifest their existence in different manners. Our country has experienced the unusual in the effects as compared to Europe, for example. I want to broaden my definition of this. It will, rest assured, affect your life in the future, as a certain problem is here to stay.

Anxious to register a point with the senator, I also sent him a more dramatic illustration of tornado frequencies as you can readily see. I had to stop though. Enough is enough. What is he were to be honest and answer in this manner, "Dear Tornado Man, It's the Economy, stupid!" I'm guilty of convoluting other details of phenomena already mentioned previously. An example of where I am at odds with so many, why no mentions of what was a rarity in Oklahoma this mid November? Nine tornadoes whipped across that state. Everybody, other than the oblivious have got the "Oops, don't disturb anyone" syndrome, a real bad case of both flaws. I take that back, just a smidgen "Atta Boy" for a hurricane expert in Florida who made a comment about Mitch battering Central America, just a peep, "This is very unusual". But meteorologists are not paid to get mixed up in the morass of politics and hypothetical explanations. A shy bunch, self conscious and wary of being discredited they stick to forecasting and warning. Can't say they belong in the "Oops" category either. I've worked alongside dozens of them- I know what their philosophy is about.

The graph below previously shown on page 7 in a less then neat form, exactly as it was sent to the Vice President and two senators, their "runaround" responses in following pages, my intent to remind hoping that they would read between the lines, such a monstrous trend, 100 tornadoes in one week just this past winter of all times, had not been brought to the public's attention. During these past few years of course, hence the reason why everyone except victims are near sedated*, most in the government agreeing with them (convenient excuse and enamored over credentialism) ultra-conservative think-tank pundits** punched out article after article flashing super prose, ridiculing environmentalists and making sure El Nino was to blame and not Earth Warming. Please note periods where we used much less energy, including other nations since OPEC had decided to squeeze us all resulting, despite the continued increase of big V8 cars on the road during the 60s to 70s, in a cooler atmosphere (not much but it's there in graph form) and therefore as it is the basis, less heat, less tornadoes.

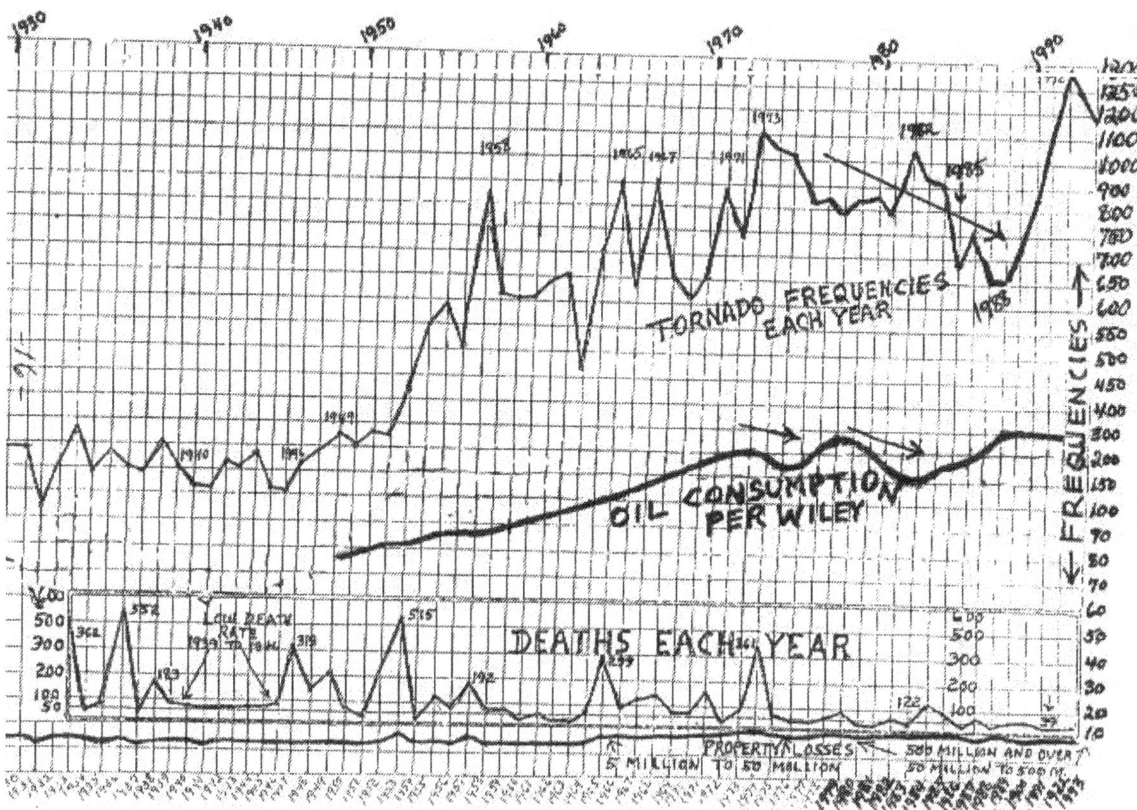

The only official to display mouth-open shock while viewing devastation with President Clinton was James Lee Witt administrator of Emergency Assistance.

**Not Weathermen!

I'm not upset over these two lady researchers whose report is briefly described in what follows – just disappointed. Same old story, "Oops". Our Providence Journal newspaper is the greatest despite being tagged as conservative – every single report by researchers, scientists and even spin-doctors is published. The chairman of the board is an avowed environmentalist. Thank you Mr. Stephen Hamblett. You are a fresh breath of air. I'm tired of so many researchers ending some interesting report with, "The evidence suggests, but…" or "Further studies are needed". In the case of these two ladies their article ended with their indicating they were not pursuing evidence of Earth warming as the cause of an unsettling summary regarding a fifty year trend of increasing humidity, so much so, despite not having indicated a fact already reported by newspapers contributing to the death of poorer and/or elderly afflicted with poor health, who could not sleep because of oppressive heat and heavy humidity. These two, and I think they merit my thanks and from all of us, obtained 50 years of recorded data from 113 weather stations and established there has been a 88% increase in the **number of heat waves** since 1949. It is an excellent article, thorough and accurate but they stopped short of making a commitment. I'm not afraid to, rest assured. The article stating the year 1949 marked the onset of change reminds me again how naïve and misinformed so many people are who, despite showing some savoir-faire in English composition write absurd letters to the editor of our paper. This man said Al Gore doesn't know what he is talking about, because earth warming ended in 1940. He didn't say **about** 1940 either. He must be young or unlike Truman's belief when he stated, "If you don't know history, you'll never know the future" he hasn't done his homework. Actually, beginning right after Pearl Harbor, gasoline was severely rationed and a coupon system lasted until the end of the war. Highways were practically empty of cars. I ought to know it took me 14 hours to hitchhike from RI to Cape May New Jersey. My uniform didn't matter much – there were few cars to wave my thumb at. Of course our atmosphere cooled down! But after the war, happiness restored, more and more vehicles and a growing economy, slowly but surely, it got warmer. I bought a

1938 LaSalle and it sucked up it's share of petrol, believe me. If I insert this portion of a letter to me from a state senator, member of the general assembly, it is only to suggest I do have knowledge of engines and in this book it is important that we integrate environmental factors with vehicular emissions. At that time of the letter there was wrangling between the state and DEM over how and when to enforce auto emission testing laws. I had suggested to DEM it was dangerous to use dynamometers unless they could find some way to keep the engine cool while stressing it at an applied equivalent load corresponding to 60 mph. I'm sure, dear reader, you realize a car at that speed needs the ramming effect of air it is hurtling through. The engine fan cannot provide sufficient cooling while a car is chained down on a treadmill device, that should be clear. My oh my, the phone rang. Some polite gentleman from DEM had decided to see what I was about and what I was up to even though I was very humble in the manner of presenting a suggestion based on my experience with dynamometers. This fine gentleman had also responded to my requesting information as to how, chemically, unhealthy ozone formation in the city of Providence during hot weather, formed from the rather dubious (I thought) theory his department had published. Here were thousands of out-of-state cars and some 6000 diesel tractor trailer trucks rolling through Providence on Route 95, big black smoke belching rigs at least 5 days of the week and lots of cars 7 days a week and DEM stated in print, "57% of Rhode Island's ozone problem is caused by RI motorists". And so, this is how you don't make friends with state of federal functioneers. I dared to mention my experience in meteorology had taught me that the prevailing summer winds were from the southwest and quite naturally the overwhelming quantity of highly populated metropolitan areas around New York city, New Jersey, it's industry and that of Pennsylvania really were a major factor in little

Mr. Andre Laferriere   January 13, 1995
30 Washington St., Apt. 501
Central Falls, RI 02863

Dear Andy:

    Thank you for your letter. I enjoyed reading your clearly stated explanation of auto emission problems and possible solutions. It is my understanding that something is being considered that is similar to your suggestion,--to have local service stations do some of the inspection process. Another suggestion similar to yours which is under consideration is that a group be formed to look for alternatives. I will apprise you of their recommendations.

    I presume you are aware that the Governor has already put the program on hold, suggesting that DEM take a hard look and evaluate the program, and the EPA seems to be willing to cooperate.

    Thank you again for your interest and time. It did not go unnoticed, and we appreciate it.

                       Sincerely,

Rhody here getting a goodly donation of pollution from somebody else. The man didn't really agree but we hung up our phones with no indication of any ruffled feathers. Sure enough, a year later, an article appeared and DEM reported that while auto testing for emissions would help the environment we were, here in this state, subject to quite a bit of pollution from NYC, NJ, and PA.

The point of all this, dear reader, please write your representatives. Here, our governor was under pressure to set up I don't know how many test sites, buy extremely expensive dynamometers, train mechanics to test and obviously this would have been very costly, along with the risk of testers not very experienced. I wrote to Governor Almond. He didn't answer. He's a Republican. I told him I had voted for him, which is very true, but no letter did I receive. Obviously, the senator mentioned previously must have talked to him. But we were all saved – I didn't want my beloved steed chained down on a treadmill and have it's heart baked. The state of Maryland, having incorporated dynamometer testing in the meantime halted the program, abandoned the idea after damaging over 100 autos. Even though it's in the realm of possibility I had saved this state several million dollars no fame, no accolades, no medals came my way, alas. Oblivion, I remain in. Such is the fate of a born loser, I tell you.

Before I return to the subject of weather and expecting your silent gratitude five or ten years from now, when you will be able to pass judgement on all things meteorological, recognize simple effective ways of anticipating good or bad long into the future. Make wise investments therefore in wheat futures of whatever. I must tell you what two gentlemen visiting my daughter's home, one an uncle to her husband had to tell me from their experience living in Honolulu some twenty years. They were a bit amused and perhaps disgusted at the wasted money thanks to the governments rather strange approach to a problem. Buoys rigged with electronic devices bob miles out to sea from the Hawaiian Islands for the purpose as best as I could ascertain, to measure various surf characteristics, currents and I have to suppose temperatures and other information. One big problem, somebody keeps stealing the radio transmitters in those buoys! And they keep being replaced. I spent several hours questioning these two chaps, both interior decorators and college graduates from long ago, one a son of a Navy Captain, the other a Lieutenant at one time and what they had to say was both fascinating and distressing. Let me save that for later pages for now however. Just one comment though, from the Captain's son. He wishes he could cap off one of the volcanoes as the air is sometimes permeated with mercury dust. Bill, the volcano hater (and I can't blame him), looked very serious about capping the source of disenchantment over his lovely island getting periodically dusted with something as toxic as mercury. Someone at the party drew some laughter when he said, "You cap that off and what happens if it pops up in your backyard?" I've yet to read about this – another one of the "Oops" deals probably.
Having told you two sad stories, one where my profound intellect was unrewarded and worse, I can't sue DEM for plagiarism after robbing technical and very original material from me, e.g., New York city and adjacent industrial sites are sharing their dirt ridden air with us. Seriously, how did DEM determine the specific percentage of pollution contributed by RI motorists? And of course, the above mercury story – how Hawaiian residents are not as enthused over earth's noisy smoky habit of relieving constipation as are tourists. Bill went into another detail after grinning over inheriting a backyard active volcano. When you catch a fish, best throw it back – it is loaded with mercury. Two stories. I do want to provide some relief here and there as I go along heading for sadder topics; hard earned revelations if I may say so. Insofar as individuals out to railroad cleaner air proponents I've a few more bits of condemnation left in me – I trust you will understand the harm they are doing.

I believe I can be rewarded with a Nobel Prize with just this one paragraph, unprecedented in it's brevity, but I shan't send it to Oslo out of sympathy for the elimination of at least fifty climate scientists even though taxpayers would benefit. When I was 17 it wasn't a very good year, just before Pearl Harbor, but I remember well skating at night every winter for weeks. A pond remained a solid unmelted layer of ice shining from not only a week of moonlight but illuminated on one side from a roadside lamp pole. There you see that's proof it was colder in the 30's than it is now. Such frigid clear weather, sufficiently cold enough to freeze ponds for a month sometimes, is a rarity now. The 30's were also a cold era of political resistance to Franklin D. Roosevelt pushing for a Social Security Act. It took five years and in 1936 he succeeded. It was too late for us. Fatherless, dad having died in 1931, my younger brother at age 6, not quite as strong as I was at 8, both of us suffering from malnutrition we had to live in a charitable institution. Mother had broken down. She remarried.

Thanks to Republicans who succeeded in dragging the issue out for a half of a decade while so much suffering was going on. I'm unreasonable in this perspective I sometimes throw at argumentative people, some conservatives have the mentality of ancient nomadic Africans who let the elderly fall behind to be devoured by hyenas, such is my contempt. While this party I shun at the voting booths has some usefulness at keeping liberals from getting carried away, I fear their influence. From a sorry childhood I can't shake off ill feelings. Too many of the far right wish to reinstate the days of Vikingry – or survival of the fittest. Let me remind everyone, never let "laisser-faire" economic come back, because there were three major economic crashes resulting from such a callous philosophy. Two disasters, the dates I forgot, some time around 1850 and 1860, occurred after each collapse had been preceded by a majority of Republicans from the President down to Congress and that bunch also in power before 1929 for ten years. Right now, and most certainly the coming center of attention in the next few years, earth warming, the same problem – ultra conservatives, and even Democrats wavering because the affluent constituents are bothering them. But I believe all will eventually yield when the cost hits home which promises a protracted period of little being done. You would be in error to think I'm a socialist if I go on with high finance flaws of the 20's. I don't go around haranguing my political point of view. But, there is animosity I want to make clear and for a very good reason, roadblocks set up by spin-doctors jamming the airwaves, pushing inaccurate nonsense, fraudulent so called experts making stupid unscientific statements, all this is going to hurt me also and my family. What lies ahead is sad.

Ah, I was going to stick my head back in the clouds. Infatuated as I am over this topic of weather and the mystery that it is not, even though the foolishness of those, at one time, enamoured over the non-existent puzzle of the Devil's Triangle, selling this juvenile crap to the eager, there is a hint of the illogic mind at work. I am amazed at what seems to be an attempt to present again, the puzzle, the elusiveness of nature's ways. It's quite beautiful and I have this habit going back five decades of looking at the clouds because certain ones, their streak or irregular designs and all that pretty stuff, has messages for me. How fast is the air moving at 20,000 feet? What does that jet's contrail tell? But, let me gripe a little more about some political emanations that just perhaps are corporate originated and well, I'll let you think over the arithmetic or the lack of it involved. A son and a daughter expressed concern over the continuous worry as to the future imperiled Social Security fund. Others I met definitely felt they would never benefit and so on. First, big and small businesses would like to get rid of social security – which is not an assumption requiring the brain of Oppenheimer. Five percent net profit for a corporation is great but without social security you are talking the fabulous! How quickly, how easily it would be to run off to some third world country and profit some more! I get a little disturbed over this issue. One time I blew my top at some guy, smaller than me of course, because he made a crack about old fogies sucking up all the money in America. He was about in his fifties, I heard he was an accountant and figured he hadn't worked his muscles into a state where it would be dangerous of me to pop off at him. "Maybe you'd like to work in some factory and some German soldier poking you in the –s with a bayonet, no?" We were sitting in a crowded section of Lincoln Downs, a dog track, lots of videos, horse racing and all that, got to have a change of scenery sometime and I allow my wife only ten minutes at the slots. There were as usual mixed in with the young, the elderly, the lonely, not many years left you know and this bespectacled senior baby-boomer angered me. Now for the arithmetic part and I'm not positive of my opinion either. They say there will be few workers and too many elderly…? Now, the mentality explaining why we have millions of immigrants can best be summed up by a businessman who called in on one of those radio talk shows I like so much when he said, "I like immigrants, the more the merrier, they are a big help for my company." That's fine, nothing wrong with that, but, millions arrived and more surging through in great numbers, changes, in my view, the arithmetic factor of "few workers and too many old people" thirty years from now. We give immigrants welfare, food stamps but they can't stay on the dole too many years. In general they are prolific, married or not, at least two babies and sometimes more from what I see. Having survived tougher conditions in their countries of origin they strike me as a pretty healthy bunch so I don't think they will have a problem with producing a family and, check your pencil and paper now, arithmetic in mind, that's a lot of workers thirty years from now. Furthermore, social security yearly raises don't compare with the rising income levels of workers and the increasing amount of Social Security money pouring into its fund. But, you see, and you will not grasp this I suppose, and I'm not prejudiced, there is a genetic factor built into the boss breed: worry, worry, keep moving, don't stop, hmmm, the stock market, watch out, got to keep the GNP rolling and seven days a year vacation as we average is plenty. Some fifty years or more I've yet to read or hear a truly optimistic view of the future. It's their nature that's all. As a result, despite their brilliance in many fields, my family has this worrisome thing gnawing away and so do millions of others, such is this negativity factor prevailing from ruling circles. I'm sitting here, an old man in a cave and I had the above message of "Fear not, you have nothing to fear but fear itself" passed on to you so that perhaps, after the arithmetic you will make somebody smile. All I need however, is some guy paying a visit to this all-knowing Guru, dissident in a way having learned there is also a streak of illogic running through the veins of those degreed pundits who direct our destiny – not all, as you shall see, but enough to bother me. Just as this guy whose eyes are glazed from computer whatever and he says, "According to the Internet our calculations of the ratio of workers to retired folks still stands – Social Security is doomed." That's all I need, really. I was talking to the chief meteorologist at a NOAA weather site, the whole place resembling NASA's Control Center in Houston – there must have been twenty screens, all kinds of computer imaging and truthfully I was quite fascinated. Maybe seven or eight young studious types, hunched over making notes and I wondered how the hell did we get along without all this when we had to worry about icing on wings, no deicers were part of our aircraft. I'll cover more on our dexterity acquired from going outside in between drawing maps, etc. I went to this amazing weather station, it's big radar ball making it easy to spot for visitors but visitors are only allowed in there one day every two years; why, I don't know. Intent on finding out one thing – how come this huge thunderstorm, maybe a half mile wide, swept into Syracuse, New York, one afternoon in October, blew trees down; pieces of roof cartwheeling down streets, three people killed while dozens at a fair ran for their lives and no warning whatsoever. Thousands of weathermen, Doppler radar scattered strategically all over America, hundreds of handsome, hairdoed men and blondes on TV weather time and this has to happen. Nothing easier to spot than a huge lightning filled thunderhead, dropping tons of water as it marches along, what with long range Doppler imaging available. I had to find out. The chief meteorologist was extremely nice. I made him laugh when I told him I wince when a TV meteorologist states his computer projects the following weather – he knew there were limits to this marvelous high-speed calculator and nature does the unexpected; the computer can't anticipate that. He told me that the army radar near Syracuse had been shut down quite some time and that furthermore this thunderstorm had moved so rapidly, 70 mph across Lake Ontario, etc.. and his radar had insufficient range to spot it. Reasonable enough, we're improving but I hope the government moves faster in its announced intention of increasing Doppler radar placements – we're going to need it. I knocked negativity but that's one area of future probabilities where I can't be optimistic. I will do my best, however, to insert something here and there hoping you smile at least, even some memories of days of old, that might be a little different although I must admit, I've set a course aimed at stirring up uncomfortable feelings about the future, assuaging bits of experiences thrown in for relief, but sometimes designed to keep you focused, as if you did not know.

There seems to be a balance between the pros and cons. One day an article lambasting environmentalists sets the cause back a few notches and the next one, and I hope to see more of them showing a little pizzazz as this one, keep my hopes up. An excerpt from our paper, on your left, comes from a much longer article by an author not associated with any particular organization. Its tone has a bite to it. In the same day's paper there was a funny cartoon, two Nostradamus types, bearded, long hair like Moses, robes, both standing apart on the same sidewalk competing for attention. One is holding a sign that says the world will end in one year and the other, in two years. I did write a guess, and that's all it is, a guess, earth warming will make itself felt again in two years. It all depends on how long it takes for the atmosphere, now freshened up a bit from two years of incredible downpours and flooding, to again reach a point where the continuously increasing permeation of it's limited confined volume with pollution, separating as I must carbon dioxide from the same category of nitrogen-oxides, sulfur, particulates, these three and some other substances as falling into the pollution class, reaching a point where re-radiated heat from the earth at night is delayed from returning, unimpeded as it should, back to the sun or

> Nevertheless, the lobbyists and spin doctors for the oil, coal and car companies are ensuring that the United States will be a global loser as this great technology race takes off. So go ahead and listen as the fossil-fuel lobbyists badmouth the president and vice president for talking to the public about what is increasingly the most obvious global threat we have ever faced. It's their right as Americans.

wherever in the universe. Hence earth warming will return. Granted this explanation is hypothetical, but it's all mine. It's new and I intend in later pages to make you understand a fact of nature: If it did not have ways and means of periodically cleaning the air we need to live, and I repeat, a very confined body of air, lidded down some 11 to 12 miles above the equator and 6 to 8 miles above the poles, we would be in dire straits trying to stay healthy. What we need to understand this is logic and some common sense. One problem, should anyone set out to prove this or that contention he or she would have to wait years, maybe 3 to 6 years, to measure or calculate a repeat performance by nature. We need therefore, to visualize the big picture, not bounce all over the universe with dozens of concepts that every form of news outlet saturates us with, "The evidence suggest, but…" or "More research is needed." It has been that way for decades and it will go on for an eternity. We're not Einsteins, but he would frequently ask himself, "If I were God, how would I do it?" Simplistic but shunned by most scientists because they are married to the idea everything is a huge mystery, an elusive puzzle, when in many instances it is not at all. Getting back to flooding – have you any idea how much water falls on our earth every year, that is, normal years? Over 375 cubic <u>miles</u>! If we have followed the news of China's six weeks of near continuous downpours this past summer, floods killing over 3000 and causing over 20 billion dollars in damage, do you have any idea how many millions of miles, cubic, horizontally up to 40,000 feet and perhaps 4 thousand miles wide, of dirty air found its way to a stagnant low pressure trough, stubbornly not moving, wedged there by two high pressure systems? Six weeks of a continuous supply of moist polluted air but vacuumed up, widespread precipitation resulting also, dragging pollution particulates down to earth. How else can nature clean its air? Do you have any other ideas?

May I introduce a piece of elementary physics now? For once, a gentleman, unidentified researcher no doubt, claimed China's flooding was caused by the tremendous pollution this country produces, what with its coal fired industry, Hong Kong is so polluted you can't see buildings a mile away. He suggested dirt in the air was absorbing solar heat. That's probably true, but I must now refer to an elementary bit of physics taught to me long ago that can be verified by any high school physics teacher. And yet, and that's my contention, this business of earth warming, meteorology and the variety of views by scientists, has a need to get back to basics. The rule is: "A column of warm air, if it is polluted, has the ability to accumulate more moisture than a comparative column of clean air." Minute particles of moisture, whether rising from ocean surfaces or available from damp nighttime landscape, are attracted and congeal to various types of particulate substances in the air. Need I review what would be redundant, old news by now, ad nauseam, the preponderance, the magnitude, the traumatic deadliest of flooding all over the world during this past decade. While the following may be boring to some and furthermore I did want to avoid high school exercises, I am compelled to wrap up as a package the essentials of minor physics as it applies to weather in a convincing formulation of <u>exactly</u> what happens when polluted humid air is drawn in, call it being vacuumed up if you wish, into an unstable area of weather, let us say a low pressure system even if it is not a powerful one or as in China this summer in 1998 an elongated trough, sandwiched in between high pressure systems. Before the recess bell rings - -, that's how self-conscious I feel right now, thus a little humor, please; let's be certain of this positive fact, an irrefutable fact: DIRT IN AIR ATTRACTS MOLECULES OF MOISTURE FROM WATERY SURFACES, which cluster about each particulate of soot, whether each invisible speck is composed of unburnt hydrocarbons (toasted), dust originating from sand and/or richer soils, different acid based substances such as nitrogen and sulphur dioxides. Stop for a moment – give what follows some quiet thinking – acid rain is just one bit of evidence that nature has but one means at its disposal to rid the air of a condition, which if it were to continue accumulating unchecked, life on earth would become unbearable. A detail of importance: Pollution increases relative humidity from approximately 85% in reasonably clean air to about 95% characterized by hazy skies.

If you prefer a higher class explanation or it's vagueness, it's characteristically dubious endings leaving one unfulfilled, some semblance to at least the questions it leaves unanswered reflected in the excerpt below, be my guest. I would like to exclaim, "You're my man!" I hope. Such an article has great merits however even though as usual I have to pick out some weaknesses. The touches of refinements are there but I don't think we need to consider a very old and well-known entity as our limited atmosphere with it's repeated acts that have been recorded, plotted, extrapolated, etc… as meriting space-age dialogue and invariably the whole subject of weather, not at all a great mystery, takes on an aura of great mysteries yet to be solved.

> Some have postulated that the cooling since 1940 is due to anthropogenic influences rather than natural climate variability. The suggestion that man has increased the amount of atmospheric dust through agricultural and industrial activities and that this is the cause of the recent cooling is not borne out in global measurements of atmospheric turbidity, although this may partially explain the observed changes in regional climates.
>
> Climatic fluctuations involve processes on a range of spatial scales that can interact in a highly complex way so as to cause climatic anomalies affecting large geographic areas. The intermediate-term anomalies are the result of persistently recurrent weather systems of the same type. The fundamental question involves how and why particular events recur during a given season or sometimes recur for several seasons or years. To cause recurrent events like prolonged droughts or cold waves would seem to require some sort of memory. It is highly unlikely that the atmosphere alone can account for this persistent recurrence because of its highly variable nature and its relatively short thermodynamic relaxation time.

While you were at recess I found time to write how I passed judgement on another radio talk show host who, two nights ago, declared he was amazed at his six year old son who spends six hours every day playing with a computer, not only Nintendo games but other more useful applications. I definitely believe children should adapt as fast as possible to a changing world, but six hours a day? There was a guest on David Brudnoy's program who represented a philosophy, a supplementary philosophy of course, more popular than I think and that I need to be reminded of, such is my apparent need to push and punctuate a scorched-earth policy against those imagined enemies who I want out of my way. Remotely, this guest was a friend to me simply because I rarely run into this type of person. But, I don't expect everyone to be interested in the peaceful manifestations of nature either, as represented in a book, "Oh Spacious Skies" that Brudnoy's guest had written some time ago. Still popular, mostly ladies telephoning their enthusiasm over what revelations they found in the book and how people from different parts of the world had written in. I can't recall his name. I don't remember simply because my memory banks sometime balk from "information overload". Pretty good excuse, don't you think? The man is a 20-year veteran of radio with a few spots on television – quite a surprise to plug what is probably regarded as banal. His suggesting we should look at the sky a little more and interpret the messages of different clouds and so on. As for the father of the six-year-old potential computer nerd, he should consider directing the child to other interests for a few hours at least. But a word of caution, surely you already know, wait till there is nobody around to look up at the sky.

I had a short period of teaching weather at the now no longer existing Norwood, Massachusetts airport after the war. If you ever have had the job of lecturing to a group of adults who you know have been working during the day and just looking at them sitting there, stifling a yawn, you know you have to find ways and means to keep them interested. Unlike naval pilots, who are interested in staying alive and also aware of crewmembers expecting as much, this Norwood bunch, about eight of them, showed little inclination to look attentive and of course, respond enthusiastically over the subject of meteorology. They didn't look like morons, but since they were striving to obtain a license to fly some flimsy light aircraft they had to pass a test in weather. I passed photos of one lollapalooza thunderhead cloud with its innocent looking anvil head protruding over a distant cloudbank. I just reminded them it specialized in shredding aircraft designed to fly in nice weather only, which is true. Nothing bothers me more than some damn airline pilot who is hell bent to maintain a perfect straight course, sees a nasty cloud ahead and tally-ho, we bounce for a while. I am not an old salt teaching kids but as great as are naval pilots, I've seen the result of mistakes a few made. Nasty to type this, but all hands turn out, go to the wreckage and pick up what's left of body parts. The Navy has this rule, respect for the deceased. Military planes do not have insulated fuselages – the sheet metal leaves dismembered bodies. While on the subject of morbidity, authors feasted on gullibility after planes and people disappeared off Florida, four Navy planes for example and a few empty undamaged boats out at sea, books sold with Devil's Triangle boldly emblazoned on the cover. This naval incident occurred during the war and while secrecy prevailed all of us realized they had blundered into a zone of violent thunderheads and a vicious downdraft had pushed their low flying aircraft into the ocean. The Gulfstream has its share of squally weather, waterspouts and major thunderstorms. I don't know what year it was, maybe 10 years ago, these four planes were found at the bottom of the Atlantic and supposedly were intact. While my strategy to alert the Norwood "fly boys" worked for about 20 minutes, I then resorted to another killer fact. I asked them what would they do if the Piper Cub ran into freezing rain. The immortals, tall on appearances but short on wisdom didn't know.

I want to go on writing a little something for everyone, young and old, avoiding the banal of weather, tailoring as I am from what is the essential deja-vu details for most and holding in abeyance how the future has to drastically change, building up fundamental facts first. If you are seriously interested then do not discount as trite every little facet of the physics involved in, for example, flooding and dirty air or you may remain as with even alleged scientists and yes, meteorologists, still convinced there is a lot more to be solved. After mulling over the "turbidity", that's correct, muddied up synoptic rendition of weather vagaries as represented with the above excerpt, I put aside any guilty feeling and intend now to express mild criticism over certain details. For one, warming ending in 1940? Ask any old timer if in the 30's one could skate on frozen ponds for weeks. Those cold winters and how well starting a car with the old six volt battery was sometimes near impossible, yes those very frigid winters are gone – a rarity in southern New England anyway. I am not assiduously dredging up what seems far out or illogical opinions scattered about in reports transmitted from mainstream sources, which in their aggregate are a study in disharmony, such are the varied theories. Am I wrong if I told you nature has to purge the air of filth and it does so in a very heavy-handed way? Have you ever heard that approach used from any degreed, doctorated or old hand weatherman? Do you know what follows after several years of death-dealing floods all over the world? I know you don't. It is not bloated ego on my part if I tell you I do know. Will climate get worse? Of course! In 1900 the world's population was 1.6 billion and some authoritative group in the United Nations estimates that close to six billion people, that figure expected to be reached before the end of 1999.

I ask myself in reviewing article upon article, are it's authors just English majors attracted to a new growing field of controversy, just in its infancy now, waiting to envelope everyone? Are they financed policy makers, several of their empirical assertions clearly biased? While I don't subscribe to knocking indiscriminately, treating the reader to repeated verbatim such as demonstrated by ridicule-oriented literary kingpins, I am motivated, not with animosity in mind as I sometimes feel towards others, to critically examine everything in print or from other sources. The excerpt in mention does not represent the workings of a spin-doctor. It is a sincere effort but randomness is connoted and that definitely does not apply to meteorological large-scale exhibitions by nature. I intend to peck away at an affliction, affecting even weathermen, indoctrinated in questionable theories even <u>as</u> I <u>was</u> a long time ago that hamstrings there not recognizing what forces are at play, albeit I just now had nothing very convincing to offer, being forced to avoid inserting something here that in my agenda belongs in later chapters. What truly stifles the imagination required to discover scientific truths, (although meteorology is more of an art form having had its earliest revelations of fronts, etc… in 1919), is the tendency, whether it's religious or not, I do not know, to separate nature from the ubiquitous omnipotent God the majority of the world's population believes in. Well after my ten year stint in naval aviation circles, temporarily married as I was to meteorological beliefs, baffled at times at some curve, unexpected, thrown at we pseudo-scientists by nature, I adopted the idea, my IQ potential short by a mile of ever reaching Einstein's cerebral capacity. Einstein had something going for him; it was common sense, better to call it logic, of believing a power was in control of everything. Why in heaven's name do all climate scientists, journalists, researchers and academics seem to be constantly embroiled with hundreds of details when first (just a suggestion) they should go back, start from scratch and recognize atmospheric circulation, confined as it is for a specific reason within a shallow layer over and around our planet, from poles to equator and back again in a perpetual recycling mode, is an extremely well organized perfect system! And that, as we should endeavor to understand, there is a balance mechanism at play. Again, let me save the true meaning of this in terms of understanding hurricanes, tornadoes and what have you for later. Let me redundantly nit-pick the issue of the stalemate, (I'm at a loss for words) in "reasoning potential" contributing to one big unnecessary conglomeration of 2400 scientists, not 500 but 2400, paid I suppose by the government. Before I support this claim of amazing duplicated efforts (analogous to 50 hounds chasing one fox, except there are dozens of other foxes being pursued by as many large groups of chasers all over the place) let me dispense with the final negative impression I have of the aforementioned gentleman author whose excerpt you have just read. The article fascinated me but it gave me the feeling that there are undertones of uncertainty and a suggestion of randomness, unintentional as it may be, but it's there. In the overall effort to understand deviations from the norm he does say some sort of memory is required. That makes me feel much better. I have that going for me! I strongly suggest to the top brass in charge of troops doing research, that they at least go back 30 to 40 years and compare weather maps of that time with those of the past several years, particularly during droughts, abnormal episodes of tornadoes occurring well past their usual season, and to some recognizable extent the rash of hurricanes and typhoons of late. You see, dear reader, I don't think this practical down to earth approach appeals to a scientist or a department head who can't stand the lack of loftiness, the complex formulations and shimmering computer imaging. Oh my, the oratory I have heard on such TV shows as Nature's Wrath, Fury or whatever, far off at times. But, I have something kind to say about the above author for a change. I can't mix this effort of mine to promote common sense and gradually present pertinent details with items of importance that must be fitted into later information, later illustrations of El Nino's wanderings for example and the difference between yesteryear's semi-permanent systems that only maps can show with today's rather different juxtapositioning of air masses. I must tell you I was delighted that a paragraph, clipped from the same author's article, attached below, supports my belief, backed by another article you will find interesting in later pages, that El Nino is one hell of a misunderstood phenomenon! Not only that, the various versions presented by a slew of experts, suggests again, showboating by non-meteorologist types. Droughts caused by intractable high-pressure systems, such as a six-week stay of a ridge of dead descending air over Texas this past summer, are now erroneously blamed on El Nino. English Majors pumping web sites for all their worth relish the titillation factor.

> The oceans play a major role in the evolution of intermediate- and medium-term climatic fluctuations. The top 3.3 yd (3 m) of the ocean contain as much heat as the entire atmosphere, and anomalies in ocean temperature have been observed to penetrate down to a few hundred meters. These temperature anomalies create vast heat reservoirs that may affect the course and behavior of storms and jet streams later on. The anomalies are in turn frequently responses to antecedent abnormal atmospheric conditions.

But in this paragraph something reasonable and believable emerges. He refers to ocean anomalies without mentioning El Nino, which no one can deny, is an anomaly of the Pacific Ocean. Because El Nino brings about in it's travels a vast area of warm waters, some 8 to 10 degrees higher than surrounding seas, it stands to reason its ability to retain most of that heat for months must require a deep massive content of unusual warmth. According to various researchers, particularly those employed or leased by NOAA, that depth has been measured to some 200 meters and I trust the author in question above has relied on such a source of information. There is a busy bunch of government employees out there in the Pacific, believe me. If, as it is argued, El Nino was an accumulation of shallow heated surface waters from solar radiation, although I'm sure this one meter deep warmer water supplements El Nino's massive pool, at least around its edges as the Spanish One migrates northward, it would soon cool from prevailing wind currents sapping up moisture.

It is evident misconceptions over this Christ-Child's source and traveling mode abound in literary circles – whenever something unusual happens of a profound nature all the English Majors come out of the woodwork and, no sarcasm meant, standby for confusion. Briefly let me explain in plain English the following technical rendition of the not too well known or even bothered with, the north and south oscillating inter-tropical convergence zone (ICTZ), quite a mouthful but requiring illustrations in later chapters. Let me stop here for a moment. Let me play the mentor. Take that child of yours outside some day, point to some fast moving high clouds and tell the child that the clouds are moving 100 miles per hour because they're in a hurry to get to the North Pole. Buy the child a large globe, the spinning kind on a stand with Bas-relief Mountains that they can feel with their fingers. Explain how an egg-shaped envelope of air is wrapped around our earth and all that. Buy the child a compass. Explain how pilots know how high they are flying because the altimeter reacts to atmospheric pressure and so on. Make that child feel what they are a part of. Doctor Spock here now, insists you devote ten minutes to familiarizing a child with the world and cut out those idiotic cartoon characters smashing each other's heads in. The Internet has to be put aside so little he and little she can go out there and develop something else besides bloodshot eyes. What do you think Road Rage is all about? Money, money got to outdo the Jones and everybody is in a hurry. Seriously, I had to depart from that ICTZ thing for a while reminds me of computer abbreviations like GIGO – garbage in, garbage out and the guy who told me that smiled; I trust he didn't think me a dullard. God, this radio talk show host, substituting for Russ Limbaugh, went on and on about Libby Dole. She can drum up 25 million like nothing, has all the connections to pile on some more and has shown her ability at understanding high finance and she has her husband's assistance in gathering greenbacks, on and on. Calls come in. The second one, an elderly sourpuss like me, who said, "You disgust me!" and hung up. Then another lady radio host says she bought a Ford Expedition so as to be safer although she regards big SUVs as politically incorrect. My view is less light-headed. She could have bought an Explorer, not very different than that air polluting 6000-pound behemoth! Who is she kidding? The affluent are out in full force, it's an image thing they are obsessed about! You see, dear reader, I want to also irritate so as to maintain interest but I'm heading back now to what this is all about or why that ITCZ zone over the equator is doing strange things and why El Nino, a factor we never heard of 50 years ago, now has a habit of frequently washing away some of California's cliffs, houses included.

The so called semi-equatorial front oscillates north and south, dropping 700 miles to the northern part of Australia during that hemisphere's summer and months later, moving northward during our hemisphere's summertime, a distance of some 1500 miles over the northern parts of the Philippines. And as for the Atlantic Ocean, the same latitudinal wandering northward between Venezuela and Puerto Rico further eastward bordering near Senegal. So much has been written about El Nino – it's source of concentrated warmth, how it comes and goes. Colorful renditions by TV shows with harsh titles, "Nature's Wrath" or "Fury and Raging Planet" of course, abound with spokesmen, some good, some bad, but all elucidating with aplomb. Two German scientists made more sense. They had this computer imaging thing going, how El Nino begins, expands and moves northward from it's home grounds. But nobody in these shows and nobody backs up their varying ideas with the only solid evidence available and that is the good old-fashioned weather map. Today's weather maps are not al all any different than those we used long ago. I wrote that we never heard of El Nino in the 40's. I must have drawn hundreds of weather maps, the last five years in the Navy and I don't recall anything unusual in California weather – **nothing**!

That's just another way of my insisting this phenomenal change is a product of earth warming and earth warming has caused deranged atmospheric entities we call high-pressure systems. Totally different and yet, has anyone explained exactly how that comes about? While the excerpt of a few pages ago has something going for it and finally, I'm giving credit to somebody, in that it logically appraises in the last sentence causes and effects. I want to show you what several years of fastidious sleuthing and drawing on memory has revealed. Tactless I may be, but it is my aim through sharing with you an abundance of information so that you too may realize the opinions of assigned experts always stop short of completing the picture. Shortly I'll give you an example of a meteorologist who seemed dismayed over what he saw while pointing to the next phase of the Weather Channel's very thorough, very instructional, very educational renditions of the elements. I have nothing but praise for this cabled gift to mankind, saving lives also I might add, therefore what I have to repeat again, does not apply to one very surprised weatherman. Invariably, and it's getting to be old hat, the habit of putting the cart before the horse prevails and I definitely know it is with deliberate intent to shroud the truth. The objective is clear – keep everyone confused over earth warming and play up El Nino as the only reason for destructive weather. Even though its deterioration from late 1996 to this winter of 1998 and early 1999 we've had record-breaking heat waves, droughts, immense flooding, tornadoes well past the season, huge hailstones, flash floods, one occurring in Kansas City, Missouri when it shouldn't, killing seven people. All of this was extremely expensive during a long sustained period of warm to hot weather for 1996, 1997 and 1998. It is not all El Nino to blame. Once our family lawyer, young about 42 and typically preferring to deal with dumb clucks so as to play the superior one, just had to let me know what he thought of my having provided him with 20 pages of details and photographs in an injury case we are litigating, quote: "All these details", followed by a grin. "You're a fanatic, you know that?" unquote. I grinned back and said, "I'm sorry but that's how we won World War II". That I am, a fanatic but I'm dealing with an immensely complicated subject with meteorological nuances of a grave nature whose eventual effects will profoundly change the placid pattern of ours, in fact, the world's economic trend. In short, I remind you, your interest should go beyond glimpsing at the weather predictors' opinions of weekend weather. In later pages I have enclosed a letter from a senator, which if you read it really slowly you should gather that just maybe the government recognizes the dollar implications and it's anticipated jeopardy. While members of Congress and the Senate, except for Al Gore, fit quite nicely in the symbolic Three Monkeys' meaning, including most of the American public, there may be slight pangs of indigestion or change of heart because of two pertinent measures they have initiated without any fanfare. One, Project Impact started in 1997 and a right to the point announcement by the administrator of Federal Disaster Assistance, James Lee Witt, the audience impressed as I am from listening to his one hour speech to the press one night late November, had to implement stricter requirements from now on, such was the waste of these past years. We'll cover all that a little later. First, I'm not all doom and gloom and if ever this public gets aroused from its deep slumber, inured to catastrophes as long as somebody else suffers and deeming it all as normalcy, soon forgotten as always, then realizing this government needs both help and support to do its job. We may be in for exciting times in the next 20 years or so. I must also include later the terrible lack of progress, inasmuch as the EPA has problems and so does DEM in some states. Considering anti-everything extremists are doing a hell of a job on public thinking resulting in constant squeaking from this or that citizen calling in and saying, "I don't want the government telling me what to do!" huffing and puffing over the phones about this and that and wild liberals yak, yak, yak! To an extent I'm not 100 percent pro-government. For example, how come alcoholics collect social security while I see them carrying cases of beer every other day? Why cultivate a class of promiscuous teenagers by furnishing them with money, food stamps and cheap apartments subsidized by the government paying landlords 1000 dollars a month and the young single mother only pays 20 dollars a month? Did you know that? But I don't want 50 or 60 million people doing what the hell they please either! That's a form of mob rule within the tenets of civil society. I wanted to continue on with that darn El Nino thing back a few pages but decided digressing somewhat would provide relief. Imagine this, Russ Limbaugh who I listen to just so I can hear his gang call in and it's quite a gang, evaluate one segment of American character. They said we have gotten along very well without government since the impeachment thing tying up Congressmen, etc., and he hopes the Senate will stay busy over the complex issue of President Clinton walking a tight rope for another year. Thus we will be free of any new laws or whatever. The idea that some 200 million people, all having different interests, encouraged to distance themselves from reasonable legislative matters, akin to happy sailors on a ship with no one at the wheel, the captain shackled below, is about as immature a concept as I have ever heard. Let me get realistic now, I have to return to weather again, "Don't fool with Mother Nature!" Let me describe her tantrums, figuratively speaking, let me draw you point by point, what She's about. To those who think we can do nothing about weather, that's hogwash. I ask you to go back to Wiley's diagram and the remarkable drop in oil consumption back in the late 70's and early 80's. Actually both forced upon us and as Americans can do, responding to appeals to conserve energy. Roughly we reduced oil consumption by approximately 12%. The economy continued on. Consider what we import every day at present, according to Paul Harvey, over 11 million gallons of crude oil every day plus our own oil well production. That amounts to a 6% increase over last year.

Just guessing as to how many barrels of oil extremely wealthy Arabs did not sell us for a while, so wealthy some come to Kentucky and bid 10 million for a yearling race horse without knowing whether it will ever make it at the race track. I would say, basing on less energy consumption of the embargo period, there being less vehicles and less new homes built, oil heat and all that, we probably deprived ourselves, weeping and gnashing of teeth not particularly evident, of about one million barrels a day. As a side mention, there is the beginning of a thorn pricking at my side, what if oil refining Texans like Governor Bush take over the government? And Louisiana's Congressman Livingston making a comeback from cigar-curious keyhole voyeur's' stupid Salem Witch hunters' abusive and unnecessary revelation? That state is also heavily involved in oil refining. I lived there nine years and let me tell you something; Baton Rouge was one smelly place, sulphur dioxide making people sick while protesters accomplished nothing. What if Livingston, astute and refined as he is, regains his influence? Enough with politics for a while – back to weather. While the ravages of El Nino excelled in 1982 and 1983, the weather's healthy improvement temporarily interrupted by that South American Christ Child, continued to a remarkable show of nature at peace with us during the mid 80's, thanks to cleaner air. But, let me explain something very carefully, in fact, it is to be copyrighted simply because it explains what seems to be baffling doctored scientists as to why weather patterns and/or cycles take unexpected turns. Writers mystified, but covering up their bewilderment with high level renditions that reveal they spend too much time in libraries and if they have ever studied weather maps, I would be surprised. The cycles, puzzling as they appear, are as follows: The atmosphere loads up with filth, how many years it takes I can't be sure but I would guess somewhere between three to six years and when it reaches a point of caking up our lungs or whatever ecological dysfunction threatens, poof, nature resorts to turn on the laundry machine. Down comes some 500 thousand cubic miles, yearly, of water here and there on our planet, drags down all that airborne filth and voila, one, two or three years of flooding having done it's cleaning job, the air is clean once again. Not perfectly of course and less so in southern regions but ample pollution is washed out of the skies sufficiently enough to at least allow the restoration of normalcy in colder regions.

In the beginning we covered the physics of dirt attracting moisture, the exacerbating of excessive vertical cloud buildup resulting, hence heavy downpours which I will elaborate on soon. The cycle again: Dirty air means anomalies in weather systems, which drive vast air currents where they were not so persistent, not so inclined to stay beyond their usual seasonal positions (to be illustrated) which in turn nudges El Nino along to areas it was not accustomed to years ago and unwelcome as it is, it stays much too long. Then not only here but also worldwide, Nature has to resort to its heavy-handed ways. When its atmosphere reaches a degree of maximized contamination, it has no other options. Flooding resumes along with El Nino's tumultuous extremes at times, one, two or even three years of disaster results, including hurricanes of the like of Mitch. So, I should rejoice at the blessings of democracy where 51 % of vehicles sold are monstrous SUVs, huge gas sucking vans and pickup trucks whose tail gates I can just about reach over and generally not used for business except to take home a dead deer. Insulting to mine and your intelligence is that they are also free from meeting any fuel consumption figures and anti-social as it is, don't even have to meet any emission requirements just because a bunch of lawyers in Washington don't know beans about internal combustion engines and don't realize the stupendous quantity of pollution, mostly carbon dioxide plus more nitrogen-oxides, as you shall read about in later pages, being pumped into our atmosphere by the cubic miles daily. But, decided from casual persuasion by Detroit's Big Three, the economy would benefit and therefore only you and I in reasonably sized cars should be subjected to emission test laws, in our case here, $47 worth. Just think, as it stands now although California, smothered as it usually is, today's trend reminiscent of the insanity of the 60's and 70's, huge v8 engines suffocating Los Angeles, hot rods all over the place and then a rising death toll. California's legislators are beginning to show some restrained hostility toward what they know from experience means a return of the bad old days. In recent articles, even though the contents of which revolved around EPA's bumping into uncooperative industrial polluters, some mention was made to the effect there would now be an effort to close loopholes exempting SUV owners from more stringent emission requirements such as everyone else is subjected to in most states. The gimmick exploited by what I brand as collusion between Detroit and the government is to classify SUVs as trucks, for two reasons. One, they are built around truck frames which is quite understandable or otherwise the sheer weight of huge engines, buildboard dimensions made of steel would run the risk of the whole vehicle bending in half. Two, they are supposed to function as towing vehicles. While I resent the potential damage to our environment from what spews from their oversized tailpipes, I really can't blame all the owners.   I doubt very much, in view of little information ever disseminated that would enlighten the public as to who is doing what in the way of abusing the environment. The probability of placing senators and congressmen in the awkward position of intentionally or most likely unintentionally being suspected of poor judgement is in itself a good reason why I choose to write about all this but without rancor.   Consider the fallacious information that was given to me over the phone by our state's Department of Environmental Management when back in 1993 and 1994 they were pushing for auto emission testing. Because several state legislators felt this would prove costly for meager wage earners, an attribute of Rhode Island going back a century or more, a battle of the minds ensued. To gain more public acceptance of what awaited them DEM informed all the naïve, who knew better, that the cost would average about $145 to replace this or that car component after failing a test. In talking to DEM, they showed either ignorance of things automotive or determination to ease the pain and I was told, for example, catalytic converters last the lifetime of a car. That is not true at all! The main reason I am passing on this trivia is the fact that eventually million of aging SUVs, pickup trucks and vans will have tired out converters and consequently what will we all be subjected to? How about a nice additional dosage of nitrogen-oxide, carbon monoxide and unburnt hydrocarbons? The sweetheart affair between the government and Detroit's Big Three, having embraced each other some years ago will sooner or later prove to be a mess, particularly for city and suburban dwellers. Ask the American Lung Association. While I can't hope to form a coalition of intelligent conscientious citizens willing to utter protest now and then, nor am I an activist, perhaps you will not hesitate to give your neighbor a signal of disgust. For example, gently swing open hands in front of your breathing orifices when some gentle breeze wafts fumes your way from some glorified tanker warming up. Remember, as best as I've read, this group of small Sherman tanks, their width expanding year by year and on a narrow street, be careful their outside mirrors don't swap with yours of cleaves the roof of your car and at night, their headlights some ten feet behind you forces you to flick down the night side of your rearview mirror, they, believe it or not, do not have to worry about meeting any emission minimums. I am compelled to review past sorry dumb thinking after visiting Los Angeles and staying there five months, akin to living inside a giant muffler, back in 1972, soot covering my car and frequently visibility varied from two miles to one hundred yards. Yes, I will tell you more because right now, advertising has resumed it's juvenile but clever psychological inducement to outperform the other guy, this very fact having resulted in a soaring death rate and pollution reaching astonishing density. As a footnote to this segment, which we can call the complaint department, Los Angeles did not issue pollution indexes as our newspapers do but the radio there called it the Eye Irritation Index. We are, believe it or not, again being nefariously encouraged to express superiority with acceleration figures, 0 to 60 mph being a privilege you are in need of, or otherwise, you can't be number one. They now have a young lady, sitting there with her jeaned legs spread apart, who after lauding the attributes of some shiny behemoth vehicle behind her, who then, to make a point, you inferior one you, you former insignificant office girl you, while moving her thumb toward her chest back and forth says, "With this baby, I'm in <u>control</u>!". Then of course, time for the men, specially the earring guys and the baseball cap on backwards; same psychological conditioning. We now must accelerate a two to two and one half-ton SUV to 100 mph in so many seconds – an imperative for the feeble-minded seeking attention. Jousting, anyone? The Knights of the Round Table are coming back. Its effects? I asked an insurance agent what was the ratio of accidents between male and female, telling him that some years ago men were mostly guilty of reckless driving. He said it is now 50-50. But, back to pollution – I'm not wandering as much as you may think, each time you briskly accelerate and I don't mean tire squealing demonstrations, the EGR valve, that device that stops combustion from producing nitrogen-oxides, is cut out. You are now spewing out quite a lot of Nox because no catalytic converter* can digest such explosive velocities of toxics. For those of you who I occasionally see on the highway, the hood up of your stopped auto, looking a little puzzled at all the plumbing, in other words those of you who are not, as Southerners call them, shady tree mechanics, the EGR valve is shut off when maximum engine power is called for. Ladies, if you don't want to replace air bags consider the fact that the heavier the vehicle the harder it is to stop. I am all too frequently surprised at displayed abandon.

I had, some pages back, but carried away with all this extraneous stuff about the growing pollution problem, intended to give you an in depth interpretation of how pollution, humid air loaded with excess moisture, creates deadly flashfloods under certain weather patterns. I gave it more thought and feel you would find it more interesting if several illustrations were included later with more appropriate related groups of atmospheric phenomena. At this time then, let us amuse ourselves by playing the psychologists and try to see through or read between the lines as to what this letter from a senator implies. Bear in mind however, I'm not inclined to play around themes bordering on what is expository, except where propaganda agents are involved. I'll let the Readers
Digest do that. Somewhere in the following pages you will find their not too flattering report on the largest governmental department. What I want to say, the senator is a nice sincere person, but he may be reflecting the tentativeness of his peers in the way in which he is not as clear as I would like him to be. Some three months ago I had sent him pages 7 to 11, which I don't think were an undiplomatic report based on reality. I wish I could have been treated to some hint he was aware of this or that aberrant weather, clearly evident these past years and that they, in Washington circles, were at least considering an all-out effort to enlighten the public, a wise approach that I suggested to another senator and Al Gore rather than Americans fighting instead of rationalizing as a community and nervous members of congress and Senate standing back while the EPA, trying hard but floundering here and there as you shall see. The public now, their support needed, is in a perplexing frame of mind, the spectrum of which ranges from downright hostility toward the government fanned by the likes of Limbaugh and several institution pundits tearing at Al Gore's integrity as a rational thinker, to pro-environmental and anti-factions and of course most of our citizenry quite oblivious and indifferent. Unless our newspaper has missed something I don't even think Governor Bush of Texas has hinted at some connection between hostile weather and increasing oil consumption.
See Senator's letter on next page

United States Senate

WASHINGTON, DC 20510-3903

December 22, 1998

Mr. Andre Laferriere
30 Washington Street Apt 514
Central Falls, Rhode Island 02863-2853

Dear Mr. Laferriere:

Thank you for contacting me regarding global climate change. I appreciate learning of your interest in this issue.

Over the past two years, an increasing number of our nation's leading scientists and economists have concluded that there is a discernable human influence on our global climate. For example, more than 2,400 prominent scientists and more than 2,500 economists, including eight Nobel Laureates, have stated that action is warranted to address the environmental, economic, and social risks of continuing accumulation of greenhouse gases in our atmosphere.

While estimates vary, many studies suggest that if we stay on our current course, average global temperatures will rise two to six degrees during the next century with potentially serious consequences. With this prospect in mind, I believe that our nation can and should play a leading role in addressing the threat of global warming. I recognize the importance of engaging developing nations in this effort to ensure that our work to combat greenhouse gases is not undercut by increased emissions in other regions of the world.

As you may know, last year's negotiations in Kyoto, Japan produced an international agreement on climate change, which the United States agreed to sign at recent climate change negotiations in Buenos Aires, Argentina. However, this treaty is still subject to approval by the Senate for ratification. I believe that the Kyoto agreement represents an important step toward a comprehensive response to global warming, and I am optimistic that our nation's continuing efforts will lead to additional progress in this area. Please be assured that I will keep your concerns in mind as the Senate considers this issue in the future.

Again, thank you for contacting me, and please do not hesitate to call or write me if you have further questions or comments on this or any other matter.

Sincerely,

United States Senator

No matter what you read, the word "controversy" is sure to be inserted, such is the quiet bedlam caused by dissenting scientists, phony researchers with no names injecting into many scientific reports their favorite expression "We are skeptical of - - -", along with the methodical attack on rational thinking by writers suddenly in possession of meteorological wisdom, plus some nonsense shown and narrated on TV shows. Above, at least one honest oilman, glory be. And then, we also have friends in the media. It's encouraging to say the least. After I nit-pick a little over the senator's letter I am delighted as to how one of my favorite radio talk show hosts, Jim Bohanan bid adieu in a not too appreciated way to one of those eloquent and highly degreed experts who, along with another associate attached to the same institution, seemingly against not only Al Gore but obsessively playing up the role of El Nino's and therefore against advocates of earth warming. Jim Bo's guest was going along fine until he made one statement that set back a notch his apparent superior wisdom. For later, this was an encouraging bit however. As for the senator the first paragraph is reasonable but "discernable human influence" is rather vague and not too convincing. Furthermore, why do we need 2400 scientists? While some have good reason to pursue what kind of acidic emissions could be damaging the ozone layer and other functions over my head I believe we could get along quite well with 500 scientists. Just what are they looking for that is already known? I don't take pleasure in agreeing with the Readers Digest report to be touched upon in later pages denouncing NOAA, a function of the Department of Commerce, as a boondoggle with a budget of 4 billion dollars, since they fail to mention lives have been saved and so on, but it is true, so many scientists and for what? I am not malicious if I tell you the experiments made by young weathermen, running around like fools, poking at the innards of tornadoes with one after the other failing experiments, real foolish nonsense that amazed me. And I will tell you, soberly and analytically because I know differently and you will also know differently or otherwise I have failed. In the second paragraph of the senator's letter, again, much to my annoyance, the dubious factor. "Many studies suggest _ _ " now that's something concrete!   And of course the nebulous part of future, not present as it really is, temperature rises of 2 to 6 degrees during the next century and without mentioning it, the favorite song over rising ocean levels which I will try to show is not likely, again, as is hinted at threatening us in the distant future. He says, "Our effort to combat greenhouse gases" and yet not a peep over the outlandish growth in huge vehicles being sold. A dilemma for me now – what to do about this transgression against nature without disturbing the economy. I'm not a fool. I told the Vice President in humble terms, nobody is worried about rising ocean levels. That it was a tepid argument. I made suggestions as to the need to repeat the effort we made in the late 70's and early 80's to cut down on oil consumption. The other night, listening to my usual radio shows, I was surprised to hear a commercial by EPA suggesting ways to cut down on energy uses in the household. There was humor added however, the EPA doesn't want to be bossy and so some mentor kept telling a resident to shut off your TV set when not in use to which the party receiving this advice replied, "Yes master". You see dear reader, EPA knows there are a lot of uptight people out there but it is the beginning of something that could prove effective although I'm afraid I am not too optimistic. What of the rest of the world? I listen to BBC at 10 pm so as to cover news all over the world – just ten minutes of my time, that's all it takes. But this one night I heard two Englishmen describe what they saw in a wide area surrounding Hong Kong and hundreds of miles north and whatever directions the towns they rattled off represented. It is amazing. Two immense international airports are being constructed, foreign investments being solicited, highways under construction, factories popping up and so on, but, they both ended their half-hour presentation on a sour note. Millions of Chinese are determined to own automobiles, bad for the environmentalists but not for auto manufacturers.   If then, according to the senator's 2nd paragraph, my having already read from a newspaper report that some senators are skeptical over earth warming, should decide as they might and I fear that, delay ratification of the Kyoto Treaty simply because China and India, for example, have yet to show they are ready to cooperate and worse, using this as an excuse to delay action here in the USA. We have one big problem. Years will go by, rest assured but nature doesn't care about our timetables. While I don't wish it upon Far Eastern nations they may have to learn the hard way. As for scientists, the good

> John Browne, group chief executive of the British oil company, said that in tackling the controversial issue of global warming, about which the oil industry has long expressed skepticism, it is time "for change and for a rethinking of corporate responsibility."

are being questioned by the bad – procrastination prevails. Best you keep better informed and be in the position to be more effective when the time comes to write or call your representatives.

Jim Bohanan, whom I often listen to, reacted to a statement made by a guest, previously mentioned which compares to a father, whose son walks in through the front door and announces the family car has just run into a tree. "Oh no" then silence. The invited one insisted that all environmental issues should be decided by the private sector. That did it. After the silence, Mr. Bohanan quite firmly, not exactly diplomatic, added, "No business for profit has that right!" Bang. It was over, the usual "Thank you" and no other words were spoken. I was almost embarrassed believe it or not, since the guest was interesting and quite a talker. Well, we have friends, that's all I can say.

The enemy is busy. Immersed in the conglomeration of scientific excursions into imagined mysteries of our atmosphere, a few reports nevertheless making sense. There are the proponents of El Nino Forever, who prefix every destructive phenomenon with this Christ Child name. Their premises are false all too often. An example, in late 1982 and early 1983, this country underwent an expansive onslaught by our visitor from the Southern Hemisphere. It was claimed by one persistent spokesman from an elite institution that the effects of El Nino lasts for years, bathing our nation in warmth, etc... There is no evidence of such lengthy lingering effects. I don't want to drag into this paragraph the multitude of recorded temperatures I have kept track of but other than two or three years (1982-1983) unusual temperature rises the mid 80's produced just the opposite of deranged weather – tornado frequencies abated and so on.

Spicing up elemental disturbances with El Nino's handle has drama! Droughts invariably caused by intractable high-pressure systems girdling the entire globe during the summer, unmoving, unchanging, as was the case in 1998 and previous two years. In fact, our south took a beating as defined by our former president Jimmy Carter, who faced bankruptcy from failing peanut crops. The ridges of these high-pressure systems being merely elongated extensions of the same squatting over the South for some seven weeks from Texas all the way to the Cape Verdean Islands off the coast of Africa, Florida drying up and plagued by forest fires- now, according to twist artists, need a more titillating identity. Spin doctors, paid to obstruct clear thinking, determined to nullify the mass acceptance of earth warming by legislative bodies all the way down to every American citizen, brand cloudless, rainless conditions, characteristic of a feature of any and all high systems, being a downward movement of air, with an inaccurate name for dry spells. It has more impact to say El Nino droughts and El Nino tornadoes. I.E.: Steady west winds from the Pacific, dragging warm surface waters toward the West Coast may create confusion (see page 103). I do not need any further confirmation from esteemed newspaper sources to convince me some members of both houses have let it be known they are still skeptical over earth warming. The reasons above explain why. And too, lobbyists are quite busy. El Nino is now being proclaimed again!!

I need to digress a while. I have to freshen up. Change the subject; that's it.

If you subscribe to a thorough newspaper there should be a long list of cities in the USA including Honolulu, San Juan and probably European sites showing yesterday's temperatures and expectations for tomorrow. Maximums and minimums are displayed and from that you can extract the means. While it may not interest you, perhaps you have a child who shows an inclination to draw, fabricate puzzles or whatever and who might jump at the opportunity of becoming an accomplished weatherperson at the ripe old age of, well it depends, seven, nine or ten. Buy some graph paper. It is only $1.00. On the following page look at that beautiful spike down to 37 degrees below zero in Fairbanks, Alaska, December 17, 1998. Six weeks have passed since I typed the above Arctic surges. I have come back to this page since a low of –55 degrees was reached February 5, 1999. On December 24,1998 that mass of frigid air from Alaska rolled in and we had a white Christmas. Now, that's not all that fascinating, but you must also realize that your child is now an accomplished scientist and could prove useful in Washington, DC, or perhaps help all those scientists employed at NOAA. On a more serious vein, I spend perhaps five minutes a day plotting different sites, but just during the winter as it is quite boring the rest of the year. I just wrote that this drop in temperature was beautiful simply because it was an announcement of nature that we were destined for a normal cold winter; albeit persistent unwinter like warm temperatures occurred in San Juan and Miami, two key indicators which can cause rather strange weather along the east coast in terms of wet snow, freezing rain, etc… I will explain how a purged atmosphere by worldwide downpours affects more northern portions of the earth and somewhat less in the Caribbean, for example, but later please. Right now, January 17, the Weather Channel meteorologists are warning Arkansas and Northern Louisiana of potential tornadoes. They concur that this is the result of an unusual warmth in the southeast quadrant of our country, but I know as I will explain later, that they do not understand how earth warming modifies the south Atlantic's mass of contained solar energy by sustaining it's voluminous power due to a combination of extra ultraviolet radiation since three decades and some greenhouse effect, whereas cleaner air in Arctic regions combined with no sun allows rapid cooling of surface air up to how many thousands of feet of altitude I am not certain of. For now, and I'm not sure how long it will last, icy air spills down over Alaska and Canada and the conflict further south is quite impressive in terms of wild storm formations.

Because of what it may suggest I had intentions of pecking away at another aspect of the senator's letter in his professing to having the services of 2500 economists, no less, but daily developments in the news compel me to comment, sparing no details inasmuch as tornadoes in mid-January is about as freakish as weather can get! Last night, January 19th, I also heard the president's State of the Nation

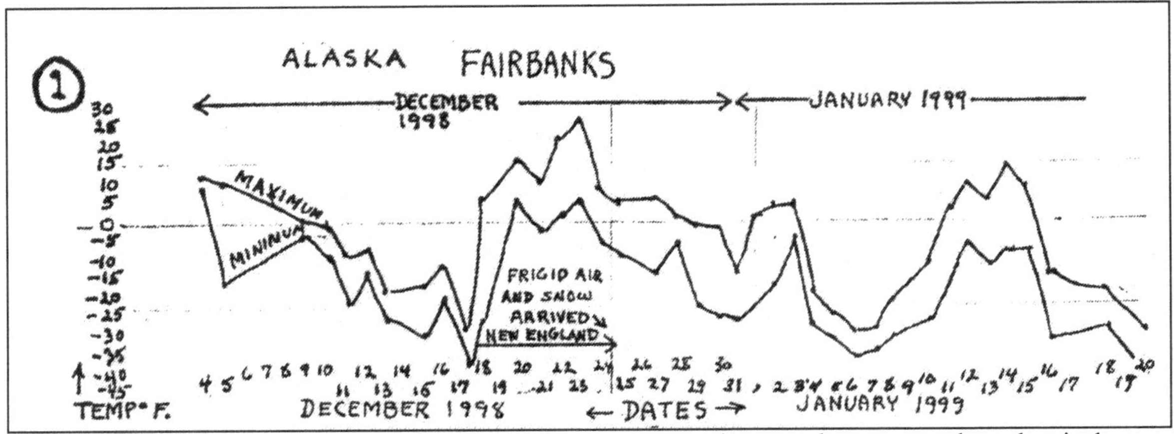

address and while it promises several decades of joyous prosperity, one has to wonder what in heaven's name does a small army or economists portend? Typing perhaps two hours a day, I stay in contact with all media, plot temperature trends of strategic sites, i.e., Oslo, Fairbanks, Alaska, Honolulu, Miami and San Juan. This offers a picture of pending arctic invasions against massive bodies of solar energy always on hand to resist or figuratively speaking, two teams at opposite ends. After each storm the cold invader will weaken and be absorbed by it's opposite at lower latitudes. Heat is power – ice is not! I stop for a moment. You will never hear such a viewpoint on one of the most crucial factors of exactly the position of power and influence represented by the rather innocuous appearance on weather maps of a very large belt of high pressure, illusionary appearance, lacking the more impressive isobar lines of high pressure systems wending their way at more northern latitudes, that girdles the entire globe on both sides of the equator. In short, I am viewing this from a very different perspective. We are talking, you and I, but never heralded by meteorologists, of an enormous storage of solar energy that is responsible not only for summers good and bad weather, but having been excessively maintained in it's heat content by ultra-violet radiation and some greenhouse effect, retains its girths and influences far too long into the winter season. In days of old, and that is where memory serves experience best, this varying belt of languid appearing semi-tropical mass usually identified at its northern borders as doldrums, the Sargasso Sea, and of course the soothing balmy breezes of the Caribbean Sea, this powerhouse as I am introducing it as, always shrank in its dimension-wise beginning in October or November and rebuilt its volume again in the Spring. On January 17, 1999 the town of Jackson, Tennessee was struck by tornadoes and they were fueled by a very active and very much alive in influence by the aforementioned powerhouse of solar energy. That is as aberrant as weather can get. That is so unusual I just had to digress from commenting on the senator's letter. Eight people died and I'm sure you have already read all the details of destruction, Al Gore awed by the power after visiting his home territory. What is my point? I want to criticize weathermen, although they did their job of warning all viewers of the Weather Channel, for years and years of restrictive thinking. They are unable to break away from traditional concepts taught to them, <u>as I have,</u> and therefore they have limited their verbal presentations as if they were insecure and dared not offer something more substantial and more definitive. No, I'm not arrogant, in fact, I do believe from hearing a few of the older and perhaps even frustrated meteorologists state they were open to suggestions. Here is the standard fare you will see and hear on TV: "Warm very moist air flows northward before the advancing cold front over the Midwest, (probably western Arkansas) and we are issuing warnings of potential tornadoes for Arkansas and northern Louisiana today, Sunday, January 17, 1999." They show this warm south to northeast flow, very prominent at higher levels, clouds barreling along viewed by my favorite modern revelation, the satellite image, in action.

Fastidiously, I would prefer to say, "The unusual presence, in terms of volume and heat content with a high humidity factor of the semi-tropical high pressure system over the south Atlantic, abnormal temperatures in the low 80's prominent in San Juan and Miami for this time of the year, is funneling a steady and rather vast stream of warm moist air up the Atlantic Seaboard states, etc." And explain why! Why is it unusual! Did Al Gore, the environmentalist, now afraid to repeat his chastisement of Congress, accusing them of having their heads in the sand simply because he may be running for President in two years and knows full well this dumb-downed public doesn't want to hear it. They want to do as they damn please and if Governor Bush gets involved in any debate he will knock Mr. Gore by emphasizing the man wants to dispossess us of our cherished ways of mobility. This country needs unanimity, now. President Clinton has again repeated "We must do something about the environment!" Like when? Ten years from now? Maybe 15? No, Al Gore, dismayed of course, his mother having been born in Jackson, Tennessee, uttered in hush tones his sympathy to relatives of the dead.   I don't believe he is going out on a limb now and cry out as he has in the past the reason for such abnormal weather. While cleaner air in northern latitudes, because that is where all surface air comes from, let us say the south Pacific around the Philippines, bordering China, India and other polluters, after torrential downpours have cleansed the air to a degree, ends up. It wends its way from roughly a southwest to northeast flow at high levels with the ultimate destination being the North Pole regions, hence the Arctic for a while has cleaner air and consequently, sunless days and lessened greenhouse effects allows a normal rapid accumulation of frigid air. A process arranged by God and/or nature wherein descending air settles over icy regions, builds up in volume and periodically spills southward during winter. See previous graph of Fairbanks for these surges. But, you have the opposite effect in the South and here we must resort to facts and apply some logical reasoning. There is no question at all about Ultra-violet radiation from the sun increasing some 5% in the Northern Hemisphere and 8% in the southern half

As a result, UV radiation on the surface of the earth would at least double in the Northern Hemisphere and quadruple in the South, compared with current increases of 5 percent in the North and 8 percent in the South.

"Furthermore, all these impacts would have continued to grow in the years beyond 2050," the report said.

1997

Greenhouse gases up sharply last year

Emissions of greenhouse gases from cars, factories and power plants in the United States rose sharply last year, according to a new Energy Department analysis that aides said further complicates administration efforts to devise a strategy for dealing with pollutants blamed for global warming. The 3.4 percent increase found in heat-trapping gases could put pressure on the Clinton administration to propose even more stringent steps to limit emissions. But many presidential advisers are concerned that tough action to limit pollutants could have serious adverse economic consequences, and they might use the Energy Department study to bolster their arguments for setting less ambitious goals for reducing emissions of greenhouse gases.

of our planet. No question at all! The American Medical Association reports a four-fold increase in skin cancer in the past decade. Australian authorities, their country bathing in the semi-tropics south of the equator, registered fear of rising cancer. Today, meteorologists generally warn beach goers to limit their sunbathing to a half hour while in my youth you could soak up sunshine for one hour and a half, unless you were very light complected and not wore tons of sticky goo as is being used on all beaches nowadays. Enough said. Jean-Pierre Mouligne, a Newport Rhode Island resident who is a prominent yachtsman involved in those crazy around the world races, always reporting in a very professional way about the weather and the seas once said, "I am proceeding slowly through a very spongy high pressure system". The semi-tropical high systems on both sides of the equator have languid characteristics, primarily it's voluminous content of heated air, and of course, the seas having absorbed a little more heat despite the cooling and neutralizing effect of evaporation.*

EXCERPT SERVES TO SHOW ESTABLISHED HIGHER UV RADIATION!

In this smaller excerpt, part of a much longer article outlining the
predictions being made by the World Meteorological Society that the ozone layer should be healing after 1999, but it is not expected to come back to it's finer state of healthy thickness of the 80's for several decades. They feel that having stopped the use of Freon and aerosol sprays has brought about what

All warm air returns to the poles.

they feel are signs of improvement. But, in other articles that I have read, there is also the belief that sulphur-dioxides from both industry, vehicular and even volcanoes may have worked its way through the atmosphere's lid (we call it the tropopause) by some penetrative chemical process. All this alchemy business is best left to others. And as you probably well know, the ozone layer as alarmingly reported some years ago by scientists interpreting Nimbus-7's monitoring instrumentation as not only had it's hole over the Antarctic or wherever greatly enlarged but worse, the ozone layer had thinned out all over the world. Insofar as the excerpt of UV radiation they were principally estimating what would have happened had they not stopped the use of freon and so on. Insofar as carbon-dioxide is effectively shown to aggravate the consequences associated with retention or preventing solar heat absorbed by earth to return to the sun, but I'll leave how that is done to the Einstein's, the process of cooling mainly at night, one has to be very aware the content of carbon-dioxide in our atmosphere is a very variable factor! Insofar as how much of it lies over the south Atlantic and both low and high levels I can never be sure. From week to week, Nimbus-7 probably senses variable amounts of carbon dioxide, more or less at mid-latitudes where high level winds may fluctuate between 30 to over 200 mph particularly during the winter. It should be also pointed out our production of carbon dioxide over America would be substantially greater than the 3.4% rise established as a new world average. That figure is a computation extracted from worldwide scanning along with numerous weather rockets periodically launched. In short, the Greenhouse Effect has to have differences in concentrations of carbon dioxide since atmospheric circulation at different levels over America for example, transports this gas from west to east. Even if we know what each city produces, the transport system of pollutants and gases at near ground levels varies from week to week and depends on weather systems. High pressure moving across Los Angeles, for example, would keep carbons dioxide concentrates at low levels, moving its content along with it spreading to Arizona, Nevada and so on. Denver's products would go to Kansas and so on, particularly in winter when green plant life is at rest. On the other hand, storm centers are but updraft entities and we have both the cleaning action of precipitation plus the benefits of carbon dioxide sent heavenward to be caught up in higher velocity air currents whose main function is to return all air back to the north pole. As for the high emission potential of the East Coast, considering also an enormous production of carbon dioxide during cold seasons by heating systems, the same rules for weather systems apply of course. But the ocean has the ability to absorb much of this ever increasing amount of carbon dioxide and unfortunately some particulate pollutants. There is enough to keep any typewriter extremely busy over atmospheric vagaries going on in our Northern Hemisphere. I'll let South Africans and Australians worry about their half of the planet even though the latter gets caught in the middle of ours and theirs when the equatorial front (ICTZ) wanders from north of the equator to south of it during our winter season. In all likelihood, those 15 yachtsmen who perished December 1998 after sailing from Australia caught in an all engulfing storm, warned no less, tossed racing vessels about was probably due to the ICTZ's usual move over northern parts of Australia and adjacent seas. The characteristics of our Northern Hemisphere's air at ocean surface levels are not much different than that of the Southern Hemisphere's equatorial zone or in this case, Australia's surrounding waters. But in arriving from our colder half of the planet in December of 1998 while on the subject of another tragedy resulting from mortals having no fear and considering the human race's bland outlook on meteorological functions, that northern Pacific invasive air had an entirely different temperature value at higher levels than that of the South Pacific which after all was undergoing the beginning of Summer. Quite a contrast at higher levels above the surface, but not much difference as measured at sea levels. Granted, this may be all very boring, but my experience with the elements could be quite a contrasting one from most folks. As a starter at age 15 I witnessed the 1938 hurricane of New England and five miles away over 400 human beings drowned. In December of 1942 the convoy of ships that brought us to Morocco ran into a storm two days out of New York City and 60 foot swells makes quite a permanent impression as to how powerful nature is. In December of 1942, 15 happy soldiers who hap participated in a brief encounter from Vichy-led (compulsory or get shot) resistance at Port Lyautey, Morocco, now named Kenitra, were allowed to go swimming in an extremely beautiful endless beach such as the like one would never find anywhere else on this earth. They drowned, pulled out by a powerful undertow. As for the Australian yachtsmen who drowned despite warnings of a storm 24 hours earlier, it served tragically to instill into the minds of those sailors who survived, ages 30 to 60, a different attitude. Quite a few decided that they would never participate in any future races again, so traumatic was their experience with the Almighty. I am gearing up for this episode of tornadoes that began this month on January the 17th around Arkansas and Memphis. While I know who is partially responsible for an indifferent public in America, I'll not be too forthright with my accusations.

If I am uncertain of carbon dioxide content over the south Atlantic, that zone's placid appearance on weather maps but what I know to be an atmospheric area vastly more powerful than the dynamic appearing surges of polar air rolling across America. Since in its circumferal year round permanence extending around the globe, I see something entirely different. Uncertain I am over its trapped carbon dioxide remaining, I am positive of the thermodynamics involved within its enormous spread latitude wise north and south of the equator and most important, it's vertical voluminous content some 11 or 12 miles high. Let me emphasize a fact and I hope some intelligent weatherman who has been wondering as I once did is reading this. The constant stream of ultra-violet corpuscles and whatever other radiation particles I have to leave to physicists, from the sun absorbed by tropical seas is rapidly returned from its receptive watery medium and transferred to overlying constantly moving air through a process we all know as evaporation. Hence, in the past 50 or 100 years (who cares) the ocean temperatures have only shown a one to one half rise in temperatures. But, bear with me, you know all about evaporation and high humidity levels in the Caribbean, the Sargasso Sea, Gulf of Mexico and on the Indian Ocean.   But has anybody explained what happens to that extra <u>five</u> percent of solar radiation that we know to be a reality for the past 30 years or more? Could I wax poetic and explain this my way? There is a lot of molecular excitation going on and therefore our semi-tropical air has expanded its dimensions and unlike previous decades of substantially larger volume of tropical air remains beyond its former scheduled autumn dimensional reduction. Or better put, a progressive contraction that once upon a time began shortly after the hurricane season. I greatly respect an elderly gentleman, who is shown a well earned deference by meteorologists, younger than he, as part of an interesting and important show revolving around an approaching hurricane, taking place I suppose at the Hurricane Center in Miami. This profession, after academic indoctrination, is purely one of piling on memories of similar weather patterns and that is it. While the younger but very bright weathermen sit by their ancient counterpart may have seen five or six hurricanes on maps, plane spotting and all that, they know that their much older professional has a brain quite busy reviewing the antics of some 20 or 30 hurricanes that he has witnessed for some 50 years or more. I hope he goes on for many more years. Poetic again: Each invisible molecule of water vapor has affixed to it a portion of solar heat. Focusing on the rather spectacular and deadly five day streak of tornadoes, at a time of the year where no such phenomena should happen, I'm very convinced a lot of thinking involved for some five years or more, there has not been any recognition of how potent, how dominant, how nearly intractably resistant a factor warm air masses containing a thousand or two thousand cubic miles, the supply of which is relentless and continuous in its content of latent heat attached to every molecule of water vapor, really are! Instead, we have been conditioned by 80 years of false premises.

Below, illustration 2. What normalcy looks like. Illustration 3. A view (satellite rendition November 21, 1998) or abnormalcy – why Washington and Oregon sustained a beating lasting three months as a consequence. The TV weathermen this 21st day of November reacted as the surprised and inexperienced would. Gesturing with both hands he repeated, "It's unbelievable!" "We've never seen such a huge high pressure system!" If most Americans far removed from the northwest coast remained indifferent, possibly yawning, residents of Washington and Oregon must have been uneasy. Surely, sales of boots and umbrellas skyrocketed. And batteries of course –

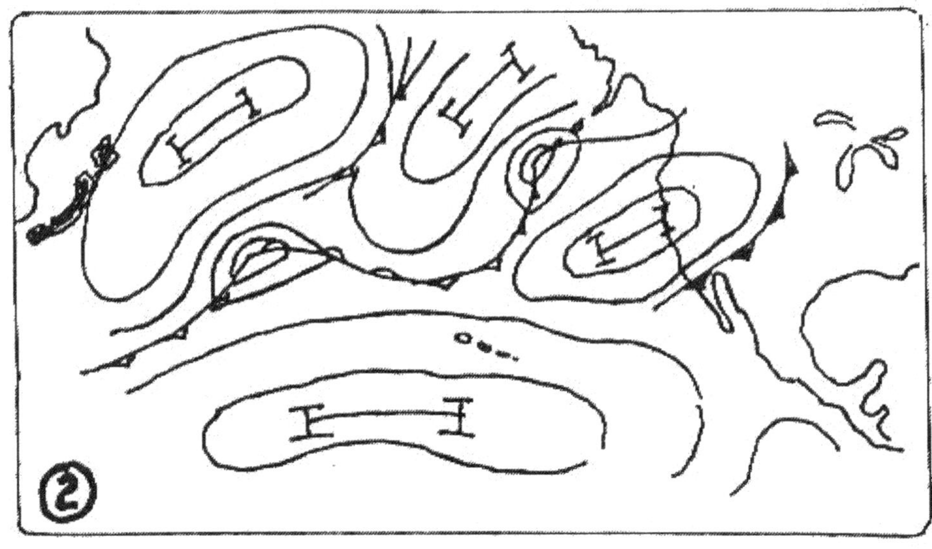

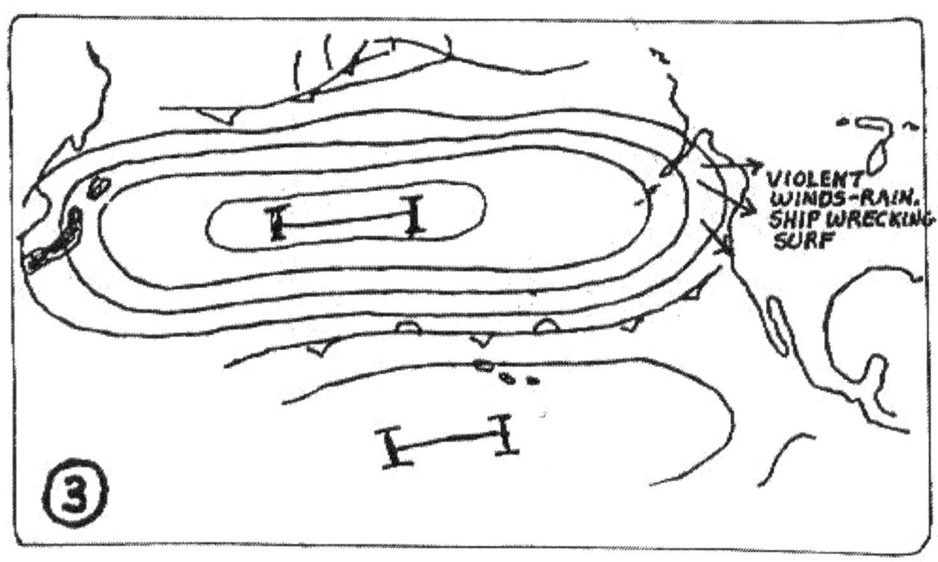

down would come trees and the inevitable loss of electricity. In my search for factual disclosures, determined to provide a definitive survey, I follow weather developments day by day. For northwest coast dwellers our emissions can be left out in the explanation. China, India and others are to blame (reasons based on physics already shared with you and more to come) but the Kyoto Treaty must wait. Trade is more important than arguing over sources of pollution. Additionally, we are not setting any good examples, EPA having been slapped down by court orders (you know all that) and Tom Donohue, President of the U.S. Chamber of Commerce announcing with some obvious satisfaction, "It's a victory for all businesses and a blow to EPA." (Published week of October 25th). I understand all this but Nature will have the last say so.

Other than an occasional ineffective squawk symptoms of a deranged climate go on skipping right over the heads of our citizenry. January 24, 1999. Beautiful warm weather across our southern tiers from Texas to Florida. We are shown scenes of couples strolling in summer attire, scantily clad joggers and golfers waiting their turn to tee off, courses busier than usual. Again, a young weatherman on TV (I have nothing against the young) comments, "It's a beautiful balmy day for late January and everyone is out enjoying it." I didn't say I would have predicted what would happen but I certainly would not smile about unusual warmth. It is a very dangerous time of year for unusually warm weather, the Gulf pumping 80 to 85 degree air loaded with tropically acquired moisture over southern states. To the northwest, sometimes delayed and prone to unexpectedly surge forward, a typical winter cold front. Oh, warnings the next day were issued but this day, the 24th, everything was rosy. 30 hours later and 150 miles to the east (Arkansas and a few miles further) there would arrive the ominous line of boiling clouds and tragically black tentacles would reach down snuffing out lives and scattering property into rubble. Blithely we sail on through uncharted waters, uncharted because there is no navigational course set by leadership, no consensus, senators at odds with each other over when and how to cope with recalcitrant industry and wary citizens preoccupied with what they consider more important.

It is now January 25th. The news disrupts my train of thought. A weatherman just said that Tennessee and Arkansas experienced 101 tornadoes, some 18 to 20 people killed. One newspaper account yesterday devoted two short columns over Paula Jones being nearby as President Clinton walked around what was left of a city block. A photo also caught him smiling, but it was very important to describe her avoiding eye contact. James Lee Witt's mouth was open however, quite shocked. Back to January 17th. For five days one after the other segments of cold air masses pushed and shoved over our Southland, petering out, dumped its deteriorating mass due to land warming into the mid and south Atlantic. Then another surge from the Arctic coming down across the USA to make another try at replacing the very dominant warm air mass the likes of which ranged from the semi-tropics all the way to New England. This condition does not produce huge tornadoes that are associated with fast moving cold fronts. Instead, a temporary delayed cold front serves only as a slanted wall upon which a very vigorous and quite endlessly loaded with latent heat column of air, climbs over…you know the rest. The results are a line or lines of major thunderstorms and many spawn smaller but vicious tornadoes. The real muscle in weather systems is obvious but I'm afraid the illusions of power weather maps convey has weathermen hooked on what I consider obsolete. It's not the vibrant looking masses of invading cold air that determine just how fast and just how far south or east they will maintain their integrity. This last word was used to relieve us of the monotony of weather jargon. As for jet stream forecasting, that's loaded with inferences. I have to, for the first time, resort to a hypothetical view: Such a stream or streams reflect what is going on at the surface and furthermore, if you will consider the need to try to understand another fact, to be explained in more detail later, there has to be a balance between pressure at the north pole and pressure and/or excess volume at the equator. Therefore, the jet stream is constantly streaming to adjust rather rapid changes from one high-pressure system to another very far away. It is, how would I say, an effort required mainly in mid latitudes but it's your prerogative out there in those weather offices to stay with your jet stream concept. Personally, I'll have none of it. Let me summarize one aspect of all of the above. If it were not for this cold air meets warm air, then replacing what it has siphoned off through the formation of lows, then stale air if I can call it that, would have no means of

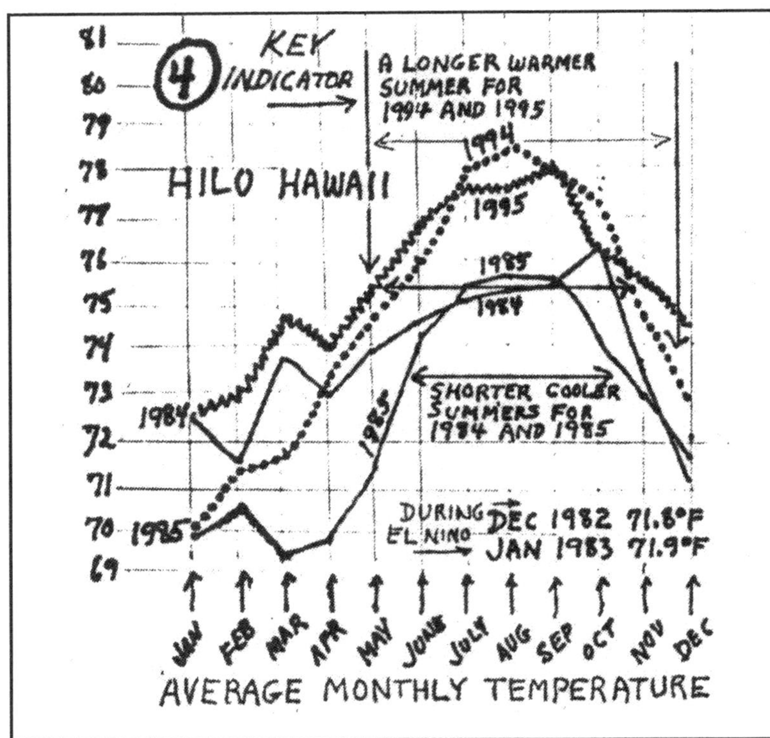

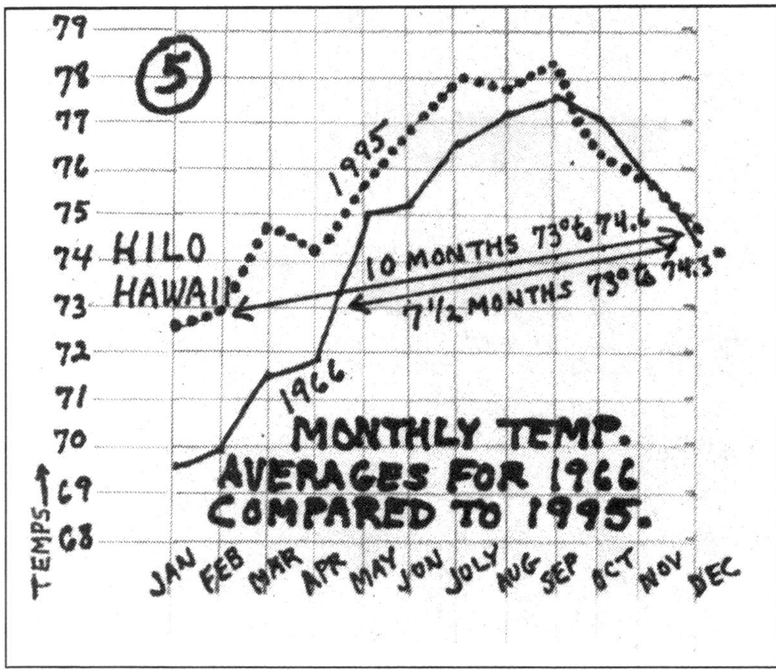

getting back to the refrigerating arrangements of the poles. Some of it would rise up over the equator as it does and wend it's way northward at high levels but the buildup in heated air would not allow normal life to go on.

I am winding my way in these pages very slowly, ponderously if I may say so, wanting to impress in your mind every component of what I know to be a very reactive nature and while titles such as "Nature's Fury" could make one wonder we must understand why we are unintentionally sinning against it. We must understand every elemental facet involved. Before I work my way back to what amounts to a case study of the effects on public thinking from repeated denunciations, however veiled, of expressed concerns made by environmentalists along with Al Gore being savaged, let me show you a bit of the Pacific again. I'm still on the topic of UV Radiation. I like graphs and so do doctors. There you are in a hospital room and there's a chart hanging in front of your bed. Hmmm, temperature variations, the doctor is seeing a record of what your body has been doing. No genius technique do I offer. I am distracted by television, watching the weather. On TV, a weatherman is raving about how nice it is to have warm weather over most of the south. A fount of information, but short on memory, as he is not seeing what is extremely dangerous now. Let us, you and I resort back to sound basic fundamentals and make some sense. Before looking at the graph, I must first redundantly emphasize that oceans maintain constants in their surface temperatures. Boringly, at least for this New Englander, two key geographical points of reference, San Juan and Hilo, Hawaii, continuously bathed in ocean breezes, never changing easterly for the former and westerly for the latter, temperatures day by day stay in a narrow range. I don't mean to be sarcastic, but why did it take 500 international members of a meteorological association five years to determine earth had warmed along with sea surfaces about 11/2 degrees Fahrenheit? While two to three degrees increase I have found in some areas doesn't seem like much it represents extra solar energy on a vast voluminous scale and meeting normal cold air over a

AVERAGE TEMPERATURE (deg. F)						KILO, HAWAII							
YEAR	JAN	FEB	MAR	APR	MAY	JUN	JUL	AUG	SEP	OCT	NOV	DEC	ANNUAL
1966	68.7	69.9	71.8	71.8	73.0	73.1	76.6	77.1	77.4	77.0	75.6	74.3	74.3
1967	71.9	73.9	74.1	74.3	77.1	77.3	77.8	77.1	77.8	77.2	73.4	73.3	75.6
1968	72.9	74.8	74.3	73.8	73.7	76.4	76.8	76.5	76.8	77.1	76.2	73.7	75.3
1969	71.4	72.2	72.5	72.8	74.5	76.2	75.8	76.2	75.0	74.5	73.3	71.5	73.8
1970	72.8	71.1	71.1	72.4	73.7	73.7	74.2	74.9	74.8	75.3	73.9	72.7	73.3
1971	71.3	71.8	69.4	70.9	72.2	75.0	76.6	76.8	76.6	75.4	73.8	70.9	73.4
1972	70.1	70.7	73.8	72.7	73.0	75.3	75.4	76.5	76.4	76.0	73.3	71.4	73.7
1973	72.2	71.8	72.5	72.2	72.9	74.6	75.7	76.3	76.3	75.8	75.4	73.8	74.1
1974	74.5	72.6	73.1	73.3	73.8	75.3	76.1	76.9	77.3	76.9	73.8	72.3	74.6
1975	71.8	71.8	71.2	72.6	73.1	74.4	74.8	75.7	75.5	74.7	73.5	72.2	73.4
1976	71.3	71.2	71.8	72.1	73.8	73.6	74.5	76.2	76.9	76.2	74.5	73.2	73.7
1977	73.9	74.0	75.3	74.2	74.7	76.2	77.8	78.1	77.3	76.9	73.2	73.7	75.6
1978	71.7	72.5	73.2	74.2	76.2	74.5	77.1	76.8	76.2	75.3	74.1	71.1	74.6
1979	68.8	70.6	71.5	73.8	73.8	74.2	74.8	73.6	76.5	76.1	73.0	72.8	73.5
1980	71.6	72.6	72.3	74.5	77.3	77.6	77.8	75.0	75.7	74.8	73.8	74.2	74.8
1981	73.3	72.7	71.6	72.8	74.2	76.0	76.1	76.1	76.2	74.6	73.9	72.0	74.1
1982	71.8	71.8	70.3	71.2	72.9	76.3	76.7	76.9	76.1	74.9	74.6	71.8	73.8
1983	71.4	71.9	72.5	71.9	72.6	74.3	74.8	75.2	74.9	74.1	73.8	72.9	73.3
1984	72.4	71.3	73.8	73.0	74.0	74.7	75.2	75.3	73.4	76.5	73.6	71.1	73.9
1985	69.8	70.5	69.4	69.8	71.4	74.4	75.4	75.7	75.7	74.3	73.0	71.6	72.6
1986	71.8	73.6	74.7	73.6	75.4	76.8	77.8	78.5	77.9	76.4	75.1	72.8	75.3
1987	71.8	70.7	71.6	72.2	72.5	75.4	76.7	77.9	77.8	76.8	74.7	73.1	74.3
1988	71.9	72.3	72.2	72.6	74.2	74.7	75.7	76.0	76.6	77.9	76.3	74.9	74.6
1989	72.2	71.8	72.4	73.1	72.7	74.7	75.2	75.8	74.6	73.6	73.4	71.3	73.3
1990	72.1	70.4	71.2	73.3	74.1	75.0	76.0	77.0	77.2	76.2	75.4	72.5	74.2
1991	72.0	73.8	70.8	72.6	74.2	74.8	76.0	76.8	76.9	76.2	73.8	72.8	74.3
1992	71.2	71.4	72.3	72.4	74.8	76.2	76.2	77.2	77.4	77.7	73.2	73.4	74.7
1993	71.8	70.1	71.4	73.5	73.3	75.4	75.8	77.0	77.1	76.0	73.4	71.7	73.8
1994	70.8	71.3	71.7	73.4	74.8	76.0	78.1	78.6	78.1	77.4	74.9	73.0	74.8
1995	72.6	72.9	74.8	74.1	73.5	76.9	77.9	77.3	78.2	76.3	75.6	74.6	75.6
Record Mean	71.9	71.2	71.5	72.3	73.4	74.9	75.6	76.1	76.0	75.4	73.8	72.0	73.6
Max	79.3	79.2	79.0	79.4	80.7	82.3	82.6	83.2	83.5	82.9	80.9	79.4	81.0
Min	63.3	63.2	64.0	65.2	66.2	67.4	68.5	68.9	68.4	67.8	66.6	64.6	66.2

wide range of our southern section of land. The abnormal UV radiation dating back several decades serving to add more heat to the already humidity charged air breezing northward. Over our Southland it is heated more. Against cold fronts moving down from the north or from the west, the slanted wall as they are, forces air upwards. That's too elementary for me to use but I am going all out to emphasize while painting a picture in your mind of rocketing air, propelled upwards by virtue of each molecule of moisture, coalescensing with others, releasing it's heat it carried all the way from the tropics to the ascending air. This extra contribution of heat accelerates vertical currents, ice crystals form, hurled outward in all directions and further increase a temperature differential conducive to violent updrafts. Given such destruction, potential existed Sunday, January 17, first manifested by looming super thunderstorms, it appears some folks had adopted a casual attitude, quite enthralled over taking pictures of lightning, according to a reporter. Leaving that aside for now, let me depart from convention strapped premises. Let me play on your innate logic as I wrote in the beginning and de-mystify what is not at all a mystery. In other words, we can put an end to eternal conflicting commentaries and in our own minds ignore partisan-promoted controversy. In the previous graph it is clear that warmth prevails longer than it use to. If we want to compare a 29-year change the above graph will do. It emphasizes what I know to be the effect of more intense rays rather than the carbon dioxide factor and it's correlation with Greenhouse influences. Both islands I have graphs for represent sunny climes. While winter in the Northern Hemisphere is subject to excess CO2 emitted from continents, a steepened increase of gasoline, diesel and heating fuels consumed, oceans as sinks absorb much of it. However, I am not claiming the Greenhouse Effect can be totally disregarded. But, let's see if you and I stick with a common sense appraisal and note in the above graph, September to December's temperatures are about the same for both 1966 and 1995. Therefore, the differences for the rest of the years have to be radiation influenced. CO2 concentrations in our atmosphere, greater over busy land activity, much less over oceans would have, if pronounced in density over Hawaii the year round, much more of a problem in the 90's than in the 60's, caused 12 months of obvious warming trends and not just for eight months. Please note what is <u>more</u> <u>important</u> <u>than</u> <u>anything</u> <u>else</u>, that warm weather lasts now much longer. Only a 5% increased solar radiation in the past 30 years or whatever can do that. That is why our Southland is being rampaged, scattered, destroyed towns resembling pieces of Hiroshima of World War II days. I have attached a typical NOAA data sheet for your verification or amusement if you are so inclined. Let me again insist a 2-degree increase is not a trivial matter. Apply heat, as solar energy does over the oceans to a body of air having in a sense, flexible, yielding walls, such as our semi-equatorial high has and you expand it's dimensions vertically to 40 or 50 thousand feet, horizontally north, south, east and west and therefore, its volume. Given that this high system around the globe has 18,000 miles going for it, from west to east or vice-versa and north and south, a width of roughly 3000 miles varying to a maximum in summer and less in winter. We are talking of many thousands of cubic miles! If it were not for landmasses, the overall equatorial zone approximates 23,000 miles. If James Lee Witt, accompanying the President inspecting what was left of several towns and exclaimed, "My God, it looks like bombs blew everything up!" that is the understatement of the year. One major hurricane and my newly defined warehouse of the sun's power can pump them out at a rate of six to ten times last year and surely the coming years and each one has the power of several hydrogen bombs don't you think somebody else besides me should get motivated. Should someone else adopt some resolve instead of hiding the truth, stalling, worrying about public reaction and start implanting at least more vigilance, a greater sense of reality? Point out the gravity while not frightening the easily frightened? It does not require momentous deliberation.

Let's go visit San Juan, Puerto Rico. Talk about a monotonous temperature. It has had highs of 82 degrees for the past 30 days. Having had the luxury for a decade of not dwelling over snapshot interpretations of varying localized weather events, that being a prerequisite, an obligation for forecasters and see as I see that our atmospheric world is a small one within which two opposing forces interplay, mainly the refrigerated poles and the heated equator, and you will realize in the ensuing pages that there is nothing really complicated about it. Having reviewed with you Hawaii's rather humdrum temperatures or analogous to a doctor having removed a thermometer from your mouth and states, "Hmmm, you do have a fever". Even though I frankly want to make you leave this book after reading it feeling a smidgen of uneasiness and frankly again, I'm more interested in our economy damaged in the next ten years while foolishly, our leaders sweat over delaying tactics afraid of disturbing a vote-getting rambunctious economy and such delays from my point of view may prove too little, too late. I submit San Juan as another patient to be analyzed in the following pages. While two ocean sites may seem inadequate as reflecting vast areas of ocean or land influences, I do record ten other locations. One, Oslo, Norway for example and of course the wild gyrations of Alaska's temperatures. Both island sites however are fairly accurate indicators of effects of UV radiation and probably accompanying pollution factors since from the standpoint of their moving position to the sun they reflect what I know to be of paramount importance. Having access to a multitude of cities, some 60 in the USA, Hi-Lo temps, expectations for the next day and 30 foreign sites, I have to credit our Providence Journal as being more intelligently put together than some that I have seen. Additionally, much space is devoted to anything environmental.

While Hawaii, being quite distant from China, that country and others pumping enormous quantities of pollution into the skies over the Pacific, then their unwanted products drifting our way, generally escapes the effects because of the west to east flow of air at high levels. Much of carbon dioxide and whatever just misses those islands. But, our West Coast further north is now being affected by dust from China's Gobi Desert. However, and it's temporary, torrential downpours all over the world even New England towns recording one and one half foot of rain in June of 1998 costing two million dollars. The whole world for the past three years has undergone death and destruction from rain alone. How else can nature purge the air and how else can I account for a definite renewal of cold weather, a product of an unrestrained cooling process over the Arctic. But, it is not so clear sailing for semi-tropical San Juan. We, as a very busy nation and thanks to vastly increasing sales of huge CO2 spewing vehicles, the fact of a 3.4% rise in that gas's presence in our atmosphere during 1997 coinciding with happy Detroit's profit figures. They are now planning bigger pick-up trucks, their best profit item they have, are really responsible for changing trends, deleteriously so in the Atlantic Ocean from roughly Bermuda to approximately areas just south of the Tropic of Cancer. However, information gathered by patrolling satellites and launched balloons are seemingly either a secret or of not much interest to the media. They are infatuated over a subject that has made a few sleaze shows very prosperous ones, therefore, we can't be positive as to exactly where and how much pollution there is. I am certain there is but in the case of San Juan, I can't differentiate between UV ray excesses and CO2 or even particulate pollutants. Therefore, resorting again to a modest form of sleuthing, it not requiring super intellect, staying focused on sound principles, let us see what has been recorded in

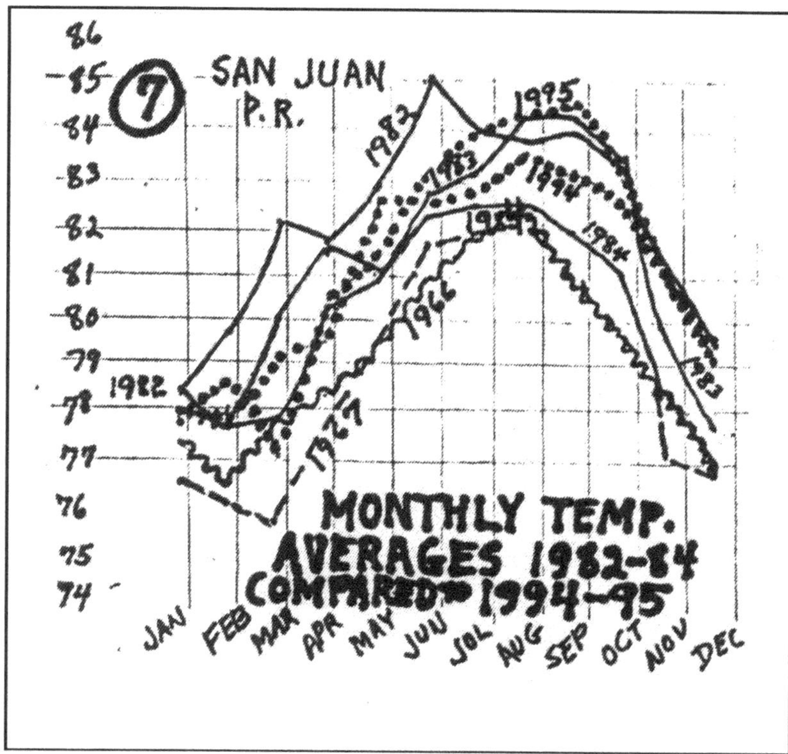

Puerto Rico. In the first graph (7) we have to allow, ruining my work of art as it does, El Nino's warming influence during it's biggest demonstration of trouble-making, to the tune of two billion dollars, during late 1982 and early 1983. Consequently, warm air rising from it's widespread presence off California up the vortexes of storms it created and maintained, was shunted eastward at high levels and by overlaying above the already warm Caribbean air contributed to Puerto Rico. In this limited synopsis of ours, no healthy dispersal of surface warm moist air, ascending upward, as only unstable characteristics moving air up, can promote. Treating this subject on a novice level troubles me but it should be recalled that cold air at high altitudes and warm air at warm levels are atmospheric features forcing air to rise. All equatorial zones are replete with rain showers, some of a violent nature and of course, rain forests of Africa and South America, well known to all of us, have extended periods of excess precipitation but, if your attention has not been fatigued by now, there is a coming and going process of wet seasons moving north then south controlled by ICTZ (equatorial front zone) for which, if I may venture into a more spiritual area of thinking, such an arrangement of distribution is a Godly design guaranteeing for now, no deluges and no deserts. Witness Africa's Wildebeest, on The Discover Channel of course. They seasonally know all about that ICTZ arrangement and what it offers and off they go by the millions. Getting back to the confusion of El Nino's influence during 1982 and 1983. Having had much of it's warmth shipped eastward via airline altitudes instability is toned down and to a degree stability is enhanced, that translates into less surface heat over San Juan propelled upwards. For this reason, an expert on hurricanes in a Colorado institute along with agreeing

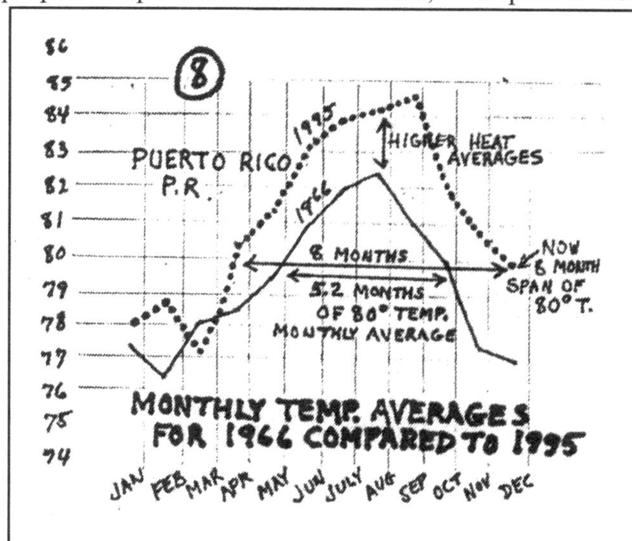

meteorologists maintains during interviews hurricanes are less apt to form during pronounced El Nino seasons since both cyclones and tornadoes require excesses in unstable air masses. Diagram (8) has something else to offer, thanks to my methodology approach. I have to blow my own trumpet of course, nobody else will. It is of paramount _importance_ you remember this salient point: Warmth, 80 degrees of it, lasts 8 months in the 1990's as opposed to 5.2 months in the 1960's.

Keep everything in perspective. While you may be enjoying winter's chill and frequent blue skies, some day you may find yourself packing up everything in your car, family and all and heading for the hills, depending on where you live or where your job has sent you and your wife. I'm not a fear-monger, but more people should buy a

NOAA weather radio that beeps a warning when some weather bureau is issuing advice on pending danger. For further information try this web site http://www.nws.noaa.gov/nwr. While the Readers Digest published an unflattering opinion of NOAA's overstaffed and widespread duplicating structure, which I will detail later, I must remind you NOAA has also saved a lot of lives and there are some brilliant people in their employ. One newly arrived scientist solved a problem that no one else could solve concerning a cruise ship having it's hull ripped open when it was passing in the Block Island Sound even though the pilot that had boarded the ship to guide the captain had assumed from experience and consequently the hull's height above the sea bottom was safely of a sufficient depth from reading charts. The captain had asked if it was all right to speed up the ship's movement from 15 to 25 knots. "No problem" thought the pilot. Fortunately, the coast guard and others saved everyone, but weeks went by and experts could not figure out what had happened. Along came this new NOAA employee and shortly came up with the solution. When you speed up a ship it tends to go into a sinking mode called squatting. In other words, in this case, the entire ship's hull was pushing through waters at a lowered eight-foot drop, apparently unnoticed. It was not as easy a calculated hypothesis as I've explained it, however. Methodology, that's what counts.

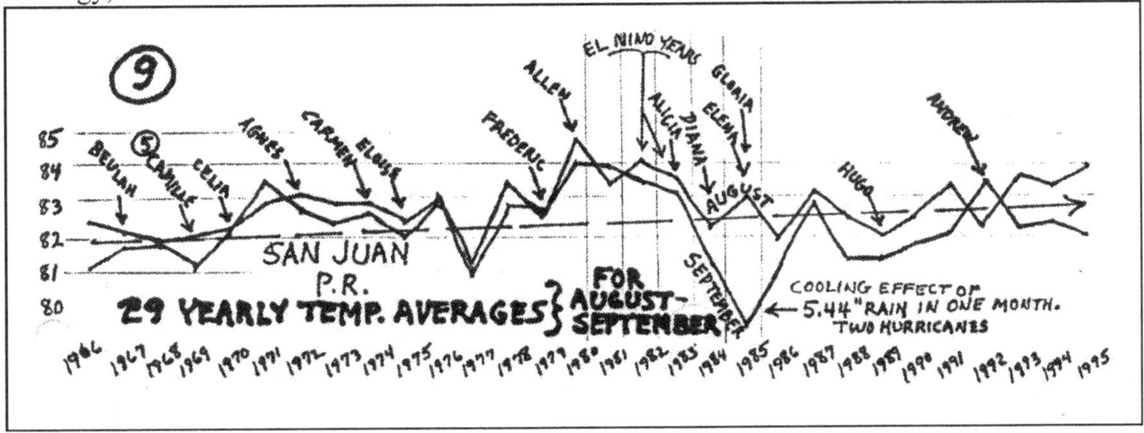

As a further study, the above graph showing a 29 year record of August and September, major hurricane months, always a threat to that region I am writing about and as shown in illustration 9. We should notice, despite ups and downs of yearly average temperatures the usual upward three-decade climb represented by the long dashed line. Not very exciting, but diagram 8 should raise a few eyebrows. You might be wondering what all this Puerto Rico business has to do with 101 tornadoes beginning on January 17th. The answer is a lot. Referring in your mind again to the temperature disparity of item 8, a prolonged heat season means a prolonged presence of the same huge and extensive air mass north and south of the equator, my warehouse of the sun as I like to call it. That in turn means a lot of warm heavily moisture-laden air pumped for weeks on end up into every part of the Gulf, Florida, the Bahamas, the Keys, Texas, Louisiana, Alabama. It was further heated before it reached, in this January deadly episode, Arkansas and Tennessee – from 80's to near 90 degrees. The equation is simple, no college degree needed – heat means volume expansion. As for hurricanes listed in diagram C, these were of category 3 and greater, three being winds of 111 to 130 mph. Nothing much occurred in 1982 and while theoretically El Nino had faded early in 1983, hurricanes started again late that year. If El Nino buffs want you to believe its influence remains long after it's innards have cooled, please note the drop in temperatures from 1984 to 1990 inclusive. But this book is about sinning against nature, who and what is responsible, what the government is doing and not doing, and the eventual outcome of it all. Don't muck up your thinking process over the idiosyncrasies, the morass, and the quagmire of El Nino's behavior. It's like trying to hold on to a slippery eel you've just caught. Don't let, what's the word, turbidity overwhelm you. I am promoting common sense and logic and with a little attention you should recognize reality is not an elusive puzzle. Five years for what? Send 15 researchers out for a few weeks over the oceans, let them dip thermometers in the water, record it all and then come home. Voila, it's finished. Oceans control vast spreads of temperatures and supply moisture for needed precipitation. What's new?

How can ocean levels rise for the next decade anyway since we have had deluges all over the world these past three years? Some 40 years ago when our climate might have been more pristine it was calculated that 360 cubic miles, not feet, but miles of water fell over this earth annually. Try to imagine how greater the increase recently when even towns in Massachusetts were partly wrecked from 15 to 16 inches of rain during one month alone, that month being June 1998. Seven people died in Kansas City, Missouri last November, caught by flash flooding. I can't have my typing the multitude of countries affected reflect a loquacious-bent. You know all about Bangladesh and worldwide torrential rains. Let those breakaway icebergs melt! Having driven across Texas twice and wondering if the car would survive unless I gave it a rest every four or five hours, you know one fifth of it covered with water last year represents a pretty good section of a melting iceberg, it causing such a fuss. God, how General Motors had cheapened their low priced cars in the late 60's. I should know. I was a service manager for one of their dealers in Louisiana. No wonder Japan, eye balling this trend, decided to take advantage of our auto markets later. My car made it from California to Florida, then to Rhode Island but it was never the same. Had it been a horse I would have to have it put down. Now, I have to highlight some good from NOAA, even though this book would rile them up. I just can't see their scientists as being a very logical bunch. Not when one degreed scientist says, "If we knew what causes tornadoes we could save more lives and property." Property? Then on TV, a highly placed NOAA administrator, who after hearing his statement, I chalked off as selling the services of his department and therefore not grimacing over it. "We feel it will be to a great advantage when we can predict El Nino five years in advance." You'll understand my attitude when I elaborate later as to what extensive sonar research probing the sea bottom near Easter Island revealed. Excerpts of course will accompany my comments. Some of the astounding findings have to do with El Nino and it's probable source of heat. As for my saying El Nino has gone away these past few years, I have been careless. It persisted but its heat content has steadily waned. This fact tripped up forecasters who expected a lot of rain but it didn't happen the winter of late 97 and early 1998 in California. The thinking processes of the human race have always had its anomalies. Years fly by before some gross misconception is corrected. A standout in paradoxes but not applicable to Americans, although I will draw an analogy from it, Hindus are shown feeding thousands of rats in the streets. Bowls of food are tenderly placed on the ground. Some rats even tug at pant legs for more. In the meantime, this practice condoned by city officials, they in turn inject some worried citizens with bubonic inoculation drugs. Reminds me of our government, other that a few Al Gore types, not really expressing anything disturbing over increasingly violent weather, sending its senators to selected cities and without much fanfare, pass out funds for the express purpose of building emergency shelters. My wife has just inserted a cassette and a wonderful rendition of "It's Impossible" by Elvis Presley has me wondering – a bad habit of mine. Only women should let lyrics play on their imaginations. But, I've seen truck drivers in one of my favorite truck stop restaurants put a coin in a record machine and out comes some silly love song.

We'll not have action soon – weather problems can wait. Money bags before body bags, that was and still is the philosophy of some. An example of an ugly fact: Just before World War II, German submarines cruised beneath ocean surfaces some 15 to 25 miles off our east coast. Nighttime suited them best for not only surfacing to recharge their batteries, but once underway beneath the waves, periscopes up, the brightly lit cities served them well. Heavily laden cargo vessels, hundreds of them destined to help England, made perfect targets. What could be better than the black silhouettes of ships against the bright background of Atlantic City, for example, periscopic cross-hairs carefully turned just ahead of each ill-fated transport. Thousands of merchant mariners died in this greatest of wars but the carnage off our coast remains a sad memory and reminder – Germanic genes of both friend and foe manifests themselves when decisions involving lives are a study in callousness. A plea was made to our Congress. Sunken ships, men drowning in oily churning foam, some, their legs broken from a torpedo impacting while walking about – that didn't matter. No, Congress would not mandate a blackout of all coastal cities because it meant sacrificing a portion of tourism money. Unbelievable, even after the war started and I was stationed in Cape May, New Jersey, then a neat but boring sea resort, this tip of land, partially immersed in the Atlantic still remained lit up as all coastal towns and cities were. Perfect replicas of video games kids love to play, a passing target on glass encased screens to shoot at. To think, one summer night in 1942 I walked over that glitzy-lighted boardwalk, out to sea, perhaps some bearded submarine skipper scanning for passing transports. Sound and feel carries a long way underwater. Our barracks were but one-quarter mile from the seashore. At night, even though it was the spring of 1942 and Germany's awful success in warfare was declining rapidly, sub after sub sunk by both planes and warships, I recall feeling the thump at least three times of torpedoes finding their mark. Always, of course, at night and probably more often than what others and I felt early in the evening lolling around our bunks reading or writing letters. Some time early in 1943 our intense anti-submarine campaign off the coast of Morocco resulted in our first face to face contact with captured German submariners. A dozen or so were escorted into our mess hall. I thought, "My God, I thought Germans according to Hollywood were all tall Aryans." This lot were just about a few inches over five feet tall*, long greased-downed sandy hair, looking innocent even though their escapades out at sea won them cruel notoriety and reverently fawning over our men handing them cigarettes. I didn't lower myself and from my vantage point looked upon them with scorn, the memories of Cape May coming back and knowing full well unlike our civilized attitude these savages, heirs of marauding plunderers of centuries ago, took pleasure in shooting at seamen floundering about after their ship had been torpedoed. Their own military films revealed that inclination of theirs, even some laughing after the "Turkey shoot" was over. Their true nature was made that much clearer if you ever saw how they laughed as marksmen plugged away at descending parachutists. In those days I was, at age 19 and 20, inclined to make my mark in history, go home a hero with some symbolic German scalps to my credit. But I am glad my excursions over the seas in our sub-hunting aircraft, droning for 12 hours, checked out in the use of a waist 50 caliber machine gun, never resulted in anything but total boredom. Oh, I have seen unusual sights out there between the Canaries and the Azores. Once we circled a ship cut in half, quite afloat and upon its deck one

sailor, waving at us some 500 feet above him. He had fashioned a clothesline and was busy hanging clothing to dry. Was it the captain? I'll never know but such a casual demeanor. Smiling, we could see that easily, our planes, PBYs, lumbering along at about 120 knots, this brave soul seemed to be saying "Hello, nice of you to drop in, see you later." Off we went, half hidden in low clouds at 2000 feet, all hands staring at the waters below. I

At that time I did not know of cramped quarters in subs.

never saw a damn periscope! Others sure did. Now submarines can fight back when trapped on the surface. A 40-millimeter bullet killed one of our crewmembers and the plane was badly shot up. It managed to come back but without wheels it flopped down in the bay of Agadir, Morocco. Of course, these were amphibian aircraft, no chance to use inoperative hydraulic systems to lower landing gear. One day, years later, I met a Portuguese fisherman who could speak a fair amount of English. I thought back to scenes that amazed me when out at seas. One hundred to maybe two hundred miles out at sea, we would spot these God-awful small fishing boats, maybe 20 feet long with what appeared to be makeshift sails. More often than not, wrapped in the unused position or whatever, manned by two men, Portuguese no doubt, as we flew by Portugal quite often. Down we would go in a mild dive, my stomach not very happy about it and wondering if the wings would hold up, the idea being to see if there were radio masts. Germans were helped a lot by such means but not these men plying their trade – not much choice in those days. They just stood there, hands by their sides and I know they were scared wondering what awaited them. You could see that. But hundreds of miles out to sea in such rigs? That's amazing! This fisherman told me in broken English how frightened he was one day. A large mass of boiling water surfaced near his boat well out to sea off the Azores. That was something told me in all sincerity 40 years ago. The sea is very deep in that area or otherwise one of our submarines, reputedly testing at a 900-foot depth*, or so the reports went, would not have ventured to try such depths. Maybe, El Nino, I've got your number. Later I want to verify my belief substantiated by a much more intelligent scientist, excerpts of his reports for your analysis of course. He being less imaginative and not prone to elevating oceanic phenomena from ancient routine rises and falls of water temperatures in Pacific equatorial waters to astonishing, enthusiastically endorsed by attention-grabbers world- shaking proportions. It is now January 31, 1999 and sure enough, the newspapers grabbed at reports! It's here! La Nina, that is. It chilled with its cold upwelling waters El Nino, the hot one, into temporarily disappearing. Something new and exciting and yet it has been a cyclical event for centuries! Soberly, let me attempt an analysis. Cold currents, both ancient, called the California and the Aleutian are now more vigorously being pushed southward and one way or another sink down and eventually are forced to rise up again in the equatorial zone from the effects of converging south and north hemisphere deep water opposing currents. This can be explained in fancier terms such as "thermocline inducements" but let's leave it be. My point, cold water

Reputedly lost.

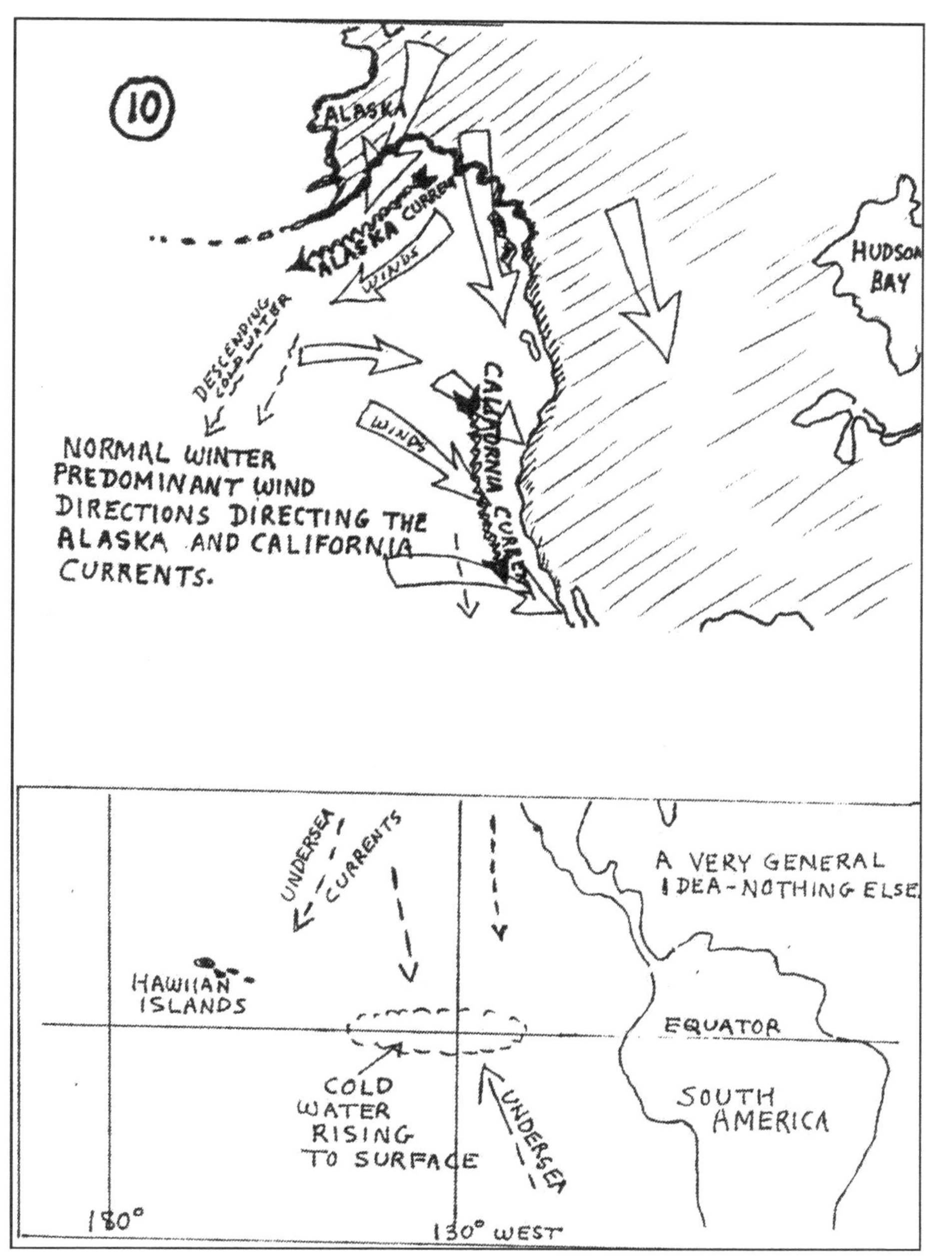

has to come from somewhere. Temporarily disarrayed by earth warming weakening cold accumulation in the Arctic, therefore lessened Arctic surges of icy air, are now reactivated since two months of normal winter weather is obviously back in business. Such a fuss made over snow in America (horrible!) hence cold air plows down across the Aleutians as it formerly did and barrels along our western coast as it formerly did, etc., etc., and so on. Yes, these currents are old hat.

A rule to keep in mind, ocean currents, on the surface and that's all we are interested in, reflect semi-permanent air currents that drive them along. Let's not get ridiculous and treat them as entities off on their own. I knew that the Gulf Stream had worked it's way further northward these past few decades, but I also found it interesting when chatting with a professional fisherman who told me he wanted to see normal winters return. "Why?" I asked. Oh he went into food chains disrupted, the right plankton gone somewhere else and the wrong creatures where they shouldn't be, etc., but a touch of seeing something more fascinating for a buff like myself with my mind's eye, how swordfish were not to be found in their usual locations. They had moved much further northward and of course he described their charging at prey – a sort of interesting Jacques Cousteau rendition if you will. Ah, but did you know this; little is known or publicized about ocean currents. The Pentagon and it's Navy, since missile-equipped submarines have complex problems in the depths of moving oceans, what they know is regarded as top secrets. There will be no revelations handed out over what happens beneath sea surfaces. But, simple common sense and logic, a shortage of which is obvious, is all that is needed to understand what is going on. Let journalists go on with nonsense, although some are to be respected. At least you can separate fact from fiction. I ventured off with a few stories simply because the subject of impending problems can be both monotonous and depressing. Weather as a topic I want to elaborate on compels me to stop for brief periods in typing so many details and insert some variety here and there.

It is money that matters most. I don't think deaths matter a hell of a lot – they are soon forgotten. Having previously emphasized our south Atlantic's atmospheric dominance, abnormally bloated all summer long by UV radiation and whatever degree of carbon dioxide effects I wish to introduce an economic factor the likes of which concerns me just as much as more weather calamities in the future. If the Senator wrote our White House has 2500 economists at their disposal and their alleged anxieties hinted at, then I'm not too farfetched in volunteering an opinion. Based on knowing exactly how populations stuck in the bottom basement levels of what I call the pyramid of any economic system, except perhaps communism where theoretically wealth is evenly distributed while a few on top soak up easy money and power. The opulent near the peak and the extremely wealthy perched on top of the symbolic structured pile of groping humanity. From my experience as a child and later a student of stock market psychology, yes I know what the threat of probable excessive wage disparities that could result from unexpected rapid rises in the cost of living created by destructive forces of nature can do. I know what angry crowds are all about but if it does happen here it will not be revolutions in the sense that we conceptualize. Before I get into this gloomy topic, let me say it's manifestations will be quiet, other than a few demonstrations. If Mr. James Lee Witt, shown with the president, looking stunned at the tornado damaged town they were inspecting and making a remark to the effect, God's name used, a bomb had struck, it is because he is wondering how much more money his Emergency Assistance Fund Administration will have to shell out. I will write later some pertinent points he made to a very chaste group of reporters, one hour of it and no challenges from members of this Meet the Press affair.

Our corporations, eager to capitalize on cheap labor readily available in the Far East, for example, leaving quite a few other sneaker manufacturers and their equivalent sweat shops aside, whose stock you may happily hold, do not envision what disrupted Pacific climate represents. How can they when our own scientists can't see the forest but for the trees. Still busy as bees or maybe they are just killing time, I can't be sure, always contradicting each other. Again, tiresome or not to ingest as reading material, let me repeat: The vastness and dangerously powerful continuous manifestations of Pacific anomalies, repeated onslaughts by typhoons, crops ruined in New Guinea, Australia by droughts, terrible monsoons and on and on. How can I not repeat China's catastrophe, all this is a future threat, wreaking havoc on lands of the very people who for now are peaceful but disrupted economies by nature's reactive wrath has the potential for social violence and government upheavals. Trade balances may go out the window. We will scurry to help motivated by trepidation rather than compassion and that can be costly. One final dab at the picture I paint of grave potential problems, the Pacific equatorial belt is but a continuum of the almighty girdle of stored heated air wrapped around our planet. Between the Tropics of Cancer and Capricorn alone, considering also steaming rain forested continents we are talking of 84 million <u>cubic miles</u> of moisture-laden air, in degrees averaging in the mid 80's. This heated worldwide zone exceeds that conservatively estimated volume. I've placed identifiable limits on its oceanic spread, assuming you might glance at a map or a globe. My point again, but not finally, it expands seasonally, doesn't contract when it should consequently and it will continue its aberrant ways for a long time.

It's your future not mine. I'm too old. My children know who to vote for. Whether or not Al Gore would be a good president is an unknown consideration. At least sanity will have a voice.

One delayed thought over arrested reasoning; has certain vehicular facts ever been brought to your attention such as the appalling quantity of carbon-dioxide being pumped out of tail pipes by huge vehicles? How does 125 cubic feet every minute at expressway speeds strike you? That's what the current "Hooray for Number One" trend is doing to nature. V 10 engines, displacing 400 cubic inches (6.6 liters) pump out this extraordinary amount. When briskly accelerated, not the juvenile tire squealing variety but rapid enough to spill some coffee in the console holder, an important nitrogen-oxide eliminating device (EGR Valve) ceases to function while the engine is being pushed. Hence at near full throttle while making a bee-line down an ingress on to an expressway, then blowing by three or four cars, toxics such as the above Nox, carbon monoxide and some unburnt hydrocarbons (soot) are released in surprising quantities out the tail pipe. Catalytic converters under these temporary circumstances cannot completely neutralize all such gases, functioning as they do when normal cruising produces less exhaust to convert everything into carbon dioxide, because of the high velocities and volume of exhaust. I don't consider this bit of information above as trite.   There are summer conditions in many cities and surrounding communities when vehicular emissions of toxics can suffocate asthmatics. Many children of poorer neighborhoods have died. The physiological effect is one of being strangled. Simply stated, the bronchial tubes close up when irritated by inhaled acids permeating the air, better known as ozone accumulation created by a combination of heat and the presence of both exhaust and industrial gases. Of course, we have here in Providence and probably many other cities, a peculiar topographical filth-collecting arrangement by virtue of not only being in a shadow valley with it's southern end tapering down to an ocean level and forming, without stretching my imagination too much, a mouth open, and somewhat in a position to inhale a lot of trash from extremely populated New York City, heavily industrialized New Jersey and Pennsylvania. Need I add an enormous number of people there also, which obviously translates into a summer pall of filth in the sky? Our gaping entrance, partially aimed at the smoky sites mentioned, is a frequent target simply because prevailing summer winds are west to southwest. Worse, an extremely busy route I-95, some eight to brief sections of six lanes, cuts a swath through the heart of Providence. For at least six miles in its valley, we are treated to some 6000 tractor truck diesel powered belching out not only toxics, but also very visible large quantities of soot. These trucks pass through every day, many at night that we have not estimated. Add the products of a million vehicles each week to that. If you believe there is nothing that can be done about it I have to differ with you. While I have in mind as a suggestion made to a senator a moderate approach to reducing what spectacular quantities of filth are produced by trucks, it is the same old story of a simmering anti-government attitude best represented by cajoling sometimes vehement far right anti-reason proponents. Therefore, I told my wife a few days ago not to play that song "It's Impossible" during the usual two hours that I type. That's how I feel what has become a stupid psyche pervasively polarizing a large segment of our population into whining non-cognizant types. It is impossible to accomplish anything such as the imperatives of dealing with a problem, that has to get worse, and yet not proclaimed with any resolve by vote-oriented politicians who know what that crowd out there is thinking. Or should I say as above, stupidly not thinking, and that is a growing yearly population, 80,000 new homes with oil heat, more and more bigger autos, trucks, vans, all part of a robust economy, mean more and more carbon dioxide. It's that simple! Getting back to diesel trucks, let me empirically elucidate step by step, knowing full well that I have departed from accepted premises particularly in weather, but also determined to reveal what is not publicized in matters revolving around combustion products. Referring to Wiley's Encyclopedia of Energy and the Environment, it not being a government two documents of amassed information, so thorough it is persuasively interesting in its scope. I want to elaborate on the enormous quantity of oil used by the trucking industry. I do not know what one barrel of oil produces in the way of diesel fuel, but we can get a picture nevertheless by what Wiley's comprehensive data reveals. In brief of course, I'll try. Freight carrying trucks use one third of all fuel (energy as it's called) consumed by every type of passenger transport systems, cars, planes, trains, etc. Somewhere around seven or eight million barrels of oil daily goes into the production of gasoline and diesel fuel. Therefore, it's reasonable to estimate diesel trucks require the use of Texas and Louisiana refining facilities to the tune of about two and one half million barrels a day. As you can see I am worried about the oil family progeny, such as Governor Bush running for President. Wiley emphasizes since speed limits have gone up it's quite common to see rigs flying along at 70 to 75 mph and consequently, fuel consumption efficiency figures have gone down. There are major reasons for speed. I listened to a lady truck owner, three of them, speaking on the radio. She pointed out the profit margin is small and therefore time is of the essence – we all know that. But, I wonder if you know factories waiting for deliveries can be harsh. If a truck driver arrives at 9 am and he was supposed to be there at 8 am it is a common reaction, according to the lady, for the factory manager to say "Go back with your load, it's too late". Here we go again, should the state or federal government have some say over this? Somebody from this country should go to Europe and see what their trucking system is all about. They now have a satellite monitoring system that keeps drivers from not exceeding speeding limits and above all, the important matter of driving only eight hours a day. I can't figure out what satellite means in this matter. I suppose the hue and cry here should such a plan be implemented, "Oh no, more government!" Better we endure several hundred cars crushed by tractor-trailers whose drivers have fallen asleep pushing their endurance behind the wheel so as to either get home sooner or meet some delivery schedule.

Wiley points out the increase in wind resistance factor rammed up against massive square front ends of tractor trucks is due to higher speed limits and consequently a rise in diesel fuel consumption in the past decade or so. One should note what few new trucks are around; measures have been taken to reduce such horsepower wasted. The fenders are even sloped quite attractively for what were once sheer air buffeting angular lines. An interesting point I want to emphasize, but certainly not an original one, is the turbulent suction or more correctly defined in aerodynamic terms, vacuum drag, at the rear of any box-designed freight hauler and very conspicuously those of SUVs, vans and what seems like unimprovable pickup trucks. As for the latter, it is surprising how a canvas cover sometimes seen in the back is very effective functioning of course to keep rain off some materials. It also deflects air by, rather than crammed against the tailgate. At the front of any vehicle moving along at high speeds we have a build-up in air pressure elevating our standard at rest 14.2 pounds per square inch atmospheric pressure to possibly 16 pounds or higher. If some genius can volunteer a calculation based on square footage of vehicular frontal area and come up with how much horsepower is wasted, that would intrigue me. Sticking my arm out the window at 65 mph while driving, no one behind me is about as far as my experimenting has gone. Consider my passion but short on hope for pursuing legislation through letters to Washington, aimed at reducing air pollution, if the above strikes you as extraneous and not interesting. Just looking at the rising popularity of poorly designed heavy vehicles, appealing to many affluent who have run out of ways to express superiority (not all) and Detroit tuned in to the quest to be different, to be outstanding out there showing their stuff, shunted aside any vestige of guilt and as one CEO said, "Give them what they want." Can you imagine this irony? An all out effort to increase fuel economy utilizing the best of engine controls at a reasonable cost and furthermore, the excellently streamlined autos. But, the benefits of improving air quality completely cast aside after promoting sales of gas hogs whose percentage of the auto market has now reached 51%. Imagine when these vehicles that could have been used in Desert Storm, having a machine gun mounted on the roof, get weary and their emission controls, not subject to testing so far, no longer do the job. Stubbornly, let me repeat again, new or used, the amount of carbon dioxide produced is shameful.

I suggested to a senator, why doesn't the government encourage, or if that does not work mandate, that all tractor trailer trucks utilize the streamlining canopy everyone is familiar with, setting on top of cabs, since too many mule-headed "Six packers" truck owners have not adopted what has proven to save fuel. I also asked if tandem trailers should be utilized in greater numbers and while dangerous at high speeds, perhaps impose speed limits. Unfortunately, there lacks incentive and united public support for the government to chip away at excessive oil consumption problems given that the insanity of "Bigger is better" of the 60's and 70's suffocating Los Angeles as it did, the death toll peaking later, rapid acceleration being hustled in commercials as it also was in the good old days. Don't think me vain if I try to broaden your perspectives. It was a matter of delegating responsibilities to several groups but without malice. Your improved visibility is needed.

What were the residents of Jackson, Tennessee thinking Sunday afternoon, January 17th? We'll never know. Episode after episode of death and destruction, but it's regarded as routine. "We've always had tornadoes", that's what outsiders say and no one to my knowledge seems inclined to inform otherwise. I heard one gentleman on a TV show go so far as to say meteorologists are nervously crying "Wolf!" every time some weather situation appears to have potential for twisters. That is unfair, particularly when certain conditions call for an alert. Then worse, the previously mentioned meteorologist that I chatted with told me that the public could ignore warnings. All I know of Sunday evening's Arkansas and Tennessee experience, nothing remaining of one town, is what I read in two newspapers. A couple there were quoted as being occupied taking photos of lightning from their porch. They heard a strange sound and ran. A gentleman, in charge of Emergency Management said, "We're accustomed to tornadoes in April and May." You figure it out. Weather, as a subject is not a prerequisite in school – not even regarded in it's true light as having a profound influence on our lives. Disaster after disaster all over the world, but case after case of total ignorance. Eighty-seven people, camping at the base of the Pyrenees Mountains on the border of Spain abutting France, drowned one evening. A cloudburst in the distance that's all it took to fill the valley their campers were parked in. There was one survivor. He described this beautiful sunset, a towering anvil shaped cloud, it's wispy wand in color, high in the northern sky, poking up behind distant hills.

Despite freak weather, such as nine tornadoes in Oklahoma during the month of November, a sure indicator there is something very wrong, but the public in general remained dumb-downed. Not a peep out of anyone, the rarity of this stirred no one.

After the invasion of North Africa in November 1942, most of the short period of fighting taking place in the Mediterranean, several divisions of soldiers were at rest in Port Lyautey, Morocco, waiting to be transferred to Tunisia. Given a day off they charged off to the beach, a spectacular stretch of sand as far as the eye can see. Twelve of them were pulled out by a vicious undertow and drowned. What do leaders from inland states know about the ocean?

Lightning means extremes in vertical cloud build up. The same muggy warmth from the Gulf, the same potential threat presented by a coming change to cooler, rolling and churning their way. I am reminded of that frequently used expression, "From the Gulf of Mexico" as being incomplete, a misnomer of sorts. That body of water is merely a transit area – the dangerous 80 degree air in January has its source further south and southeast where solar radiation has been in a state of storage for too long. And the ocean's innocuous appearance? How beautiful the beaches of West Africa – miles and miles of white sand. But, as with all west coasts of continents, waves have relentless for milleniums excavated the land. Beneath the surf or its absence the sandy bottom below drops off abruptly. After the peak of high tide and it begins to wane the current below rolls downhill while the breezes drive the surface waters toward the shore. Deceptive, what lies deeper is not seen.

It doesn't matter if its mid January and you've been told by the inexperienced what wonderful weather you're having for winter. There was instead, a spring condition, the same ingredients that fuel tornadoes during their once-upon-a-time season, transported from in-depth equatorial boundaries and steadily being funneled over one third of eastern United States. When will financed anti-reason pundits stop suggesting earth warming has yet to be proven?

One tabloid usually found in supermarkets is now promoting secret Bible predictions of terror in the form of insane weather will annihilate certain parts of our country. New Orleans is going to be flattened by tornadoes, and so on. Theatrical performance must be a supplementary course taken by certain writers, even the more subdued Anti-Gore bunch. A tornado in New Orleans, Biloxi, Mississippi or any landmass whose shores are bathed by Gulf waters is very unlikely to experience such violence. Further north of course, the land-heated air presents a different problem. Oh, they occur near the coast all right but they're not the twisters roaring along, cutting a wide swath of destruction. They are what I call the uncertain or undecided kind, skipping along or dipping down for a moment. We had one sometime in the 60's when we lived in Slidell, Louisiana, about 30 miles from New Orleans as the crow flies. It pulled somebody's modest shack of a home up into the wherever, spreading boards and tar paper in nearby pine groves, but left a refrigerator standing intact on what was left of the floor. Because I commuted nearly 80 miles a day working in New Orleans, I bought a gasoline stingy English Morris. One day, with my wife, returning from a ride there being a five mile bridge across the eastern end of Lake Ponchartrain connecting Louisiana's main attraction with its annual Mardi Gras and the town we lived in. We spotted a waterspout about a quarter of a mile to our right. Seeing a few cars stopped we did the same. Stepping out of our unpretentious import we decided to jump right back in. The spout seemed to be interested in the bridge and headed our way. It wasn't exactly noisy but the sound it was making sufficed to initiate a hurried departure. The cars ahead pulled away. My metal steed had reached its limits of 80 mph and allowing for speedometer error, probably a real 75 mph. My wife was frightened and said the spout was catching up, as it seemed to be intent on moving parallel with the bridge. She panicked. Beating on the dash with her hands she kept yelling, "Hurry up, hurry up!!" Later, we had a good laugh. Lake Pontchartrain, really a vast body of water, there being a 30 mile long bridge across its center, is but a depository of the Gulf of Mexico, gentle tides moving Gulf waters through a narrow channel called the Rigolets, a spot where a few tarpons have been caught. Hurricanes, of course, are a different story for southern Louisiana's land mass, the part nearest the gulf gradually tilting down and 40 miles north building up, the result of enormous quantities of soil dropped by the Mississippi but taking away at its widening exit. All of New Orleans and its suburbs lie below sea level. While we lived in that state of Spanish moss and Live Oaks, there were three hurricanes, Betsy, Camille and Hilda. Sharing losses with Florida, Betsy cost seven billion dollars. Lives lost in Chalmette below sea level numbered 38 but the total deaths for three hurricanes exceeded 250, factoring in Florida and Mississippi. While everybody in Slidell stayed home, my other car, an Oldsmobile, transported us 50 miles north where we stayed in a motel. This happened twice. I'll let those more stout-hearted brave the elements and going along with the main theme of this book, the inclined to self-destruct who, minority or majority, do not comprehend or even visualize how the immediate cost of disasters has a built-in multiplication factor, coming under the heading of increasing cost of living, some aspects of the economic ramifications to be touched on later. As for the immediate costs of NOAA's simple graph below is clear enough but that is only for hurricanes.

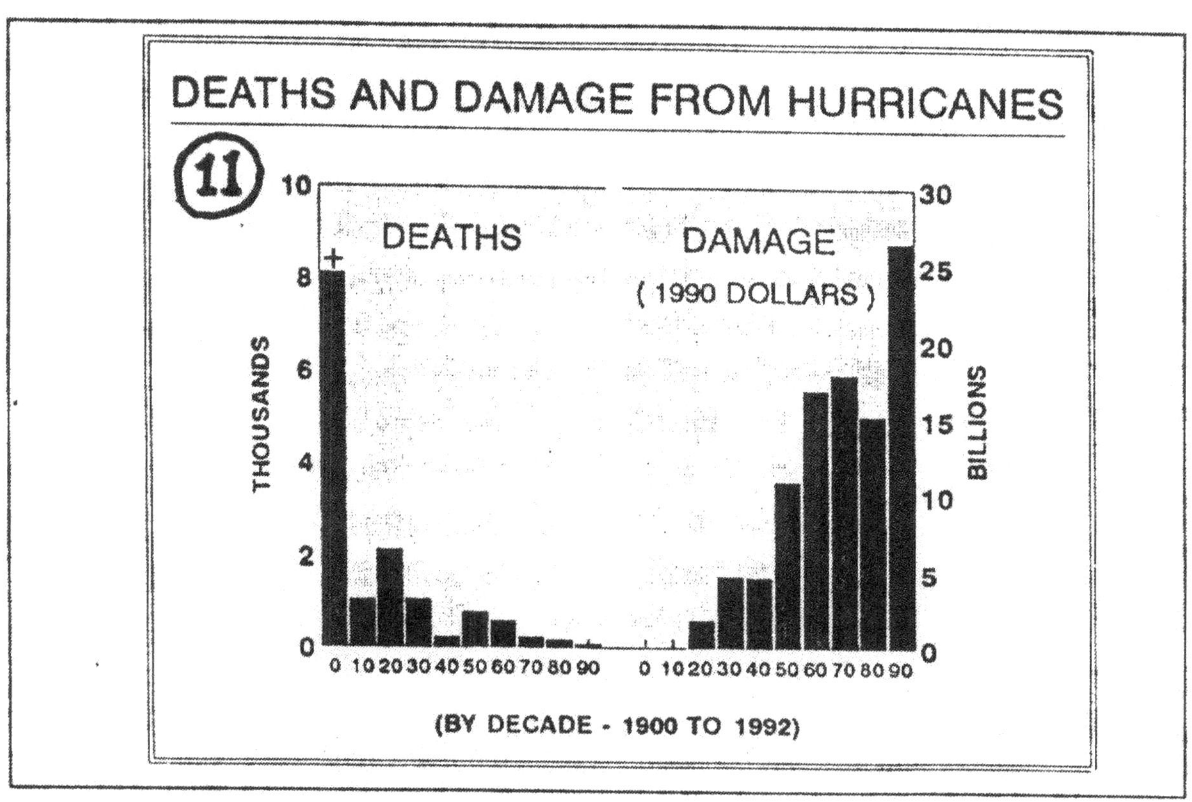

This Environmental Whacko doesn't think in terms of present cost following some destructive incident. Russ Limbaugh used that derogatory expression five times in two minutes the other day. Once his zealous infatuation over the impeachment affair is over, I know what he is gearing up for. As I wrote before, I listen to him as it gives me an insight as what ails his followers who reverently flatter him over the phone. So conceited is this man that invariably he finishes off a conversation by showing he's right and the caller, well, he or she doesn't quite have it right. So much for the Missourian who, no sympathy for the have-nots looks like he was well fed in his childhood. Let me get away from the ordinary.

While I'm not that fond anymore of President Clinton, I wish they would have chosen a better photo of him in the company of James Lee Witt, administrator of the Federal Emergency Assistance, standing in the wreckage of what was once a town some time during the week of January 17th. The president was smiling. Mr. Witt was not and his mouth was open. He looked stunned. The press quoted a statement but the first two words are "Enough said". "My God!"

Getting back to the graph before I continue as to how Mr. Witt plans to handle future funding. Please note the decade of the 80's. First we had El Nino at it's worse in 1982 to early 1983 and as an accepted fact, there are less hurricanes while this phenomena churns up storms over our west coast and it's seas. This has to do with its warm air being pulled up into high altitudes and then transported eastward over the Atlantic. High level chilled air is required to create the necessary unstable arrangement that both hurricanes and tornadoes thrive on. Warm air therefore, cancels this high level promoter of powerful updrafts. Note, I also included tornadoes, something that escaped being recognized by a showboating doctorated expert who, much to my surprise, said tornadoes over Alabama and Georgia killing over twenty people was definitely caused by El Nino. The second reality that contributed to the 80's experiencing less hurricanes and definitely less tornadoes is the Wiley reported and recorded fact that we used a hell of a lot less oil.

One night in November, I listened to a meet the press session over my radio, of course Mr. Witt being the guest. While he never said so, I know darn well that the federal funding barrel has been seriously dipped into of late! 2500 economists implies serious consequences can be expected also in terms as to the effects of continuing earth warming on future economics. Mr. Witt made it clear that the government would not spend money more than once if a home owner, having had his abode slide into the Pacific, requested monetary assistance to rebuild on the same cliff just because the scenery is beautiful. He also nixed any thought of some homeowners living in flood prone areas or their home parked on river's edge because the catfishing is great, entertaining the notion that they could receive help to rebuild in the same locations. While he never used catfishing in his presentation, that being my feelings on the matter of previously wasted taxpayer money, he emphasized help would only be rendered if applicants chose higher elevated areas and a safer distance from rivers that flood every ten years. The questioning reporters at the end were content to ask for a few more details. There was nothing that suggested in their tones of voice that they were not totally in accord.

Two Senators came to Pawtucket in November also. Without any fanfare they met with city officials to tell them one million dollars was on its way from Washington to refurbish and reinforce what buildings here will serve as Disaster Shelters. Project Impact was passed in 1997. Every state will select a city each year and either build a shelter or improve on an older one. Initially one million dollars will be issued but additional sums for repairs and what have you are available. In other words, some initial sums bordering in the low 50 millions are to be spent on shelters every year from now on. If Mr. Witt keeps his position for the next decade or so, I'm certain he'll be saying "My God" quite often.

Is Project Impact an indication of anticipated problems or a response to complaints? Perhaps numerous hard hit communities of late let it be known through their Representatives in Washington, DC, facilities to shelter citizens before or after so many calamities were inadequate or did not exist. Little news on a national level having its source from the White House that could remunerate in clear terms is forthcoming, that is obvious. Is it fear of disturbing the stock market? What is so frightening as to the effects on investors should America as a whole cooperate and adopt at least ten measures to reduce pollution, that I suggested to a senior senator no originality on my part can I boast of, the nature of which in it's aggregate would be painless? Is the public that numb-downed or that intent on resisting governmental guidance, constantly expressing a wish to be left alone? I have my convictions of industrialists pulling strings, their ad-men convincing the naïve including wishy-washy members of both houses. As a result, from what I have read, the crucial importance of earth warming is being kicked around and indecision results.

I see two major weather disturbing pollution factors, distinctly separate and yet not viewed that way. Leave health hazards aside for a moment; that being the low-level contaminant issue. If EPA would be regarded more as a necessary function of government rather than a foe to be resisted, much could be accomplished. First, abnormal warmth during the months of January and February in particular over the Deep South has to have a reason. For one, ultra-violet radiation, it's increased intensity discovered several decades ago, has not diminished at all. A cold front however modified by land heat but dropping Miami's temperature a few degrees after disappearing east of Florida has little residual effects. What cooling it brought to eastern coastlines from Key West to Savannah or for that matter most of the southeast, rapidly changes back to abnormal warmer temperatures within 24 hours. There is no mysterious force at work out there in the Atlantic or the Gulf of Mexico that some imaginative writer would like to get his hands on. Summarizing, a cold front passes through, the sun shines and the temperature jumps back to 80 degrees. The ozone layer in it's present somewhat debilitated state and the cause in dire need of remedial application but stalled by recalcitrant industry's CEOs whose smoke stacks belch out sulphur dioxides and other contaminants, suspected as finding their way to the ozone layer, is the main reason for maintaining what weather people refer to as unseasonable warmth. That is not exciting in the way it is expressed, is it. But what does this mean in bringing about a much earlier start of the tornado season? Just think, here it is February 7th and the Weather Channel is reporting a tornado has touched down in Arkansas. To the nonchalant, a twister this time of the year means nothing but that is why the human race is very vulnerable. I want to belabor every facet of changing climate. What I have to offer in later pages is a departure from conventional thinking; therefore, I painstakingly lay the groundwork. I can't expect believers if I don't apply a meticulous approach replete with all available information. Read on. Let us go back to Miami and see it as I see it. After that, the second problem, the insidious increase of carbon dioxide. There are three divisions of pollution.

If we consider Miami, or San Juan or Jamaica as true thermometric representatives of thousands of miles east of these sites, since they bathe in warm air continually flowing from one end of the South Atlantic (Africa) to the other (Gulf), allowing for brief interruptions already mentioned, what is more significant as Miami's unusual maximum temperature rise, hovering at 80 degrees in January and February of this year <u>when</u> <u>no</u> <u>such</u> <u>abnormality</u> has occurred in 29 years going back from 1995 to 1966! The top max was 76.7 degrees for February. The maximum temperature now, in Miami exceeds that by 4.5 degrees! I know the sun's effects and possibly that of CO2 when I see it. There are no Devil's Triangle mysteries going on. The Gulf Stream is really but an extension of the air current influence driving it coming all the way from the South Atlantic. Starting as a mass of warm tropical water moving westward, known as the Caribbean Current, moves between Cuba and the Yucatan Peninsula, driven on into the Gulf of Mexico and then having no place to go. There being inertia involved, finds an escape route, and reverses to an eastward movement between Florida and Cuba.   From there it switches direction toward a path northward, brushing by Block Island, driven by prevailing southerly winds and swings toward Europe. That reminds me. Thanks to the Gulfstream Europe enjoys a maritime climate but they having accused our country of producing too much pollution some years ago. I want to suggest they consider the fact that their pollution driven eastward frees them of much accumulation but it finds its way to nations east of the Alps. However, gasoline prices running up to four dollars a gallon and the imperatives of small cars, I know we did not make a fuss. But, even though we sin against nature our position as roving sheriffs all over the world is very costly – only a major heavily industrialized nation such as ours is in a position to defend liberty or even economic advantages. Therefore, a little chastisement from Europe is acceptable while they, in the old country, keep in mind the realities of 220 million Americans who, I believe, eventually will see the light. I expect they will rally around the government as they have done before. Please evaluate the graphs on the following page showing 29 yearly (1966 to 1995) temperature averages for the month of February only. I am moved to such a depiction simply because a gross abnormality now exists and yet, weathermen make no fuss over the fact that now, just in the first week of February, heat records are being broken from Texas to Florida and up to the Carolinas. Be reminded that in 29 years the highest single maximum temperature ever recorded in Miami was 76.7 degrees in February! We have had 80 fairly consistently in that city and even now, New Orleans is also high and San Antonio, today, the 8th, 86 degrees. The sun shines today over most of the south but it is not as gentle as it was perhaps 40 years ago. However, I hope I am not reflecting any anxiety.   My main interest is to explain as best as I can in positive believable terms what is going on. If there are any alarm bells to be rung I'll let somebody else do it. One year before it happened I had written to both Clinton and Gore that sooner or later pieces of American cities would be blown away by tornadoes. They never answered. Probably some dingbat clerk opening mail
thought I might be deranged and decided not to pass it on. It happened, first in Arkadelphia and then in Nashville. Gore's hometown got hit again this year. I do not feel vindicated. It is simply the odds favoring such calamities. But, as the writer of a fine book, "Agendas, Alternatives and Public Policies" namely John w. Kingdon points out, a blow on the side of the head is required in American politics before action is taken.   So far, rhetoric with it's usual years before implementation.   We have a terrible atmospheric arrangement right now with normal cold outbreaks from the Arctic pushing down across America but abnormal warmth waiting to be engaged. That is why I prepared these graphs as to reiterate the latter condition. The West Coast is being battered continuously by powerful winds and tremendous rainfall. They certainly will not have water shortages for a while and that, ironically, is a blessing! No mention is made as to the enormous high-pressure systems out there in the Pacific, maintaining their

volume and power through the more intense sun's influence now that Earth is positioning it's northern half for a better share of it's rays. From the Atlantic to the Pacific, quite a switch in my analysis but the battered and saturated West Coast offered a more dramatic case in point. Henceforth, I'll avoid the cumbersome; let's settle for a moving sun rather than waste words over orbital angulations of a spinning Earth relative to a stationary sun. Yesterday, February 9, temps in the Caribbean and southern USA extremities jumped to the 80's.

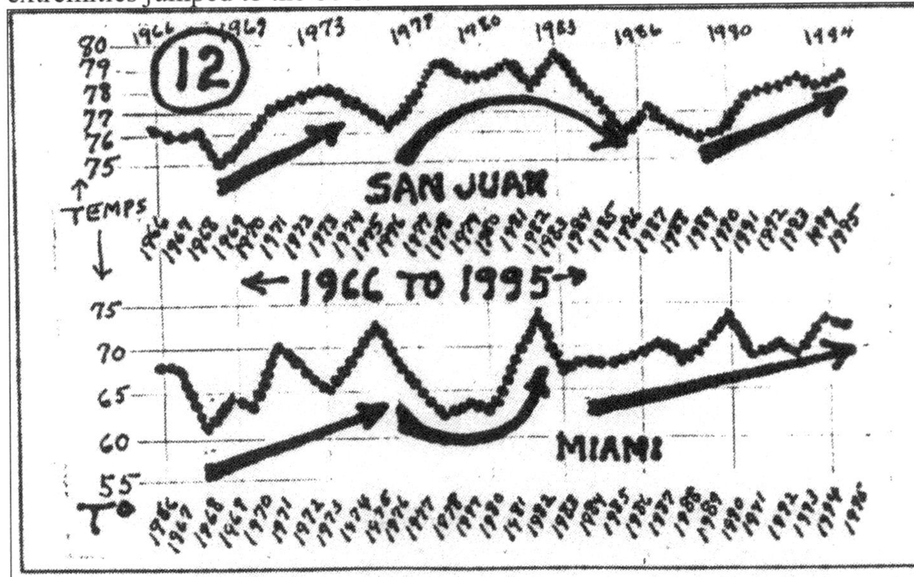

San Antonio jumped up to 86 degrees. Records were broken over dozens of sites. Nothing good can come of it even though it's fine for tourists. I am mainly interested in temperature trends over years as indicated by heavy arrows. San Juan is 1300 miles from the equator. Its daily temperature range not showing any invigorating fluctuations reflects that. Further north, some 1800 miles from the equator, Miami is subject to more pronounced daily and seasonal variations. Both locations however, are under the same oceanic and therefore atmospheric influences that create and drive hurricanes in their direction. Both sites also experience the same warm humid air that fuels tornadoes in America. One does not have to be a wizard to expect an increasing frequency of both violent phenomena since the very condition and ingredients that initiate and powers them is steadily on the rise. We could also go on and on with other destructive forces, such as droughts and floods, but this year 1999, presents the possibility of not duplicating 1998. The question: Will the Arctic produce its normal seasonal vigorous invasions of summer's cooling air or will it again, as it has for quite a few years, remain more or less impotent due to the Greenhouse Effect and pollution at high levels shipped in in ever increasing amounts from China, Japan, India and other countries. In normal times, summer and some of our autumn season was blessed with occasional showers and minor precipitation bearing storms triggered off by a healthier Arctic region. I have no way of knowing whether the present state of normalcy there in icy regions, attributing this to widespread and continuous rain all over the world last year will continue on for very long. Lulled by the lack of information, the public and the media not very interested anyway, I am hardpressed as to what degree of CO2 and dirty air will come back toward the poles. It depends entirely on the capacity of the spacious atmosphere to reach a point in its collection process where normal nighttime cooling is inhibited. I know darn well that scientists don't know either. I don't believe they even realize the cleansing effect of rain as a major worldwide factor! If I could compare the extent of increasing pollution over northern regions with that of previous years, I would have something to work with. Information is not to be had because I am quite convinced that their approach to understanding this or that in atmospheric anomalies has with it some shortcomings. Do not capitulate to ignorance. You have to resort to logical reasoning. You have to emulate the "Thinking Man" and not swallow everything that is thrown at you. Not only are there degrees of ineptness in every major organization, but there are also politics to be considered and the usual in fighting between various department heads. For the sake of thoroughness, let me go back to the graph. Look over the deviation between Miami and San Juan, this latter city showing remarkable higher temperatures from the mid 1970's to 1983. In this period, eight major hurricanes reached and affected Florida. The relation between rises in temperatures sustained over a series of years over the Caribbean and violent destructive weather is obvious. I wanted to avoid excessive details and better still keep major causes and effects relative to an ailing nature for later pages where I intend to illustrate and advance my convictions in more detail. But having brought up the point of San Juan's six years or so of peak warm temperatures and very expensive hurricanes and of course a lot of victims, I have to explain why. The south Atlantic high pressure system, even though such a title short changes it's position as being a major part of a worldwide belt squatting over both sides of the equator, during such a series of hurricane years has in it's volume and late summer intensity, stretched it's eastward dimensions and influence over Africa. It pumps that continent's super heat and the adjacent warmer ocean's contribution of excessive moisture rather forcefully westward toward the Caribbean. There is also the connection of the equatorial front (remember ICTZ?) having migrated northward and the force behind it packing up against our Atlantic High (to be explained later). Rather persistent and fairly strong easterly wind result propelling the values already described in one beeline toward our part of the world. Believe it or not, 1983 and 1987, according to oceanographers taking samples of ailing corral reefs off Florida, dust from Africa was analyzed as containing some kind of bacteria, those years being the worst years for strong steady easterly winds. Despite that remarkable research effort, it never received great publicity. The usual yawning. The usual airhead mentality as if dying corral reef was not serious. If I remember right the Learning Channel ran an hour show on dying reefs but nothing much else came of it. I spent five days in Freeport, Bahamas about 10 years ago. My family chipped in and paid for everything for both my wife and I. A little disappointing in a way. A glass bottom boat took tourists out to see fish and if they saw two, looking for mates that no longer exist, I thought them lucky. We didn't try our luck but it was quite a trip. We left Boston in a snowstorm. While parked on the runway in Freeport, the pilot started a run; we were heading back home. After a very short run, the plane came to an abrupt halt. The pilot announced that there was an alarm going off but he wasn't sure what was wrong. We waited about 15 minutes. Memories of World War II came back to me – scattered pieces of bodies and aluminum sheet metal, a piece of wing sticking up toward the sky. I admit my nerve ends were firing stark messages. My wife, always scared of flying anyway, kept looking at me for reassurance. "We'll be all right". Off we went. I always sweat take-off, having flown some 1100 hours in the war and also to 1950 witnessing so many crashes, particularly on that engine straining climb heavenward. I confess that I am not inclined to read anything on that run down the runway. No sir, my hands are rather set on holding on to whatever is available. But this pilot got to me. Something might have been wrong, but off he went anyway. Boston was socked in with snow again. Too much traffic and so we went round and round over Providence for an hour. Couldn't see anything. Do you want to know what I was thinking when we finally headed for a landing, not a peep out of anyone? "I hope that the radar operator guiding us in is skilled and sober." All you younger buffs might be laughing at this, but what I have seen in 10 years of military duty would take quite a few pages to describe. Maybe I will later. The subject of weather can be monotonous, but be reminded. The subject in the future will switch from what does the weekend hold, to where is all my money going. Already flood insurance is either hard to get or very expensive. Medical insurance premiums last year jumped 20%. Guess why. Somebody has to pay for all the sickness and deaths from heat and other calamities that we have experienced. Property insurance? Wait and see, you like SUVs, you pay. The government is not going to tell you that.

Getting back to the graph which by now is a few pages back. I have to go off on tangents to provide relief from my dismal feelings that I hope doesn't get you down. Miami, in contrast with San Juan as you can see took a nose-dive in temperatures from 1978 to 1980 inclusive. There has to be a reason. Subject to prevailing westerly winds at higher altitudes, where no doubt emission products from the west find their way over the southeast quadrant of America, such as diffused and/or diluted CO_2, being in a settled mode and locked in one stratum or another going for a ride on its way back to the North Pole, the route which I will show not being a direct one. Miami experienced the results of cleaner air from 1978 to 1980. Less CO_2 and other heat retaining contaminants means heat collected on surfaces during the day is allowed to flow heavenward at night. First, nothing much can be proven in weather, therefore, scientists are on a merry-go-round. An example: I could claim Los Angeles pollution, although reduced by stringent rules, finds its way to a corner of Nevada, which, despite its aridity had a serious flashflood some years ago, or neighboring Arizona, but it would take years to verify what I know to be factual.

From Los Angeles to well into Arizona, there lies a rather wide valley over which Route 10 runs. The very passage of a high pressure system over southern California collects and hold pollutants gathered over what is a heavily populated area while disturbing them in all directions. The essentials to form thunderheads are three. One, the impetus to propel currents upward requires an unstable condition that only very cold air arriving at high altitudes can provide. Two, Pacific humid air, its humidity content maintained by the presence of pollutant particles to which moisture molecules cling to. These two factors might suffice, but let such a forming rising cloud run into a ridge or a hilly region and the worse can happen by virtue of the whole potential thunderhead system given a boost upward. Before I go back to a special atmospheric condition experienced by Miamians let me complete a pertinent and very deadly arrangement lying in wait for the unexpecting since I had above presented conditions over southern California which have lead to very violent flash flooding. If you note on the following page the clusters of such incidents, some deadly, just east of Los Angeles, you have an idea of what I am talking about. Additionally, other concentrated spots lie over busy areas of Texas. The rest of the country represents the pitfalls of living near major rivers, although, here and there cloudbursts have been experienced that are not connected with river flooding.

In a cumulonimbus (thunderhead) cloud buildup on the top of a ridge or hill the worse happens at the peaks. On the other side of ridges or low lying mountains there is the usual gully carved out by previous cloudbursts so powerful, autos seem like toys and are carried away half submerged. Authorities, from whom I extract information, claim that one half of deaths occur with occupants trapped in their cars. That is one very good reason why everyone should spend some time becoming familiar with this or that aspect of nature, in particular the messages clouds convey. The problem with such a scenario of the obvious dry riverbed or gullies, called Arroyos in Arizona, is that the bottom where a million tons of water (one cubic foot weighs 55 pounds) roared downhill then fanned out leaving what might innocently look as the ideal spot for setting up a camping stay. From the few I have seen, since I traveled out west and stayed in North Hollywood five months (no, not in the movies), the allure of a flattened area, perhaps attractive plant life, some tall cacti which had a good dose of water perhaps a year before, ample room for autos and other tents and mobile trailers, can be a fatal attraction.

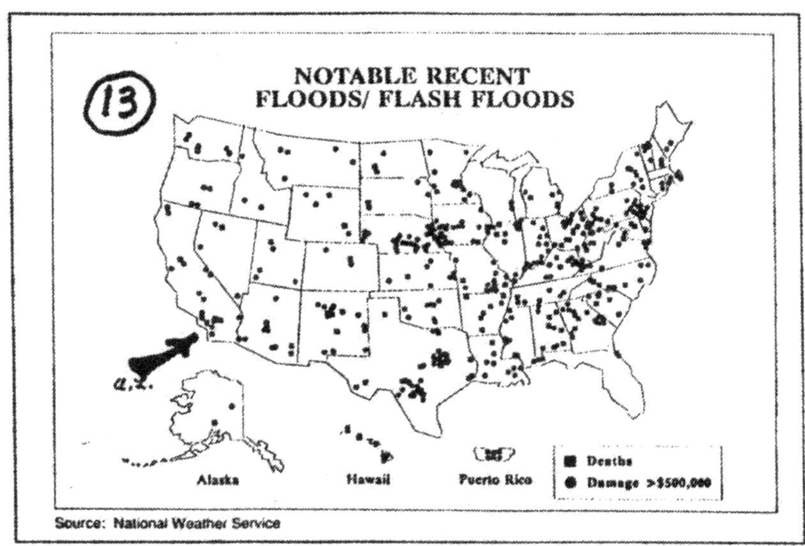

As for Miami, finally, and again I want to recycle events and effect already outlined in the beginning of this book, of an era spanning some ten years or so where thoughts of leaving cars at home and walking to the store (My God) transformed our citizenry into cooperative partners with the government. With gasoline at $2.00 a gallon, teenagers lost much of their enthusiasm for tire screeching exercises at night. Shades of World War II gasoline rationing coupons haunted many.

Following the Oil Embargo of 1973, President Carter prevailed on the Department of Transportation, EPA and others in the power structure, to mandate ways and means to reduce energy consumption. Detroit's Big Three had their arms twisted and were compelled to meet certain fuel economy requirements for auto fleets. The mid 70's saw the beginning of emission controls the state of California at the forefront of demands made on auto manufacturers. This state got tough enough to ban certain new autos if they tested positive for a disease now afflicting 1999 vehicles, namely gas guzzlingitis. "Don't ship that car to us, Detroit!" That's how the streamlined Sherman Tank builders got the message. After 1973, gasoline prices rose gradually. You don't think the Bush family suffered do you? However, a major serious jump in energy costs took place in 1979; lines of motorists waited their turn into gas stations. Later, Japanese cars arrived in droves and this nightmarish scene prompted Detroit to build economy oriented Colts, Chevettes and Escorts, etc. Insofar as pollution beginning out West, there is a domino affect as the accumulation of cities east adds to the total, obviously. Miami's drop in temperature

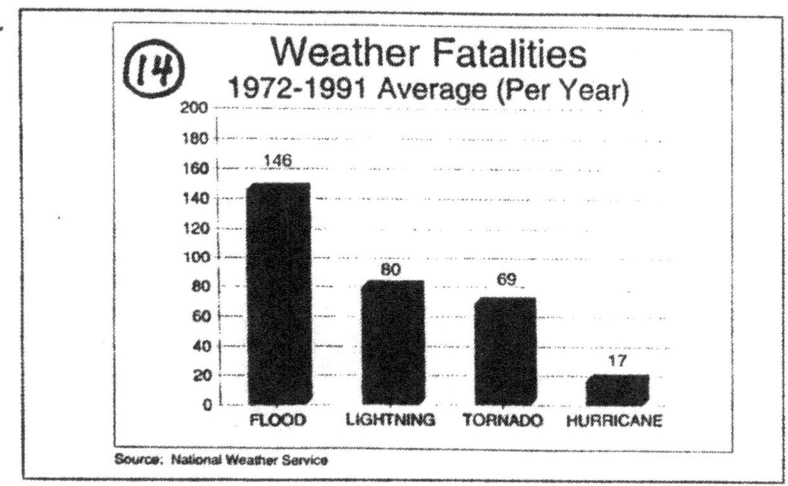

was due to cleaner air but no way in hell will authorities ever concede that fact, especially since they have condoned for the past decade the return of "Bigger is Better:" Whatever states west of Florida produce in the way of emissions, particularly loads of sulphur dioxide from scores of oil refineries in Louisiana, but more so along coastal Texas, along with other climate altering combustion products from heavily populated and industrialized southwest regions. These emissions rise upward to high levels, settling at different stratums and are transported in an easterly direction. Later, I will show you why Hurricane Mitch, killing over 10,000 was dubbed Hurricane Exxon by one of the more intelligent authorities. If you have never driven in Louisiana or Texas or even a few other bordering states, you will see hundreds of tall pipes, perhaps 50 or so feet high, belching plumes of flames, fed by natural gas being released from deep underground sources for what reason I really don't know.

Quite a sight at night, a panorama of candles viewed from a distance. The trouble with this, and I don't really think much can be done about it, it is a constant source of carbon dioxide. While I don't want to treat any reader as not being bright, it is night cooling that is hampered by both CO2 and particulate

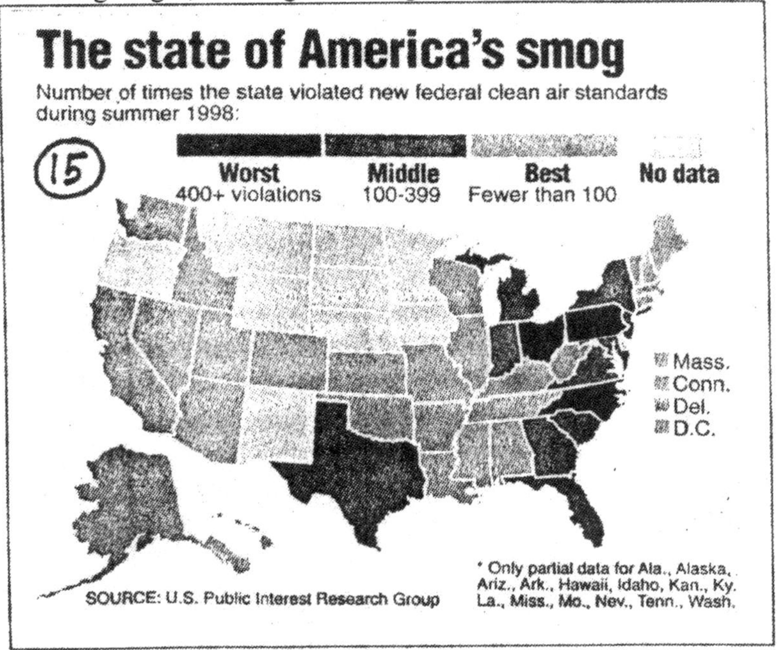

pollutants. Earth's daytime absorption of solar heat, more intense now, cannot return as long wavelengths of energy to the heavens above. It does so at an abnormally slower rate. Notice the illustrations below. They do not represent the maximum polluting states but rather states that are delinquent in cooperating with the EPA. That is a major problem confronting the EPA, maligned by the stupid and selfish and the greedy. You can see from these illustrations how Florida would be affected as is much of the rest of the country. A little humor now. We were driving through Baton Rouge and this day the odor of sulphur dioxide was strong. Our eight-year-old son, then, kept sniffing and finally said, "I smell something but I don't know if it's good or bad." It has an odor of sulphur but it is very toxic.

If all this bores you, wouldn't you like to live an extra two or three years? Some states in New England prove to have its residents blessed with remarkable longevity because they are out* of the firing range of airborne trash.

*I thought this – now I read where Vermont has problems.

Not all lung cancers are caused by second hand smoke. The metal etching power of nitrogen oxide was clearly recognized years ago as accelerating the wearing out of engine components when cars of old did not have PC valves that allow combustion gases in the crankcase to be sucked back into the fresh fuel mixture to be burnt away. I definitely want to bother you because from talking to many people for so many years some don't seem to have the slightest idea of what toxics and acids can do, including breathing in soot. There are two kinds of lung cancer. One creates a necrotic state, figuratively speaking, a degree of tissues dissolving and the other, the proliferation of cells growing wild or tumorous. I am bothered and find such people completely out of touch with reality when I see joggers slogging along in a city such as Providence, or wherever, busy with autos buzzing by and an occasional bus or truck belching black smoke. There they are, stopped by a red light, dancing and all sweated up, sucking in garbage at three times the rate of citizens walking about. Here I drive a Ford company van and inside my dash soot collects because I spend 3 hours of stop and go city and suburban driving, the windshield collecting greasy black unburnt hydrocarbons. Gas masks anyone? Naturally, if you only drive to work, say 15 minutes to half an hour, and go back home, you'll not appreciate what I experience. Can we do anything about it? Of course we can. Environmentalism doesn't mean sacrificing too much of our lifestyle. It promises all kinds of innovative products, inventions, and businesses to develop different and exciting vehicles. The problems is oil companies don't want to hear it and as you shall see if you have not already read, the latest battle looming ahead, then you will see why too many years will go by before concessions will be made between combatants.   Excerpts to follow later as I need to back up everything I am espousing. My cause is an intense one.   Clear up the fog over climate disruptions, make cognizance of ocean currents now kicked around to a point where it has become preposterous, encourage you to be heads above the rest, know who is responsible for tumultuous deadly weather and who, through narrow minded thinking or paid off, will damage our economy, especially our mainstay, the power of our agricultural advantages already beginning a downhill slide, slow but steady nevertheless. Last year our cotton crop lost 27% of its normal production, and so on, due to excessive rain. But I cannot go through weather vagaries by leaps and bounds, I must build up

U.S. gasoline not all the same

Average amounts of sulfur — a major cause of smog — in some cities' gasoline:

City	Parts per million
Chicago	580
St. Louis	
Billings, Mont.	
Detroit	
New Orleans	
Atlanta	
Denver	31
Dallas	3
Philadelphia	30
Miami	28
Pittsburgh	220
Kansas City, Mo.	190
New York	170
Minneapolis-St. Paul	80
San Francisco	40
Los Angeles	40

NOTE: Figures are averages; fuel sold in one city varies widely (from 300 to 820 parts per million in Chicago, for example) SOURCE: American Automobile Manufacturers Association

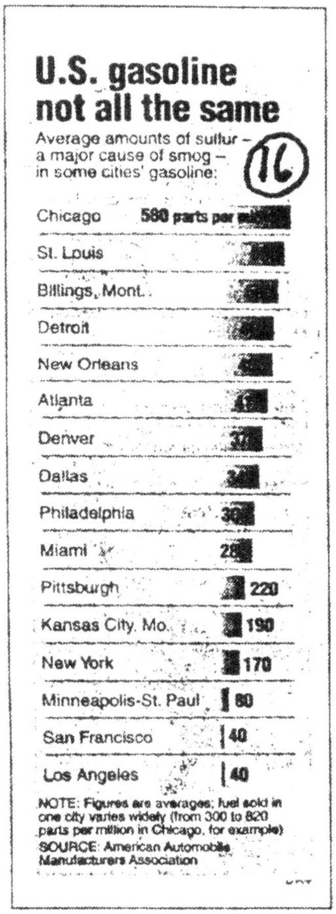

Becker's group calculated that over the next 20 years the other proposed rules "will achieve equal or greater emission reductions than low sulfur fuels."

GM's Leonard said, "You've got conflicting demands on the manufacturer. Do you want fuel economy or do you want (lower) emissions?"

ing the sulfur content by half, to about 150 parts-per million, in 22 eastern states with the worst smog problems.

"Why should motorists who enjoy wonderful air quality in Wyoming pay an additional nickel a gallon?" Meteyer asked.

This may not be the right way to clean up the air, said Jerry Taylor, natural resources director of the Cato Institute, a conservative think tank in Washington. These types of rules force people to keep older and dirtier cars because newer ones are getting more expensive, he said.

"They slow down auto fleet turnover, and auto fleet turnover is the main means we (use to) clean up urban air in this country," Taylor said.

After EPA issues its rules, get ready for a fight, Becker predicted: "These are very strong lobbies and we are taking on two giants."

piece by piece, repeat this or that in varying ways so as to build up conviction. My first ten years in the Navy were but the usual experience that all weathermen accumulate. This book is not about forecasting, although I will offer suggestions should by chance a meteorologist, who as most are inclined to, think over the possibility of something new is to be pursued. Instead, I am taking advantage of a decade or more, ample time available to do some serious detective work and pass it on. Below are revealing illustrations that accompanied the lengthy article I mentioned detailing EPA efforts and problems, the fight between oil companies and auto manufacturers, the usual butting in by acclaimed organizations who artfully find ways and means to distract attention from realistic appraisals of climate problems. The growing resentment against gigantic SUVs exempt from emission testing and so on. Below, one guy is complaining about five cents. The auto manufacturers want oil companies to cut down on sulfur in gasoline. The oil company wants the auto industry to improve on auto emissions. Twenty years? Already Washington and Oregon states have been taking a beating by nature for two months! Below, the expected battle. On top of that a think tanker says new cars have helped to clean the air. To a small degree, yes, but CO2 rises fast!

I didn't know SUVs, vans and trucks, new ones
that is, use to clean the air?

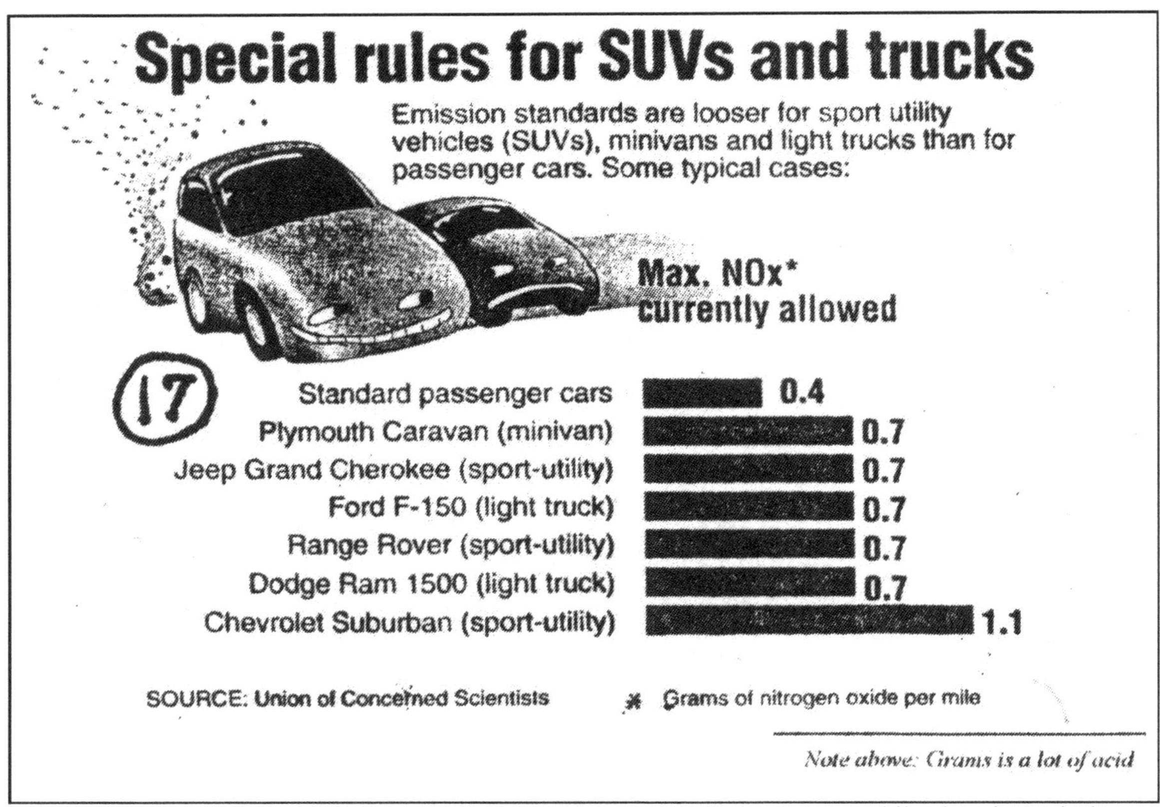

I walked in to a major car dealer and looked over a huge 1999 SUV. While the above indexes, for example 0.5 for my car, is boldly shown on price and detail sheets stuck to new car windows, you have to look real hard to find what the SUV manufacturer has to say about it. In extremely fine print they claim that they have achieved a 40% improvement in emission reduction. A magnifying glass would help if by chance your bifocals have limits.

If you are a jogger, head for the hills and only do so when the wind is blowing. Don't do so in the morning during calm conditions. Meteorologically speaking (I've got to leave autos behind) all dirt settles to surface levels during the night and does not waft away until late forenoon. One of our more eccentric drivers with a perverse sense of humor has a decal on the rear of his truck: EAT WELL, EXERCISE AND DIE ANYWAY. I'm not sure if that's so funny – I just know what is going on with our atmosphere primarily.

I wanted to return to the element I'm most comfortable with but let me take another shot at rather unprofound thinking. Owners' of old dirtier cars are poor. Their cars are usually parked all day behind some grimy New England red-bricked sweatshop. The poor for obvious reasons do little travelling, one being the unreliability of old autos. It appears from an expert, locked in his office with two computers on his desk thinks all new vehicles never get old. Has he ever bent over and kneeled down as I have and seen how small the catalytic converters on huge SUVs really are? Cost, you know. Now look at older V6 and V8 (now rare, the V8) engines and see how much larger converters are when manufacturers were more conscientious.

On the assumption that the young may read this, let me treat the subject on an elementary level first. I had defined Florida as both a surface and high level recipient of pollution or the absence of it for a short period by virtue of the prevailing west to east direction of an expressway system in the upper atmosphere shown here that propels either dirty air or clean air back to the poles. The illustration serves only to show the general direction.

Of course these flow pattern have a serpentine characteristic and imposed within this moving winding mass is the high speed conveyor belt, sometimes tow of them, called the Jet Stream. I have my own

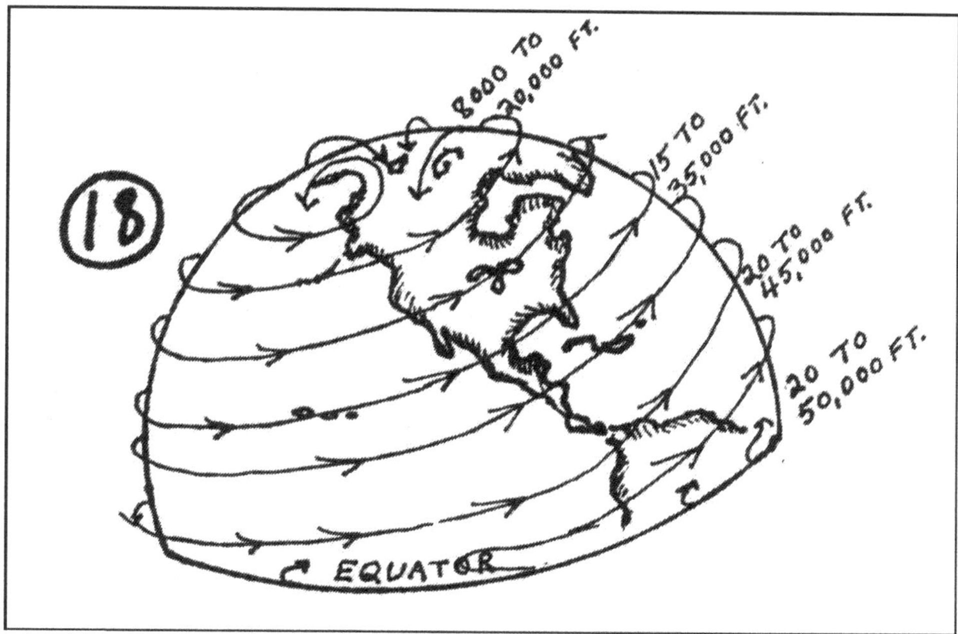

theorem over this high velocity stream but it is near impossible to prove.

Every day I have heard the same- two months of non-committal mindsets. Meteorologists, repeatedly, day by day, explain how very humid air coming in from the Pacific in powerful waves over our northwestern coast has maximized recorded rainfall. One town in northern California saturated by seven inches in a 54-hour period. Daily TV scenes of walkers leaning against drenching winds, an occasional umbrella deciding to flip inside out, autos plowing through hubcap deep water and a swinging stop light. While saying this or that record has been broken, none venture beyond that point. Couldn't they condition Americans with a suggestion that all is not well? We do know China, most prominent as a producer of pollution, along with others on the far side of the Pacific, has filled Pacific air with trash, which has already been established by observers and reported by the media. It appears a physics fact, namely that dirt definitely increases the capacity of air to amass greater quantities of moisture from the ocean surfaces, some 5000 miles of it, doesn't enter the minds of professionals who should know better. And just as limpid, the usual "Unseasonable warmth" when they should be saying, "Such a trend portends serious problems." They are leaving Americans with the feeling that soon things will be better. I can't advocate fear mongering. At least try an instructional approach so that when the day arrives, or politicians get hit on the side of the head with a few crises and decided that their bonding with auto manufacturers, oil lobbyists, utilities and industrialists is an ill fated marriage. Therefore, the public might be receptive to governmental intrusion, as promoters of free-for-all mob rule call it. I am glad however that the Weather Channel group does not venture into the morass of El Nino as some have. Even Paul Harvey, who specializes in brief but interesting reports, having, I suppose, read the Science magazine covering the surprising behavior of El Nino during 1997 and 1998, quite distant from our shores, painstakingly elaborated upon by a respected oceanographer furnishing ultra-scientific details worthy of great respect, concluded with an opinion. "They haven't a clue" I studied the article. I like what I learned, but it can be misread.

This being February 14, 1999 I want to consolidate what I have presented in many previous pages in a scattered fashion so far, those pertinent details revolving around causes and effects pertaining to our beleaguered northwest coastal regions. Let me again emphasize (I suppose before I go on with this sentence you are thinking it's more of a jackhammer style of making a point) the dramatic conflict of a temporary re-invigorated, therefore a much

more colder normal Arctic region with vast warmer systems over the Pacific Ocean. Systems that are not to be normalized as is evident in their extensive unaltered dimensions and power.   Power in this perspective, bucking invading cold air and creating huge storms. Fortunately for us, we are not at sea somewhere between Washington State and the furthest tip of the Aleutian Islands. As you probably know an oil cargo ship was driven aground and the Navy decided to set fire to it rather than let leaking oil again cause major damage to the ecology. In short, there is normal cold up north but abnormal warmth south. That is a very troublesome combination and only dreamers imagine it could change. We have a sun whose increased radiation profoundly expands all; I repeat all mid-latitude and tropical high-pressure systems girdling the globe. This belt of stored warmth, each invisible (sometimes visible) particle of moisture carrying it's share of solar energy, has reached much further north in our hemisphere. While I will avoid the added burden of elaborating over the other half of our planet, I hasten to add the Southern Hemisphere has also a similar problem. Unless the aging process has tragically limited one's brain there is one advantage to being old. I remember what normalcy in meteorological functions looked like and how the configurations of weather systems were different in terms of dimensional influences and so forth. In normal times less powerful warm highs would systematically have much of their volume drained off, pulled up by frequent low-pressure systems. Now, huge storms, and I mean they are huge, between the Aleutians and our northwest do not accomplish that – no significant contraction of expansive warm high systems across all of the Pacific. This area our topic for now, is occurring as it should if it were not for the sun's unusual effects. Let me remind you again of a precise arrangement. Storms function not only to spread precipitation over our planet, but also must siphon off warm blocking systems, get them out of the way if you will. Once that rising column of air has shed it's rain or snow, off to the North Pole it goes via a high altitude expressway.

Continuing a lengthy review, we familiarized ourselves with the findings of two lady researchers who through an arduous effort digging up weather records going back to 1949 from over 100 weather stations, revealing not only how much more heat waves have affected us but the increase in humidity on a year round basis. The only disappointment I registered with you is that they also were noncommittal as to the reasons. An example of how, besides the northwest, this steady humidity, sometimes less of course, affects California in an area that really shouldn't have so much fog as it does. A truck driver, phoning in to Jim Bohanan's talk show whose guest was a lady truck owner, said the most dangerous highway in the world for weather lay in a valley, and I'm trying to remember precisely where, was I-5, a stretch of probably 200 miles, somewhere between Los Angeles and that distance northward. The humidity rolling over coastal hills (they're called mountains out there) and fog stubbornly persists. Accident after accident to the point where this driver suggested the possibility of driving trucks up on train beds that ply along a railroad also part of that valley. Now, if you are a New Englander, or an East coast dweller, all this might be banal but a Californian might think things over should he finally be educated as to why this mess. Getting back to the northeastern Pacific, so powerful is this relatively warm high system over much of the Pacific that it held at bay a huge mass of frigid Arctic air bulging at the seams for two days before it finally broke through and pushed southward. It did, this being a bit unusual, get a rise out of weathermen. Further south, the party over El Nino is being revived, but I'll get to that later even though the topic of pacific weather calls for some more evaluation. Let me finalize the above by punctuating specific abnormalities in one tropical region substantiating the effect of the sun in late January no less. By now, you might view me as a dentist bearing down with his drill, pulling it away and here he comes again, or when am I going to let up! Honolulu has not cooled down at all, except for a day or two. San Juan hangs in there with an eternal 82 degrees, Jamaica receiving a piling up effect all the way from somewhere near Africa plays around with mid 80's when it should be mid 70's according to previous statistics. One day or so of sunny days and Texas temps loom up to the 80's. Looking back at EPA resisting Texas do you suppose Republican senators and Congress folks really care about the probability of more record breaking deadly performances by Nature? Or is it oil that matters most? Let me now dwell a little further into politics or why we are not getting anywhere, even though President Clinton in his Address to the Nation said he will try to help industry cut down on some of the sorriest, damnedest pollution, suspected along with auto emissions, of being responsible for the sun meeting less filtering as it's UV rays go through the Ozone Layer. That protective umbrella is in a sorry state. I don't expect much cooperation from politicians leaning to the right – I suspect a small army of lobbyists are waving greenbacks around. Let's look at the record of the esteemed Ronald Reagan, who is credited with bankrupting Russia. That's nonsense! He had not the slightest idea years before he began to push Star Wars that Russia would spend their way into military oblivion. In fact, for a while that country built up its forces. War with Afghanistan devastated Russia's economy also. One reason for such a system entailing

By JOHN CASTELLUCCI
Journal Staff Writer

PAWTUCKET — Ronald Reagan's two terms as president may have been "morning in America" for many people, but they were dark days for the solar power industry.

In 1985, when Washington eliminated a set of tax credits for solar energy, the impact on the rooftop solar-energy industry was devastating, according to Robert W. Chew, former owner of a small, Rhode Island-based solar energy company and current director of Project SunRise a state-subsidized for-profit solar energy effort.

"Over 90 percent of the solar companies went out of business," Chew recalled.

The tax credits Reagan eliminated dated back to the 1978 energy crisis, when solar energy was seen as an alternative to dependence on foreign oil. The tax credits led to the installation of more than a half-million household solar energy systems.

Solar power may never again become as big.

PAWTUCKET

laser tracking and shooting down missiles from some space platforms that probably turned Mr. Reagan on, even though I know we have bastards all over the world who want power, is Silicon Valley of California, loaded with electronic firms, would greatly benefit by such a, well, a Buck Rogers scheme. But for someone to suggest Reagan's features should be added to the four presidents carved out of a mountain, which is too much. The man had his merits but he was strictly a Republican, chop away at the good and build up the wealthy, so that they can run off now to foreign lands for cheap labor. He cut down the FDA and sure enough, here comes rotten meat, that's a cheering thought. I told my wife, don't she dare buy hamburgers or hot dogs! Now read this below and see how a substantial amount of energy (oil) would have been saved – a horrible thought for those clutching stocks. Just think, solar heating and so many states could have benefited. The biggest items using electricity are air conditioners and refrigerators. Natural gas or oil, that mattered more – that had priority. As long as Reagan's beloved state seemed to have more control of emissions and the Pacific blowing in what he, I suppose, thought was pure clean air. Why encourage solar energy? Chop, chop. I wonder if stocks of Sun Shielding Goo are worthwhile investing in? I can't watch our TV now without some excited Fifth Avenue salesman or saleslady, eyes all aglow, pumping out enthusiasm over this or that mutual fund, busting up a show every 8 or 10 minutes.

There is a lack of balance in our priorities and phones-in-faces people drive around in such a hurry you would think it had to be done before some anticipated end of the world. I take a slap at this crowd because that is who will oppose any action against pollution. So much emphasis of dollars per second in advertising that I have to put up with some guy on the radio rattling off in auctioneering frenetic style the merits of this or that product or service. No wonder psychiatrists are busier than ever!

And the EPA. Its pursuit is a much needed one. I'm not in agreement with them over the use of dynamometers to test auto emissions, but if I have to I'll sue the state if they overheat my engine. All it takes is a rookie mechanic tester and he stresses your engine too long at high revolutions*. Here comes the steam! That's the end of the piston rings. Oh sure, your car will run all right. Some slight loss of

By JOHN J. FIALKA
Staff Reporter of THE WALL STREET JOURNAL.

WASHINGTON – Environmental Protection Agency investigators have found fundamental weaknesses in a major national air-pollution control program that have allowed many "significant violators" of federal and state regulations to escape inspections and sanctions for years.

Recent reorganizations within the EPA helped generate a serious lack of federal oversight over the program that monitors factories, utilities and stationary sources of pollution, according to the EPA's Office of Inspector General. The program, started in 1970 under the Clean Air Act, is one of the nation's most complex and far-reaching environmental regulatory efforts.

EPA auditors put some of the blame on a 1994 agency reorganization that separated the enforcement branch from one that provides grant money to the states. That has left federal enforcement officials with little leverage to push states into compliance.

William Becker, who heads the State and Territorial Air Pollution Program Administrators association, based in Washington, blamed the EPA's 10 regional offices, which, he said, were making an "inconsistent application of the program" in their dealings with states. Some states are "livid" about the inspector general's findings, he added.

power hardly noticeable because the top compression ring will be cooked and will allow excess combustion carbon to pile up behind the two oil rings below it. Then you've got a premature oil burner. Happens all the time. Do you expect some bureaucratic willy-nilly, sold on this dangerous idea by dynamometer manufacturers to accept effective and safer alternatives? But the EPA is bucking not only resisting industry, particularly in states that are under the influence of right wingers, but also the difficulty of maintaining law and order in a vast organization there being not only some to be expected failings but state departments managing environmental issues, so many of them yet to be resolved. Our chief DEM manager here lasted one year and quit in a fit of anger over a barrage of criticism for certain violations never having been attended to. Having talked to one gentleman myself back in 1994 I learned also that they are not loaded with weather knowledge or experts on auto engines. The main problem with this world of bureaucracy is here and there; the male brain has this residual Cro-Magnon cluster of cells. To dominate, to push around, to strut his stuff like the peacock, that's another reason for dissention between DEM and EPA in some districts. If you wish you can extract detailed reporting of conflicts between states and Washington, some livid comments by this or that official from a library. The Wall Street Journal, quite tuned in to the future of economics published an extensive article over EPA's problems December 28, 1998. Here a few excerpts plus the central framed piece that says it all. However, I am not nor am I in any position to encourage more criticism. We need the EPA badly, but it may be too late at the rate we are going.

60 mph

Below, an article has EPA administrators, conscious of the free-for-all mob and rightists viewing them as Big Brothers, selling their accomplishments in cleaning the air. That is a moot question. I limit this offhand new to two excerpts. One boasting of cleaner air, which I can't agree with because CO2 content in the atmosphere has jumped up and apparently it doesn't seem to worry EPA. Another paragraph suggests it is not so clean. Oh well.

WASHINGTON — America's air keeps getting cleaner, and a new pollution report from the Environmental Protection Agency shows that in 1997 the nation's air was the cleanest since the federal government started measuring six major pollutants.

It is not yet time to breathe easily, other experts agree. The nation's nitrogen oxide emissions — one of the key ingredients in smog — have been rising. Independent environmental groups say this year, with its record heat, was a particularly bad year for smog.

I am afflicted with creeping pessimism but it shouldn't erode anyone's happiness – I didn't say the end of the world was just around the corner. In fact, given the possibility of a rude awakening, we might see very exciting innovative developments in the future. The EPA's enforcement setbacks, many dragging on for years, simply reflect a degree of antagonism against regulations and that is catchy and it is spreading. State governors and their political entourage, holding hands with needed industries may be condoning stalling tactics. I don't see governors, for example, making any particular fuss over poor air quality, do you? And the head honchos in Washington? What are they telling you besides the theory that ocean levels may rise? Tell me if you can, indicative as 150 tornadoes in November is of something being very wrong, was there anything proclaimed other than the equivalent of Oh My? Clinton smiled and Al Gore, intimidated, said nothing. An author, interviewed by WBZ's Dave Brudnoy, explained how the affluent detach themselves from national interest. With each step upward in the acquisition of money their former community spirit on a national level evaporates. In business or as most working for corporations they are inclined to see EPA as governmental intrusion rather than pursuing a noble goal. Today's Providence Journal published a lengthy article by the former head of RI's besieged DEM. Even accusations of transgressions, failure to do something about barrels of toxic chemicals on a former naval site, for example, led Mr. Andrew McLeod to resign even though a multitude of failures occurred before his appointment as DEM director. The leading paragraph, the excerpt on the next page, is all that I needed. We need total public cooperation but we don't have it. Following this excerpt, another very brief one by Froma Harrop, the lady journalist and prominent commentator for the Journal. A few sentences shown representing the embryonic malady beginning to manifest itself amongst the "Leave us Alone" affluent as to what book publishers are concluding from the lack of enthusiasm over very good books. One, the Agenda, a thorough and interesting analysis of how government works - it's problems and the enormous time consumed in the passing of bills. After reading it, the reasons being very interesting as to Washington's need for it's politicians to have major crises or the equivalent of a blow to the side of their heads before acting, my glum attitude remains. Harrop's article merely substantiates the exodus of the affluent to Xurbia, leaving behind critical issues as being someone else's problem that I am writing about. They want no part of social "Inclusion". I'm afraid, unintentionally; a coalition of anti-government advocates grows.

ONE YEAR AGO, I wrote in these pages of my optimism about the work of the Department of Environmental Management and its future direction. Having recently completed a year and a half as the director of DEM, I am no less encouraged about DEM and its future — as long as the people of Rhode Island are serious about and supportive of this vital agency of state government.

It is your natural environment, Rhode Islanders, and your DEM that protects it. The department that you get is the department that you want and are willing to fight for.

I never thought a wrestler would begin to intrigue me as an interesting political figure, but Governor Jesse Ventura made waves with just a few statements he made to a crowd of reporters while in Washington, DC. Obviously he wanted to get the attention of what a local radioman called "What a flock of lawyers need to hear in both houses", when he said Americans want Common Sense, Honesty and Straightforward Plain English.

Since Tom Brokaw's book "The Greatest Generation" referring to we old duffers, has become a best seller I feel important enough now to herald myself as being Mister Common Sense and Honesty. With a diary mode of writing in mind, daily news catching my attention, the weeks of tremendous snow storms in Europe, which killed dozens, I want to remind you, never having seen so much bad weather in 50 years, that the main ingredient is excessive moisture and you need not my hammering away again at the relation of pollution.

Even though changing the subject matter to other areas that are of importance to me seems unrelated to the main topic of pollution, and it may be judged as an inability to maintain continuity of thought and my lack of journalistic experience, I simply am revealing the permanently imbedded image I harbor over political philosophies. That is important to we "Psycho-babbling environmentalists" as we are called by the Great One from Missouri. Given that fifty to sixty years of avid interest and searching for truth plus some sorry experiences, I have no difficulty recognizing wolves in sheep's clothing. Hence, on a take it or leave it basis, I review how today's partisan political orientations reflect yesteryear's parental influences plus a few inherited genes. If Pop was a bastard and also a high roller in an amoral way, then his male progeny most likely is also a bastard. A zebra does not change his stripes. What suits the rich best has priority for ultra-conservatives. Today, Russ Limbaugh made no bones of how he feels about President Clinton's agenda focusing on saving Medicare and fortifying Social Security. The man said that these two items were simply another form of welfare and that Clinton has simply expanded on it. Yesterday, the 17th of February was a peach of specious material-earth warming is not caused by humans and following that, a complex repertoire over glacial cycles that I am sure his mob advocates have gratefully swallowed. I am by nature an optimist, but the problem of earth warming and our walking on thin ice over the consequences is not being addressed with much resolve. I'm afraid I'm losing it – I just can't see anything to be cheerful about. We are now hearing cries from Republicans to cut taxes – having their way would only steam up the economy and take away budget surpluses that could be used for much better purposes. An example of their largess was and is corporate welfare*. Thanks to our government doling out tax money to companies enthusiastically endorsed by conservatives, the extra profits derived served to facilitate their running off to other countries. Let me quote Reverend Jesse Jackson whom, although not a favorite of mine occasionally makes an excellent point with his caustic wisdom. Facing four reporters, he accused corporations of stealing our tax money and running off to other countries. The reporters seemed dumbfounded – I don't believe that they ever thought of it that way. Already plants closing down have ruined the lives of many families in New England cities. But conservatives whose families are busy accumulating stocks and bonds don't care at all. The growing wage disparity which has to worsen as I expect it will, exacerbated by rises in the cost of living following repeated bouts of destructive weather has to create serious social problems. Despite glowing GNP figures, the affluent and the wealthy reveling in good times, some signs of discontentment and anger are evident already. There are jobs of course, but you can't pay the rent of minimum wages. We have here the consequences of disenfranchised parents passing on their frustrations to children. Last week vandals in a Pawtucket cemetery tipped over 50 tombstones. Tires are slashed, fenders kicked in, stolen autos, graffiti, swaggering youth standing

Corporate Welfare: 51 billion annually direct subsidies and 53 billion from tax breaks.

on streets blocking advancing autos, defiant, increased drug trafficking and according to the Mayor of Central Falls the court is jammed with defendants every day. Only 22,000 people in Central Falls, but our police overtime pay amounted to $200,000 last year. Young punk burglars are now looking for more lucrative markets and robberies are being reported in suburban areas. That is just the beginning. Why am I emphasizing such a trend, which surely will worsen in the next decade? I understand it - that is why. I experienced it and furthermore went to great pains to find out the reasons for the Great Depression. A society becomes dysfunctional and then violent. Former Governor Weld of Massachusetts has been quoted as advocating opening the doors to more immigrants. Some friends of his must own sweatshops. That's all we need – more disenfranchised minimum wage earners who can't speak English.

I don't know much about Republican Senator Lott, but at least he was honest in a statement he made last week. "Our party has alienated almost every segment of our society but the rich and businesses like us – that makes us happy." That's what I call stylish honesty or in other words, he doesn't regard most citizens as worthy of concern. So, while today's affluent fondle mutual funds and company annual financial reports, they will experience a different future, a costly and troublesome future.

Mr. John Gailbraith wrote a three-page article explaining the 1929 Stock Market Crash. The title, "When the Money Ran Out." Only one catch – he left out something and he's supposed to be socialistic oriented. A lot of that money didn't disappear into thin air, in fact, many Wall Street <u>insiders,</u> walked off with fortunes after that 1929 debacle. Because America was and still is the economic Kingpin the whole world as we knew it then collapsed. For ten years before the crash Republicans presided. The Roaring Twenties, technically was a lawless period. Insiders did not have to meet any more than 15% equity to borrow stocks and pyramided their paper fortunes to higher levels. But, as I learned from books, they also knew when to "short sell" to a frenzied crowd of investors and although next to impossible today, that method reaped them quick fortunes. A cute expression identifies this peaking stage of rising markets, known as a Period of Distribution. Following that <u>induced</u> collapse, ten years of nearly worthless stocks, for example, US Steel from 200 dollars to a low of 20. The insiders always maintaining a low profile slowly picked up stocks and businesses for a song, and of course, this was known as a Period of Accumulation. Even today, ultra-conservatives still feel it is their amoral right to have free rein in matters of finance.

The richest country in the world and my mother would receive but 12 potatoes a week from some charitable function doled out by city hall. After her walking two miles to work and two miles back for over a year she broke down. Her frequent crying over nothing bewildered me. Behind in the rent, I heard her asking the landlord to wait. My younger brother suffered from malnutrition and later would spend one year in some recuperative hospital of sorts. As for my physical ailments, it is not important here. Finally, we were sent to an orphan asylum but later we would be freed of this lonely life and bad treatment. Mother remarried. I absolutely do not seek a sympathetic reaction from anyone. It explains my choice of political parties. Additionally, I perfectly understand just how much the disenfranchised will tolerate till they react – I'm making that clear. Remember Watts? A party that insists on a small government as long as they run it their way is to be feared – history proves that. We need conservatives but not as an overwhelming majority.

I now read where a new movement may be underfoot. Environmental issues and the corrective measures that will be necessary, has proponents for increasing state control while reducing federal authority, drumming up approval. Being familiar with a mixture of state legislators, some lawyers intent on graduating to Washington, landlords, friends of car dealers, one man here running for representative while he earns a living driving his own heating oil truck, couldn't even speak respectable English, most claiming to be the epitome of good citizenship, "I was on the school committee for three years." "I'm married and have three children." In all, a variety of sincere and not so sincere candidates. This idea does not set well with me. Let the EPA be the whip boys and work in conjunction with the Department of Environmental Management, a state function, even though temporarily DEM is in disarray.

Today, Russ Limbaugh, some caller probably intent on or induced to initiate an anti-environment tirade by a rather influential radio talk show host, Republicans quite enamored over this agenda, rattled off in a steady style, accustomed as I am to his hesitant speech, looking for impact words, suggesting he had already prepared for this a very complete and scholastic review of ice ages. The warm periods hundreds of thousands of years ago, following up with a convincing "Human beings are not responsible for earth warming!" I'm sure many of his followers were impressed. That is what we are up against. But as with others, there is no Earth Warming – it's a myth – so they say.

According to polls, 70% of Americans are concerned about the environment but I'm not convinced that they'll accept some blame for it's deterioration. For one, I know why Al Gore has been silent over the matter lately. Yesterday Dan Rather announced we had broken a record for gasoline consumption in 1998 but he doesn't dare broach the subject of a possible connection with that year's horrendous weather. General Motors cannot keep up with demands for huge V8 powered pickup trucks, most on the road already quite polished and showing no evidence of being used for business. For a polled majority of concerned Americans this doesn't strike me as being ready to cooperate. Now the EPA is turning their guns on two-cycle engines, snowmobiles and other popular units of work or pleasure, my knowing as many others that this is a filth-producing polluter, spewing out one third of every gallon of fuel, unburnt and loaded with oil. I find it hard to believe but supposedly, such engines represent 25% of all emissions produced by cars and trucks. It has to be or we let nature beat us into the ground. Granted I am being a pain, but we are floating on the perimeter of dangerous times.

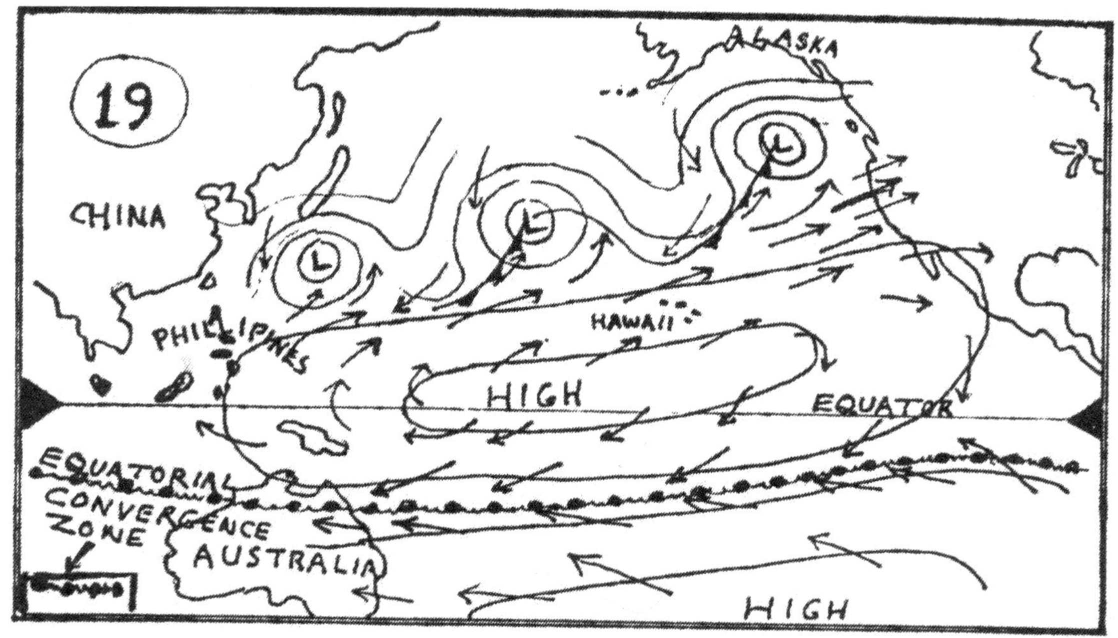

Am I revealing some teeth-gritting attitude about the past and showing too much disrespect? Actually, I'm quite relaxed as I type. The past is the past, but for me it is lessons to remember and apply. In fact, our TV set is on; I watch the Weather Channel or TLC. A meteorologist is explaining the continuous massive surges of high humidity laden air at all levels, clouds barreling over the west coast, muddying up two states and one half of a third, California of course. The satellite view of weather in motion is more than impressive; it is strikingly powerful and far-reaching! Over 80 days of almost steady rain and tree-bending winds brings out the usual explanation. Every day, the same limited synopsis I've listened to and watched – the man apparently not cognizant of what this augurs for the near future does say, " A tremendous amount of moisture is coming <u>all</u> <u>the</u> <u>way</u> from the Philippines" while raising his eyebrows. Viewers must have sensed that he recognized the unusual. That's 7900 miles! Remember the big surprises for a weatherman when he, at least, did a mental somersault November 21, 1998 after seeing the immensity of a high system scanned by Nimbus 7? Well the one on the next page is the same Jim Dandy except it has drifted south. That's what the ICTZ zone we covered is all about. The equatorial line of two hemispheric highs migrates about five degrees south of the equator. I can't be exact as to its irregular line but Australia is affected. This gradual push southward occurs during our winter. I have a theory, one of the few. Over the Atlantic this invasion by a high system that belongs to us reaches down into the southern hemisphere during winter when our green life no longer can produce oxygen, and I suspect that we are borrowing some from South America and Africa's greenery which is in full bloom during their summer. Not a theory this one, rain seasons shift southward in both continents above. Although I believe in God, I am not associated too fervently with the religion I was raised on. Still, a check a lady gave to me inspired me to let loose a little. In the corner was a brief bit, I suppose from the Bible, "Consider the wondrous works of God" Job 37:14. I've seen so much in my life that I am afraid I'm quite convinced that some entity is in control. Of course the check meant a lot to me also.

If I seem to think of myself as a paragon, which I am not or a study in ostentatiousness, so be it. I know if I were in charge of TV's Weather Channel programming I would instill into the weather folks to make a little more effort to teach and condition the public. In an effort to turn off the heat on the reader I have to poke fun at myself. For example, you would find me an amazing TV manager. I just want to let you know I am not ranting and raving to myself even though I am dissatisfied and elaborate too long in a critical manner. Why don't that darn weatherman add a little more emphasis to his rendition, altogether duty-like except there is a permanent smile, of 25 foot waves and enough drenching rain to make the Sahara desert livable again. At least one minute of what he believes might be the cause of such tempestuous weather that doesn't seem to end. The illusionary returning sun, now mid February, having maximized it's power south of the equator, but still abnormally strong enough to keep the equatorial zone heated up all winter long is now inflating our power system as seen in the previous page's illustration. I wouldn't expect that approach from any weatherman (means ladies too) but at least a raised eyebrow and if brave enough, raise a voice and say, "My oh my, that is unusual!" Of importance to me, Jamaica now has climbed to 89 degrees, seven degrees above normal. No need to repeat other sites reported. Of course you know all about heat records in the south. In the meantime, all the King's Men in the government and the media are masticating to a molecular pulp each and every word that can be reconstructed again and chewed on again and again over the impeachment affair. My radio is not being used too much at night now for that one reason. Now a woman is claiming according to Russ Limbaugh that the governor of Arkansas had bitten her lips – another avalanche of excess verbiage is about to descend upon us.

I sit here, fingers resting, not typing, and looking for that TV tube to reveal something more enlightening. These meteorologists, what makes them tick anyway? We are told how to drive, wear the proper attire and button up, and on and on with reports of temperatures and chill factors for too many cities. "Drive slowly" that's another one or "Watch out for icy roads". I am annoyed because so much more potentially educational is left aside. For one thing, they should advise viewers to buy sturdier umbrellas, so many scenes of leaning pedestrians hanging on to what remains of such a useful item, bent rods sticking through flapping fabric, which makes my wife laugh and I make believe it is funny but my mind is elsewhere. Especially some longhaired person, hairs sticking westward and he or she is heading south. Am I too sardonic? I wait and wait, sometimes 45 minutes just to glimpse at the most wondrous of technology, the God-like view of what a satellite camera has to show. All that churning weather down there but flip, flip, the Weather Channel allows me a lousy one quarter of a second to see a view of Europe, sometimes Africa or South America but that's all – flip, flip. More important to go on and on with chill factors and take the dog in. It's a wonderful show and I suppose I'm unreasonable. I want you to believe this; we were just as good in forecasting. When lives are at stake, particularly with aircraft filled with a crew and no radar assistance, you become sharper week by week – you had better! I could tell you how six B26's, two engine bombers, nicknamed the Flying Prostitute because of it's short wings, had no visible means of support, disappeared on the way from Morocco to England. I don't want to rile up a certain entity should this be publicized, but the pilots were told they had to wait for favorable winds, such was their limited fuel supply. Well, some esteemed high-class meteorologist radioed it was perfectly all right to attempt the flight, some 30 men aboard these unique planes. Who told Eisenhower to invade on a certain day and a rough sea practically made landing barges sitting ducks? A day later the weather was perfect. The man, we were told, had three gold stripes and you don't get them usually coming up through the ranks. A good forecaster has to be reasonably intelligent, work overtime to study why something went wrong and have a sharp memory or the years are a waste of time no matter how much schooling.

Of late, truth is coming forth because the real men who were there are telling different stories. Truck loads of dead soldiers, their stacked four deep, legs dangling over the hinged tailgate, rolled into our air station from Tunisia, from there they were flown to the USA. I don't know how many thousands died in Kasserine Pass, but some West Point graduate blundered. I believe he was a Colonel. He stretched out a long column of tanks, trucks and troops for miles between mountain ranges. General Rommel simply knocked off the lead tanks and the tail group and rained down artillery on soldiers unable to make any retreat or move forward. Oh we had some fine naval officers and generals but there were a few lulus. I'll never forget those trucks.

The good old days; I could write another book. Hard times sharpen your wits and you want to break out. You also become very resourceful when disaster looms. I think this is hilarious, but a rather thin friend we had in our gang before World War II was privileged. He owned an old 1930 Ford four door sedan. Four of us would pile in after chipping in for gas – very expensive, 8 gallons for a dollar and this one time we decided to test it's top speed. The decision was ours and not Gil's, the owner. Now, if the timing was not right and you were squeezing 60 miles per hour, the exhaust pipe and muffler would get red hot, but we didn't realize that. The floorboards caught on fire but we managed to extinguish it. That's not the really funny experience we had with Gil, who ended up being the personal yeoman for Admiral Halsey in the Pacific. Two months before Pearl Harbor, so many desperate young men trying to join the armed forces, no jobs and all that, the Navy had rigid physical requirements. Off we went to Boston, the Fargo Building I think, and we waited several hours while Gil made his bid. He walked out toward us sitting in his auto looking totally shattered – yes to him it was a disaster, failing the physical. "I'm too thin, I was 4 pounds short of 128 pounds" or whatever the minimum was. Sharp, our gang

showed the cultivated acumen of hard times. We bought 6 pounds of bananas after waiting three days before Gil made another try and forced him, I mean the last banana was literally jammed down his throat, to eat exactly five pounds worth. In he went for another physical and passed. Talk about a nauseated but happy Gil! I remember laughing when he said, "I keep tasting lacquer". I become self-conscious as I write unable to avoid the somber. The future is inevitable for some, insidious for many, troubling changes, I resort to a form of relief; a few stories are called for.

However slow its success rate as implied by the Wall Street Journal, I know that the EPA is progressing but I can't totally agree with their claim that the air is cleaner. Why such a major protest by residents in and around Atlanta last year demanding something be done about pollution? I also know that Providence isn't exactly an example of arrested emissions. The extent of emission densities varies of course – one year, particularly summer, the medical associations report excesses throwing in hospitalization figures, some deaths of children and the next year there is peace and quiet. It's a matter of weather and wind directions, if there is enough of it. I know there is constancy to emission output, considering that Georgia city and surrounding countries was highlighted in a newspaper article, two of them, one mocking the propensity of those citizens having some 1,700,000 registered SUVs but quite often such dirt (meaning variety) finds its way eastward where it is in the realm of not possibility but definitely reality, South Carolina has to be affected. From there it spreads over the Atlantic. I want to draw a few illustrations to follow, to imprint into your minds what I know are the mechanics involved governing the distribution characteristics of passing high pressure systems, particularly in colder months but nevertheless to a certain extent the rest of the year. Winter means more homes using oil or natural gas heat and of course, more gasoline is consumed. In addition to some soot, or otherwise my window panes would not collect so much dark stains surrounded as we are by thousands of apartments and houses plus quite a few factories. The reality of a constant supply of carbon dioxide is a continuous and increasing factor. Although the amount of carbon monoxide is tolerable as measured and reported daily, that too continues its upward climb.

Late in the 40's apprehension mounted over the potential threat of atomic bombing in view of Russia flexing its muscles, the Cold War not as yet reaching a publicity stage. The Navy Department directed us to prepare and send them radioactive particle fall out charts so as to determine how wide of an area would be affected. My guess then was they wanted to see just how much time the advance of such a dangerous invisible cloud potentially menacing citizens could be counted on for evacuation. Frankly, it didn't bother me feeling as I do that God is not about to destroy us but I'm a great believer in keeping the old rifle clean, just in case. Our need to maintain a powerful defense is clear – I didn't say God will protect us to that extent that we ignore the menace of insane leaders scattered throughout the world. But for us to be destroyed, that didn't and still does not fit in to my view of the scheme of things. Our naval air station, the only military unit near Boston worthy of note had been selected to prepare estimates of radioactive distribution compiled from balloon records of winds aloft. Descriptively, such charts show the speed of particulate distribution and estimated densities given to us by, I suppose, physicists working in Washington and the distances involved from a west to east trajectory. There is not enough room on these sheets so I'll reduce the compactness of lines, figures and what have you and furthermore let me transfer a

Boston study to New York City and kill two birds with one stone. One, an idea of what radioactivity traveling ways looked like and two, how New York and the vastly populated and industrialized areas within and around that area manage to send us their unwanted goods, gratis, no charge. The very problem that I wanted badly to demonstrate to RI's DEM but I thought better of it. They were determined to stick with their inaccurate assumption, rather than a scientific report they publicized. I already described this very minor incident in my life. Another reason for illustrating is to give you a hint that I have varied experiences.

Let me interrupt our train of thought. As I type I also keep glancing at the Weather Channel's continuous description of weather nationally. While the Atlantic powerhouse as I will from now on call both the Pacific and Atlantic belt of semi and equatorial high pressure, is finally yielding and contracting, thus allowing cold surges of air to punch southward, the Pacific is far from changing it's ways. Continuously the northwest coast is being barraged and drenched simply because of the atmospheric fixture, unchanging, I have been hopping on. As a side note, a small article in today's newspaper, this unrelenting sleuth never letting up, a group of ice fishermen must now realize there is earth warming going on despite a normal cold winter this season. Quote: "We have never experienced this in 30 years". Now, that makes me very happy! Guess what these poor devils did and I don't know if their insurance covers such a different kind of accident. Traditionally they drive pickup trucks over a frozen Lake Champlain to do some ice fishing. Six trucks broke through and if you need spare parts for yours, since everything according to Reagan's Supply and Demand (which is OK I guess) is sky high, except for spark plugs, you could talk a diver into grabbing a few items. There is an assortment to be had. Maine residents shouldn't be called Mainiacs but I do believe a few merit being called just that.

Back to the illustrations. If a passing low system or unstable weather (vertical cloud build up) happens the distribution of pollution takes on a different behavior. Particulates are pulled upward and any

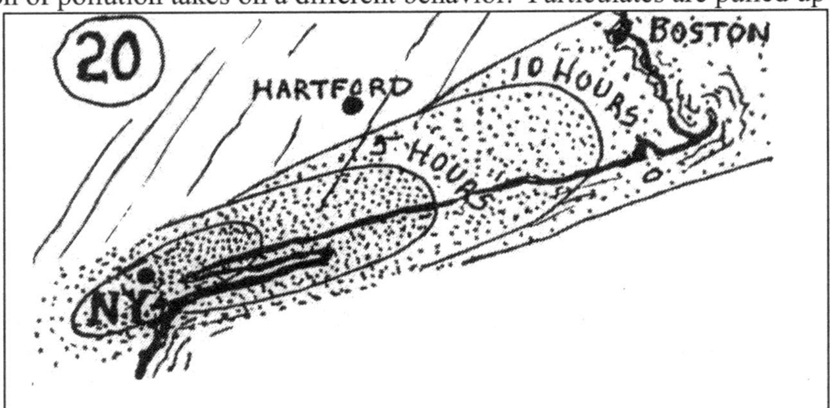

area east is temporarily relieved of emitted products. But, stable conditions, typical of high pressure and much of it is diametrically spread holds down emissions and in its usual eastward movement spreads pollution more or less at low levels between the surface and approximately 2 to 3000 feet. The graph below is based on balloon measured air speed at 2000 feet and higher gives one a good idea of how pollution transports as it diffuses. If such illustrations seem extraneous you should first think over where you live. Then watch TV's weather report and if a high is to pass over some smoke-filled area west of you and is expected to cross your own neighborhood, reach for your gas mask. Seriously, we will all be affected, sooner or later, directly or indirectly. All citizens will find the cost of everything to rise in the next ten years but those affected directly if they live in the wrong states have a double burden to live with. Destroyed property and deaths and top that off with higher prices for everything that is sure to follow is something to give some serious thought to. Reality is not for everyone, most have their heads in the sand, I'm sorry to say, and liars are busy seeing to it that they stay in that psyche mode.

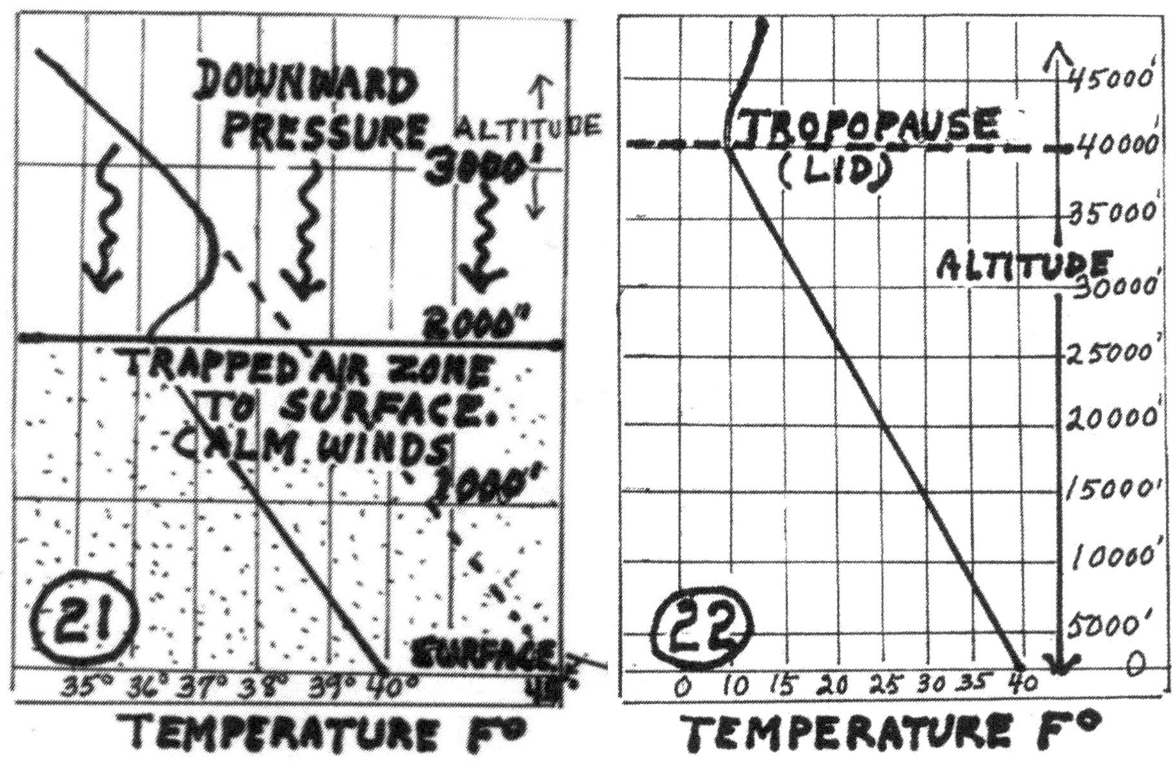

For the student, an introduction to temperatures aloft as determined through ascending balloons carrying weather instruments, referred to as Radiosonde. Illustration (22) is a sample graph of what an instrumented radio transmitter hauled to 50 to 60,000 feet altitude reveals in the way of weather data, but I've limited it to just temperatures. Early in 1943, Port Lyautey, just 100 miles south of Gibraltar where a vast aviation complex had grown from a few planes to hundreds, placed us in the lower end of mid latitudes. Therefore, the tropopause altitude was in the neighborhood of 40,000 feet. Slightly rising temperatures as shown always characterize the tropopause, which identified itself as we calculated radiosonde signals by a change in temperatures. It is an invisible lid and without waxing poetic or excessively religious, that is God's way of confining our entire atmosphere in order to facilitate some regimented order of a continuous recycling process between the poles and the equator. As yet, the balance between equator and pole is unknown*. With illustration 21, the effect of calm winds, cooling effects of clear nights, particularly when stable conditions prevail. In a city, particularly urban sections, the air contains more dirt than usual. If the EPA insists that there are 3000 cases of cancer annually

caused by pollution it should be pointed out urban dwellers are predominantly the victims. The purpose of

*_For others, anyway._

illustration 21 should not be regarded as trivial. I don't remember the exact year, perhaps in the 50's, smog in London killed hundreds! I guarantee you that the inversion condition described had to be a major factor. Most assuredly a high-pressure system was around somewhere or otherwise concentrated oxides, soot and both dioxides would have been blown away or lifted up to higher altitudes in a diluted state. In countries whose governments pay for socialized medicine there is the advantage of centralized record keeping of every ailment, causes and effects, death rates from this or that, but in our more or less private sector medicine, records are scattered widely. Therefore, I'm willing to wager EPA's claim of 3000 cancer deaths falls short of the entire harm done by pollution. Deaths from asthma, particularly children of poor families have increased over 100%. That is just another figure to add on an incomplete summation.

Previously I had explained, however elementary, the ignored factor of high-pressure systems acting as delivery conveyances of dirt (meaning everything pollution wise). Referring to illustration 21 the effect of calm winds, cool clear nights due to a stable condition is evident. Overnight cooling taking place, more rapidly if winds become negligible, there is a downward pressure typical of any high system and consequently, the temperature trace from the surface to perhaps 2000 feet drops off normally but is arrested as shown. Pressure prevails above that elevation. Dirt therefore is trapped throughout the night and into early forenoon and is not blown away or wafted upward until late forenoon warming occurs. The dashed line represents what surface temperature is required before air can ascend.

Stationed in Cape May, New Jersey and while functioning as a simple weather observer and recorder, map entering and all that, a naval officer tutored me in things more advanced. I was fascinated. It was early 1942. Years later, should someone ask me if the Navy was good to me I would reply "Yes" but I would resist adding, "I was also good to the Navy". You can spend all your recreation time off boozing and looking for girls, which by the way were few and far between in that quiet unpretentious summer resort. Certainly I didn't stay in the barracks all the time but study I did. However, the Navy doesn't force you to be ambitious. Damn if I was going to stay an observer and map enterer! This attitude caused me trouble overseas.   What follows is a simple story of a few episodes during my two-year stay in Morocco. Of course, there is not only relief it offers to the reader
should he or she feel lessons in meteorology not as stimulating as I would like them to be, but I try to connect some entertaining aspects of experiences during World War II with the weather topic at hand, in this case how temperature inversions created other phenomena besides trapping dirt. Referring again to illustration 21 the dashed line represents to what degree surface temperatures must rise to before vertical currents can be set in motion. Dust devils over arid fields, doing their insane dance spewing dried grass and twigs in all directions, manifest this sudden onset of surface air free to bound and spin upward.

I was shipped from Cape May, via a stay in Long Island for amphibious landing training lasting a month, jumped into icy waters and running like hell one mile back to a hotel the Navy leased, then packed aboard a transport mid December 1942, crossed the Atlantic in an incredibly huge convoy and 13 days later landed in Casablanca. Then indignities of all indignities, seated into what were once cattle trains, open air box cars with improvised wooden benches to sit on. We clunked along a railroad, some tired ancient engine ahead belching black smoke and we arrived in Port Lyautey. Tents awaited us, five men per canvas home and add to our discomfort, a horrendously noisy wall of clouds moved in, lightning and torrential rain making us a little more miserable as we tried to build up a small dike made of mud around the tent but to no avail. Here came the water. The next day, to my amazement, not just clear blue skies, what was one arid dusty looking valley, one north-south runway and a dangerous sandy hill for planes to climb over after taking off, turned into a blazing red carpet from horizons to horizons. In 24 hours flowers had popped out of the sand. I thought this spectacular but slogging through mud and standing in line with hundreds of others waiting for powdered egg and milk breakfast, we thought that we were condemned to live in misery. How lucky we really were. So much suffering all over the world, rows of white crosses in testimony to men who never came back home. The second morning, tension popped out of the line ahead of me. One single black youngster and in those days, I think I saw a total of three during all of my travels, such was the prejudice, was fighting with a white southerner. Flailing away and we, sort of enjoying this distraction, decided that the white youngster was going to lose and several guys broke it up. As a parting gesture of hate, the southerner spouted loudly, "I can't do it now but I'll get your ass in Alabama after the war!" Cautious this tough guy might hear our snickering we nevertheless burst into spontaneous laughter, all in concert. From Port Lyautey I would be shipped to Agadir, due to not only dissatisfaction but also an unpleasant encounter with a very unNavy type officer. This officer was a pot bellied bank president from somewhere in New Jersey who, accustomed to keeping bank clerks and assistants confined to cubicles no doubt, the Great Depression allowing free rein to slave masters (believe me I could tell you stories), did not like this lowly "White Hat" kid at age 19 to fraternize, if I can call it that, with one Lieutenant named Jorgensen and a Warrant Officer, Mr. Sheldon from Virginia who always had a ready smile. This newly arrived Boss-man, who we desecratingly named "Pus-gut" when we were in our barracks, was one of the few who managed to acquire higher ranks by virtue of age, college and social status, and I suppose some patriotic urge. While these two older men treated me kindly and aware the Navy was always in need of more trained men, there not being many college graduates available, the commander made it clear that he didn't want me around them. While they drew maps, spoiled as I was from tutoring by a MIT graduate back in Cape May, I would pull up a stool next to either one and show great interest in what they had to say. That lasted about three days. I heard this "Ahem" bit, saw his disapproving look and that ended that. On top of that he was the total fraud and was not at all skilled in meteorology but did a good job of making believe even though everyone knew where his expertise lay. However, that wasn't the Coup de Grace that sent me packing. He got rid of me, the almighty one, but it happened during a beach party. I guess that it was funny but at the time I was one worried youngster. This is about drinking and I must build up this true story. First however, the strange sights of Morocco! There was little or no gasoline available for Arabs and Europeans and so they ingeniously mounted stoves, about the size of 50-gallon drums in the trunk, the lid of course having been removed. A small curved pipe functioning as the chimney so to speak ran under the frame and was attached to the carburetor. Lo and behold, the fumes from burning charcoal powered vehicles along at turtle speeds, including small crowded buses. So much for my interest in autos since prior to Pearl Harbor, autos were gazed upon with envy, so few of them around. Unhappy in NAS Port Lyautey, I requested transfer to any aircraft carrier but was turned down, reminded that the Navy had trained me for other duties. One day destiny changed everything. If beer doesn't strike you as playing a role in how this beverage changed my life I don't blame you. Now, there were some 4 to 5000 men stationed in "Lyautey" and because the town was small we were only allowed one afternoon's liberty every fourth day. Can't have hundreds of military personnel with various objectives in mind swarming all over the streets and sidewalks and upsetting it's residents. We had to return to the base by sunset. Blackouts were in effect since German bombers were quite able to reach Moroccan cities.

Our army personnel had learned a nasty lesson when arriving in Casablanca and wandering about. Some became inebriated and were lured into an alley by raggedy native children, who managed to learn a few choice words that you now hear on TV, to the effect that a mother or a sister had something to offer to appease raging hormones. Quite a few soldiers were killed and their naked bodies were found stripped of everything. Therefore, our authorities were not about to let we youths stay out into the late hours. Extremes in poverty existed as the whole world had economically collapsed after the 1929 crash. One Arab, years later, thought transmission oil would make good cooking oil and sold it to unsuspecting natives with the results of dozens dying. This was in the news a few years after the war. And so, not much in the way of recreation was available and what's new, we gravitated to bars. I think some enterprising natives had soaked weeds and made beer but it was terrible – undrinkable. I tried cognac and got sick. Even though none of us were not that infatuated over imbibing the announcement that we would have a beach party for all hands and all the beer from America (talk about enthusiasm) that you could drink really put us in a happy frame of mind. Not just a few hours off, but the whole day! The sun shone African bright, the heat and the cavorting around in bathing trunks holding on to bottles of beer made us tipsy. In fact, we got silly and all of us began to shake bottles and spray each other. The commander and two officers sat on a blanket and seemed to be in a pleased stupor, gracefully tipping up bottles to their lips. I don't know for sure if it was an accident or subconsciously wanting to bother the boss-man, I splattered a spray of beer on the back of his neck while running around as all us were busy trying to dodge spray coming from all directions. He just looked at me but said nothing. The next day, trepidation set in as I walked toward our office, part of an operational setup built on top of a hangar. I had made second class petty officer and I dreaded the idea that I might be demoted and once that happens your career has roadblocks. Just think, your record, in my imagination for the whole military world to see. "Enlisted man, while under the influence of alcohol willfully assaulted a commissioned officer, etc." The commander said nothing to me for three days and I began to relax. On the fourth day I was handed orders to pack up, board a truck, part of a caravan and off we went to Agadir, 400 miles south, to open up an advance base. After arriving six planes plopped down, one by one and I knew that 90 to 100 men and wonderful pilots was a lot better than the organized mayhem of 5000 at Lyautey. The commander probably thought he had banished me to the wilds, but it changed my life for the better.

Six miles inland from Agadir, a town built on the edge of the ocean, it's winding narrow streets and clusters of small houses stacked on top of each other extending halfway up a 900-foot mountain. The last and smallest peak of the Atlas ranges seemingly sunk in the Atlantic Ocean, the Seabees built us livable quarters in a few days. A short runway, part of a French airport then no longer in use, allowed our aircraft to make their daily flights out to sea, scanning its surface with radar from the Canaries, halfway to the Azores and north near Portugal. Imagine how stunned I was when the news broke out some 20 years later that Agadir had been swallowed up and crushed in a violent earthquake – 20,000 people died – many nations hurried to help. I have so many memories, including the morbid, but let me again go back to weather. Illustration 21 gave us an idea of what a temperature inversion looks like on paper. This condition frequently existed in Agadir, our base on the side of a flat arid valley some 3 miles wide before the bases of the Atlas mountains, shrouded in morning mist, framed the limits of ideal terrain for an airport. One day, a newly arrived lieutenant, who I would soon learn was a hard core alcoholic, strolled out to where I was standing, waiting for the "Dance of the Devils" as an Arab called them. It was quite a novel show while it lasted, whirling vortexes, some fifty feet high, swaying side to side, then scurrying along in different directions. The officer, one of the few late comers the Navy accepted without the usual rigid requirements, motivated by patriotism and having political connections, came from a very wealthy family in Chicago. I told him what I was waiting for. When the dusty pirouetting began he was dumbfounded and wanted to know how I knew when such a phenomenon could be expected. Another day, fog rolled in from the sea and blanketed the entire valley. I had posted a warning however, it not being difficult to assess, knowing that a sea breeze would kick in about 1 PM, the same conditions existing in San Francisco, there being similar cold water upwelling along the coast. The lieutenant thought this feat, which of course I didn't let anyone know, was tied into the routine onsets of sea breezes, was remarkable and exclaimed, "How do you do that?" I was both pleased and amused. The man, a real gentleman, bloodshot eyes no less and feeling some sympathy for him, kindly and well bred, nevertheless showed signs of mental deterioration from years of drinking. Dignity went out the window. An officer in charge would send out an enlisted man in a jeep and pick him up in town, so drunk he had to be carried off. He lasted two more weeks but was sent away, to where I don't know. My only admirer and he was gone. There were better days to come, but weathermen never achieve celebrity status you know – at least I had a moment of being recognized as someone with unusual powers. I'm poking a little fun at myself.

Weather systems, the anticipated conditions at sea, the source of information, all of this was a highly guarded secret after Pearl Harbor. German submarines powered by batteries were forced to surface every 12 hours to recharge and obviously stormy or foggy conditions and their locations were sites they anxiously sought. New in the Navy, just a month or so stationed at Cape May, the secrecy of our work having been made clear to all of us, the sudden barracks incident before my eyes made me realize how we were watched somehow. The death toll of merchant marines just before Pearl Harbor and for about two years or more thereafter was astonishing. Therefore, an all out campaign to stop so many torpedoing of cargo ships was launched by hordes of aircraft and warships, the idea to sink as many subs as possible but to keep them submerged where I was told they would loiter along, saving electricity and then surface only at night. Can you imagine the slaughter? Sixty ships unprotected before our defenses took effect by sheer numbers, moving slowly in the north Atlantic and waiting, a minimum of eight submarines. In a matter of a half hour 27 cargo ships were sunk. And the dead? I don't recall these hard working civilians, struggling to make a living for their wives and families, ever being honored despite thousands lost at sea.

Tired as usual, working 8 on and 8 off shifts, I was half dozing away in my bunk. Next to me a rather peculiar new man, who worked also in our office but was quite an eccentric, a loner. Propped up in his bunk he usually was no distraction at all, busy reading a very thick book and off to sleep I would go. Then, one day he turned to me and said, "You ought to read this book, it's sensational." Feigning interest, wanting to be cordial to this oddball and wondering how he got into the Navy I asked him what the title was, "Mien Kamph". And then just like that, we no longer talked. Hell, I didn't know anything about Hitler's book. Two civilians walked in, looking like FBI agents, grabbed him by the arms, hoisted him out of that bunk and that was it. A quartermaster told me that he was a spy. Spy? Why advertise it by reading a madman's ambitions, our enemy? I was never sure he was able to be just that. Yet, one weekend, two weeks before, I met him on a train going to New York where he supposedly lived, my destination being RI and there he was, holding a briefcase. Who ever heard of an enlisted man carrying that. Later, after his disappearance I worried a bit as to my being suspect travelling with him for a few hours. Was this a ploy? A means to scare we new weathermen or was he in the espionage business after all? Quite the dummy if he was. Now, jumping across the ocean again in my writings the secret of it all reached more serious proportions. The English had broken German codes and now, we had all of the weather information for Europe. Prior to that, our weather maps were a mixture of guesswork and postulating from a few weather sites. It was 1943, probably March or so and we got the task, but little old unappreciated me, since I had not as yet wet the commander's neck, of tracing on transparent paper the secret weather data of a map showing all of western Europe and all other areas that we had been using, everything in the way of a completely drawn map by someone else had drawn and because I could draw well, there in full view on thin paper just the typical millibar lines, lows and highs, fronts but nothing else. Then an officer would take the data laden original and burn it. Fine. One problem though. Two years later I went home on leave and mother asked me what I had done wrong. Instead of showing elation after hugs she backed off and said, "What did you do overseas?" She described how agents, two of them no less (gosh, what an important man I was) who knocked on all the neighbor's doors and asked about my character and whatever else. Sure enough, at least three neighbors ran to my mother and said her son must have done something terribly wrong and were just itching to find out all the juicy details. It's a good thing the neighbors like us or else they could have told the agents that I was a bad one, stealing bicycles, bothering girls and arrested a few times. It took me a while to figure out why I had been investigated as over two years had elapsed since the big secret came into my life.

Agadir had some interesting characters. Bear Bryant, the legendary Alabama football coach who died some years ago, arrived to take over as our executive officer. Actually, the commanding aviator ran the whole show and we got along quite well. Heyward was his name and those flying men were my types of people. I tried to be a pilot while in America. The Navy was so desperate they even did not require high school graduates although I was. I flunked the physical. When I was but seven years old I bumped into my mother who was carrying a large pan full of hot water. Naked, waiting for the tub I trotted out to see if she was coming. I got cooked right there and then from one side of my chin down to my stomach. My father, ill, fainted because I screamed so loudly. I spent a month in bed but what applies in this attempt at becoming an aviator, the burn had left the left side of my mouth slightly curled down. It became hardly discernible in my more mature years, but at age 18, having this doctor examining me I heard him say, softly, to a yeoman sitting there and filling out a form, "Atrophy of the facial muscles" something like that, and then the teeth, too much overbite. Just prior to oxygen masks for pilots one's teeth clenched a tube. I soon appreciated not becoming a hero as I thought I would be, wearing a beautiful uniform and worshiping town citizens marveling at my success, at least 10 enemy planes downed. The wreckage and bodies that I have seen cured me. Well, Lieutenant Bear Bryant didn't know me from anywhere and his main interest was to make football players out of us. All 6 foot three of him and maybe 220 pounds, a football crouched in one arm, he relished charging through a dozen of us 140 to 150 pounders that he had lined up in formation exhorting us to tackle him. We'd pick ourselves up and he would just laugh and insist we try again. This last time I decided that I would tackle him in the ankles and trip this character. Fifteen feet later, my shoulder almost broken, still hanging on after his high knee action did me in, he stopped and said, "Good!". He liked me after that and although he meant well he called me "Frenchy" and after a while I asked him to call me Andy, which he did. Because I could speak French he would take me out to various villages to buy eggs and vegetables. One Arab merchant looked like Jesus, that's how regal he appeared. It was a custom to never refuse an invitation to have tea after a transaction. I knew what this guy was going to say. "Nah, we gotta go". Discreetly, very embarrassed, I told this impressive person dressed in pure white that the officer didn't understand the importance of customs there. Bear Bryant got his name when, as a college football player attending a fair, he climbed into a ring and faced a trained bear. Rumor has it that he won some money doing just that. But, the "Bear" was very likable and although I was but his interpreter he always made me feel good about myself. Next to our encampment there was this hangar that nobody seemed to use. An elderly Arab, looking very gentlemanly, stood around there most of the time and I would find out he was retired and paid a pension by some French aviation company. "What did you do when you were working?" He explained flights south to Dakar in rather dangerous old time planes required he go along on each flight, just in case of a forced landing. There were dangers from hostile natives and although he didn't identify who these Arabs were, my guess was that they were Tuaregs. His job was to talk everybody out of trouble. The Tuaregs were a bizarre looking lot. They painted their faces blue and dressed in solid black from turban down to sandals and always on horseback. Poised for action, even though I doubt if they could hit the side of a barn door, they held a long barreled musket at a ready. With a black shawl or whatever across their faces and two jet black eyes and furrowed eyebrows is all you could see, generally glaring at anyone different. One day, three Foreign Legionnaires eased up on the edge of our buildings and soon we were bartering for knives, old muskets and whatever for the cigarettes that they had come after. One was an American from San Francisco but I didn't get nosy as to why he had joined this outfit. At that stage of their careers, disheartened by Germany sweeping over France, simply putting in time at a nearby encampment, they, with an apologetic look, considering a reputation garnered decades ago were quick to tell us they couldn't do anything against Stuka dive bombers with just a rifle. One was a hyperactive Russian and I sensed that he wasn't all there – probably killed someone and ran off and found a home. I saw him in a bar one day and he was jumping around in a crazy dance while repeatedly stomping on his hat. While forecasting was a routine, the sea presenting some problems, some interesting things happened in the way of wild and wooly nature. On a sunny day, off in the distance a low lying dark cloud across the eastern horizon, inland of course. It seemed strange on such a day but soon a noisy cloud of locusts moved over us. Huge grasshoppers all heading out to sea, why, I can't imagine. I saw Arabs catching them, running them on a wire and barbecuing over a fire. From a respectable distance I watched these wretched thin Arabs eat locust. One invited me over but I politely declined. Along came one of our boys from town who had been drinking and he decided he would try eating one grasshopper. He walked back and we asked him what it was like. "Tastes like shrimp" but he didn't go back for more. Then we got hit with a sandstorm, the Sahara not being too far away. Another chap and I were crossing the field to see what a group of British aviators had to offer when we got caught. We lay on the ground with our hats tightly pulled over our faces but the stinging sand made quite an impression on me. We had sad moments also. One morning at 4 am we were awakened by loud shouts. One of our planes had crashed two miles away after taking off with seven men aboard. We had to wait for the flames to subside as there were four depth charges threatening to blow up. One did and the concussion jolted us back. I need not tell you what the Navy requires out of respect for the dead. As I told you, scattered aluminum and wings sticking out at crazy angles chilled my interest in becoming a pilot. One time in Port Lyautey a PBY, four depth charges under its long rectangular wings, two 50 caliber waist guns and one 30 up in the nose, seven men aboard, roared down the runway intent on a routine submarine hunt over the Atlantic. For some reason it ran out of runway and plummeted in to a hill, always a potential danger for heavily loaded aircraft. In those military days there was no such thing as fire fighting equipment. Everybody waited as the heat was firing off the bullets; it sounded like Fourth of July's spasmodic popping fireworks. One depth charge exploded. Finally everyone moved in expecting the usual lack of survivors or what was left of them. Lo and behold, one of the crewmembers was sitting, stunned, his head hanging between his knees some 150 feet behind the wreckage, dazed but not seriously hurt. We never could figure out such a miraculous thing – it mystified us all. It was not the policy of naval aviation higher ups to disclose any information. Later in my military career I would find out from a maintenance chief that sometimes as great as naval pilots were and still are, they would forget to shut off the carburetor heating system after revving up the engines to start their take off. Detonation (pinging in autos) would immediately set in and eat a hole in the center of the pistons. How many times I flew in those noisy vibrating planes, there being no insulation and sitting back there in the fuselage in full view of the cockpit, much of the instrument panel and numerous switches along with the pilot's right hand nursing the throttles forward. Once in the air, I would breathe a sigh of relief and see the man in control of my destiny deftly move his right hand over various switches and toggles. "Don't screw up." That thought always ran through my mind. I'm not a coward. I just saw enough of torn bodies spread around wreckage – and picked up the parts – threw them on a canvass sheet. We were young and tough and strange as it may seem, other than sadness, we would go along with daily routine not feeling ill or anything like that. My little stories, mostly nostalgia, faces of men I still vividly remember, serves here to add a little variety and relief from the subject of weather which can be humdrum. Also, what makes me tick, where I'm from and how the young of yesteryear coped and adapted and learned. One pilot in the USA that I flew in, a two engine SNB, just routine training for him made me feel very secure on that take off run I'm not crazy about. He would practically shout one, two, and three all the way to 12 as he went through the routine of adjusting various controls. He wasn't about to slip up and in my mind I happily granted him several "Atta boys". I've changed my mind about avoiding morbid details having just briefed over a pilot's apparent poor judgement, crashing several miles from our Agadir base. As I have already written, off we went in two trucks after being awakened at 4 am, we could see the glow of a fire in the darkness ahead. The reason for my change of heart stems from an opinion made by a radio host who I listen to between 9 am and 10 am. "Nothing interests anyone nowadays unless it has shock value – they've seen everything on TV, etc…" So be it, forgive me then, but my thoughts in those days of an occasional disaster were those of a very young man with no ladies to share them with. I found a severed hand hanging in a bush and there was a wedding ring on the finger. I kept looking at it, not moved at all because there was much worse than that all around us. Flying sheet metal in a 150-mph crash severs everything. I kept thinking, most of the crew was in the age range of 19 to 22 and in my mind I associated this body part with an older person and thought it to be that of the pilot who was generally older. On one hand you are a human being, sensitive to the slightest painful experience of others close to you and then, at that moment what is left of someone you didn't know doesn't affect you the way it should. You are, without realizing it, at that age conditioned by poverty, military training and believing glory lay in killing someone. Quite cold even though for a few hours there is quiet remorse amongst all hands. "Whose hand is it?" that was on my mind mostly. I thought about those years later and asked myself what did it all mean? Our preoccupation later in the day was as follows: On take off, if an engine quits, which in this case one of the two did, you do not bank over to get back to the airstrip. You try to keep that plane going straight ahead, no matter what, because on a frantic turn the wing angles down and loses most of its lift. That's what we talked about that afternoon. One time, one of our waist gunners on a PBY that had damaged a submarine and its crew were floundering about, got carried away and was firing 50 caliber bullets at those Germans down there. He was young and as I said before, you leave your heart at home in war, but in this incident another crewmember knocked him down and raised holy hell. He was furious and kept shouting "Do you think those bastards wouldn't do the same to us!" You become unfeeling. Tell me something, what effect does natural disasters have on you? So many times I've seen TV announcers, looking grim while reviewing scenes of death and destruction and one second later, turn on a smile for the next topic if it is of a more cheerful nature, not that he or she should reach for a handkerchief. I'm trying to extricate myself from having gone perhaps a little too far, a little too morbid. Reminds me of an annoyed SUV owner who phoned in to a radio show about the growing resentment of high level bumpers on these behemoth vehicles ramming you just about seat high in an average sized car. His comment, "Oh those poor little people in their poor little cars, my heart bleeds for them." Two thousand people killed by pickup trucks with their weight, their high bumpers along with SUVs contributing to the toll and this guy is going to let you know that he doesn't give a damn. Two thousand die yearly and how many crippled for life? Lately a commercial keeps droning away, "I'm a hired killer, I'm second-hand smoke." In the meantime Detroit is probably considering a 6500-pound bigger vehicle knowing full well that they can sell millions of them. I know who is doing the most harm. It's not some guy puffing a butt on a San Francisco sidewalk. Give me a break. The break from weather is over, back to why the oceans can't rise since more and more humidity is being extracted from it – millions of tons I might add. Amazing! As I type I keep glancing at the

Weather Channel and an ear tuned in to a smiling weatherman. Over three months of it, 30 foot waves off the west coast, even two surfers drowning, a ship breaking up while being towed out to sea so as to be sunk, but so far without success so rough are the seas and the man doesn't give us a clue while pointing to wind figures and so much rain – 90 to 100 miles per hour battering Washington and Oregon. Oh well, I have to refrain from getting critical, such is my views of what ails too many. On top of that, more and more do I hear uncertain forecasters saying, "The computer model predicts…" instead of being honest as I would be since I never said that I could do a better job of predicting. In this way the computer can be blamed you see. You will see what I mean shortly if computers sooner or later will be subject to, how would I say, given a good talking to, no? We had a moderate northeaster Thursday February 25 and I think somebody has no dignity to be quoted, as they were by our newspaper the next day – unbelievable to feed that to the public! Now they're talking about damages in Virginia and South Carolina, crushed cars and all that. The beat goes on – nature on a rampage. We have always experienced bouts of rough weather but not so incessant! But, no worry, we have 2400 scientists working on this. It must be La Nina out there in the Pacific, 5000 miles away. El Nino died May 1998 and without it the rest of the year we experienced extremely bad and destructive weather. But, we need a new devil don't we? It's more exciting for public consumption. Nothing like the titillation aspect to attract imaginative writers.

developed in the upper Alps in recent days carries a secondary threat: <u>Massive</u> <u>floods</u> <u>can be</u> <u>expected</u> in April when the snows begin to thaw and the swollen mountain headwaters rush into the Rhine, the Po and the other major rivers that originate in the Alps.

In France, a 26-year-old hiker died early yesterday, shortly after rescue teams plucked him and his wife out of a snowstorm in the Pyrenees mountains, police said. The French couple were a part of a group of six who went missing during a snowstorm while on a hike Sunday.

In Romania, melting snow and heavy rains swamped about 4,000 homes in northern and western Romania, officials reported yesterday. About 110 towns and villages in the country, including the cities of Arad and Oradea, have been flooded.

Europe these last two weeks of February has experienced abnormal weather, particularly in the Alps and as far as Romania, 600 miles east of that fabulous ridge of mountains Alexander the Great clambered over. So far the principal west European nations, industrialized and heavily populated have escaped having their roads dissolved nor cargo ships shoved ashore by huge waves, such as the liquefying effects of endless rain occurring in Oregon and Washington. Blessed by a continuous flow of partially purged Atlantic Ocean air, although more damp than usual, those nations are but a narrow strip of land compared to our 2900 mile wide land mass, hence it takes the lifting effects of hills and mountains to create serious problems. Mountainous regions have always been subject to avalanches – I do not use this fact as an index of anything special. The death toll in the Alps is unusually high but not all died from the sudden cascading crush of so much snow rolling down overloaded peaks. Inaccessibility created by two weeks of violent snowstorms that hasn't been experienced in a half of a century delayed rescue teams from reaching freezing victims. All storms require moisture and it is now excessive all over the world primarily due to the effects of particulate pollution that I need not reiterate. But Europe will have problems. Europe sends their pollution and I'm sure it is much less than ours, to the Alps, then over the hump to regions east. Some years ago they protested our contribution which, we're being the greatest, no pun meant, simply because it is obvious that the prevailing winds at higher levels and sporadically at surface levels being west to east during the winter and southwest to northeast during warmer months and on target lies Europe. Imagine our millions of heated homes in cold weather, chimneys pumping out tons of C02 and whatever! The problem is what can we do about it? And the world? Supposedly, I am told the world's population increased by 86 millions every year and while it may prove profitable for auto manufacturers, it seems everybody in more prosperous foreign countries are anxiously looking forward to getting rid of their bicycles and get behind the wheel of a car.

Off the subject for a moment but related to office bound writers pumping the Internet, one came up with some real bull excrement. If Buffalo, New York is getting heavily dosed with snow it is because of

La Nina. Now, I lived there for five years and I'm a weatherman, obviously not a writer, and I know cold air from the Arctic, and the colder it is, crossing a slightly warmer Lake Erie, the more snow falls over a strip of New York land on the edge of this lake. Think of it as a miniature El Nino, if you will. In mentioning the Internet I heard a fascinating business lady on Jim Bohanan's show. Her business, the name of which I believe is called Information Plus or something like that, specializes in advising companies anxious to take advantage of web sites but apprehensive over reliability. She made it clear that the belief by too many Internet enthusiast that this information electronic paragon is the Beginning and the End All, is a foolish one. She added, they still contact human beings for supplementary advice and that this will always be true. Her name is Sawyer for anyone who is interested. I mention this because now, being a member of the Greatest Generation and the epitome of the evolved man, "non pareil" plus the exemplary wise old man, since in this high flying times were not regarded as such, I think you should take me seriously. In other words, a sobering suggestion: Don't think you are going to find out everything there is to know about meteorology and air pollution gaping at that green and sometimes blue screen, please. But, if you want to find yourself confused by the variety of information available just go right ahead and keep believing you have the Beginning and End All and experienced human beings really don't matter anymore. On the other hand, I didn't say computer information wasn't a spectacular development of the 20th century. Watch out for turbidity though. I'm beginning to like that word.

I was reading where Ford Motors is planning to outdo their competitors for the biggest SUV ever. Sooner or later Europe will resume their protest of the past over air pollution but for now we're quite busy fighting some kind of banana war. Let me redundantly show a segment of the senator's letter and see if you can expect some governmental effort to stop this nonsense of promoting voluminous three ton vehicles that have already killed 2000 people every year and with smaller SUVs* no less. But my main interest lies in the effects on our atmosphere.

"Social risks of continuing accumulation of <u>greenhouse</u> <u>gases</u> in our atmosphere. While estimates vary, many studies <u>suggest</u> that if we stay on current course, average global temperatures will rise two to six degrees during the next century with potentially serious consequences."

"The more things change the more they remain the same." I don't know who came up with that one but it applies. "Nero fiddled while Rome burned." The Dust Bowl of the 30's that turned some southwest states into Sahara like conditions, sand dunes partially burying farm houses. President Roosevelt had trees planted north to south across this country and governmental experts taught farmers how to plow and plant crops on an alternating basis from one year to the other. And of course, the height of folly, the roaring twenties and the worldwide disaster that followed. I'm all for freedom, in fact I actually resent using seat belts because I maintain 150 feet behind vehicles at expressway speeds; and before I charge through a light just turned green, I wring my neck left and right and make sure nobody is in a hurry – it happens all the time. But, laws are needed, that's that. But, egregiously is how I describe the potential menace of a major affluent group, who do most of the voting and may succeed in rendering our government nearly powerless with their attitude of "I do what I want!" Furthermore, the thought of a supposedly compassionate governor, Mr. Bush of Texas whose family is in the oil business and Texas a predominant oil refining state, becoming President bothers me. In using the word, powerless, that may be too strong, but surely the progress we are making against an ever-increasing amount of CO2, for example, is discouraging. EPA is bucking exactly a negative reaction from entrenched businesses and of course their higher salaried employees who predominantly make up that politically influential class I refer to as the affluent who are
*Includes pickup trucks

not quite ready to applaud what they deem as intrusive. But that is the nature of the human race, change is an uncomfortable thought.

Radio is a wonderful source of information. Unlike television with "quickie" renditions of important developments and opinions of various authorities, the 150 thousand dollars a pop for commercials squelching the slightest fair analysis on the part of viewers, the patient and general calmness of two people discussing in reasonable depth or even debating important issues coming out of a radio is far superior and definitely more instructive! Additionally, BBC from England, just 15 minutes after 10 PM, takes me to far places. Last night, more enlightenment over earth warming effects by four doctors that I'll pass on shortly. It does not mean, despite my continuously bringing up impending problems, that I am obsessed emulating some modern Nostradamus.   Instead, let us, you and I think as cool-headed technicians and access everything available. Furthermore, thinking dollars before lives, obviously still a priority, has grave pitfalls for the cavalier obsessed with wealth that, if he leads, or influences public opinion, is really but a shortsighted Pied Piper.

I don't think the guest Jim Bohanan had one night several weeks ago fits the description of cavalier. But as an expert economist and his name of course forgotten by yours truly as usual, there being so many I listen to I thought him a bit too optimistic. Jim Bohanan asked him if the new consolidated Europe would prove, in ten years or so, a very tough competitor. Jim Bo actually sounded concerned. "I don't think so" replied the gentleman guest. "We have a flexible labor market and a floating expertise while Europe has always stuck with more rigid and inflexible traditions, etc…" I wonder if he has kept abreast of pending changes, agricultural, medical, new diseases moving north (to follow) all due to weather trends, very obvious ones and of particular significance, why 2500 economists standing by? Is he of that school of thought and theory, really, which by its very nature allows us to relax since rising ocean levels are many years away and perhaps actually believing as I have seen hypothetically presented on TV, letting me now as I type add a little sarcasm, that gondolas will serve as taxis in our New York City some fifty years from now becoming another Venice? I wonder if this theory is a very valid one? Don't they realize how worldwide flooding, the consequences of tropical air expanding it's territorial reaches northward and its seasonal duration which normally shrunk in winter now does not, derives its source of water from the oceans? Just for the fun of it, we have a world population growing at the rate of 86 million every year and how much more water will be needed if we understand 6 billion human beings and their economic needs represents one hell of a drain on water supplies. Does this economist really know 1200 to 1300 tornadoes a year and probably more than that in the next ten years, along with heat waves, droughts and all that gloom I'm selling has to disrupt economics? Apparently he visualizes clear sailing for the next decade.   Is it vain of me to say, having observed and studied passionately everything scientists and experts have to offer, that too many are afflicted with "Tunnel Vision" and that they just don't see the "Big Picture"? I watched, enthralled, even my wife who is usually bored by science, experts struggling against Antarctica's weather, cutting huge tunnels down into solid ice and from the textures and grains evaluate climate some 15,000 years ago up to today. There in this white cave were horizontal streaks of brown and I jumped, yes, my mind said, "Tell me, tell me" surely that has to be something in the way of dirt or pollution or dust from recent times or centuries ago. Do you thin that scientist was even interested? No, he has devoted his life, days and days, hours and hours, grinding out bits of information for just one specific aim – ice moves at the rate of so many inches, so many feet annually, depending on what area of Antarctica. Overwhelmingly, the equipment used to bore a mile deep into packed snow then ice then mud at the bottom and so on, days and days of struggling. There is no leisure time for the avid scientist. He must bury himself mentally into the very molecular structure of his field. Nothing else matters.

Four doctors, their opinions taped for BBC, one from Harvard, Doctor Epstein, another from Canada and two in England, had nothing but sorry news to report from their analyzing world trends in mosquito borne diseases, Malaria, Yellow Fever and Dengue. Even Sweden was mentioned as having ticks moving further northward and spreading Dengue, which though not fatal is very painful and epidemic, sometimes resulting in hemorrhaging, because their weather has warmed. South Africa, whose rainfall resulted in 40 more inches above normal last year, now is reporting mosquitoes expanding their vortex as they call it has doubled the Malaria and Yellow Fever death toll. These doctors went on and on with disturbing statistics from various parts of the world and did not hesitate to say, unlike some fluent turkeys moving in on Washington DC, earth warming is responsible. The Harvard professor indicated increasing cases of the same diseases might reach epidemic proportions in the next decade. Even myself, how can I forget the Fire Ants of Louisiana, when my family and I lived there? They're all over the deep south and uniquely, are armed with an injecting needle type tail. If you just walk by one of their mounds, the very sound of your footsteps is enough to signal a mass attack, rapidly swarming out, up your leg they go injecting venom. Welts the size of a quarter result and it is very painful. Cattle have died; they can kill an infant. I don't care how good your immune system is, when they get through should you make the mistake of swatting away rather than running like hell, you may get sick. They migrated from South America and so will a lot of other unpleasant insects. My backyard at first had dozens of fire ant mounds and the only way I could fight them off was to pour gasoline down their one entrance, throw the bottle away and light a match. Peace for about a month but they would come back. There is another quaint insect in Guatemala, the name a long Latin affair, which results in a fatal heart disease. Give it time, let us all swing and sway, mouth open, entranced by the magic spell mutual fund salesmen and saleswomen on TV or the sweet voiced over radio, create with their pitch of quick wealth. "Learn how to get rich on the Internet." That's another one. Shh, don't disturb the market with this whacko environmental stuff. After I get finished with economics, for example, will utility bills stay reasonable if 50,000 repairmen must always be repairing destroyed electric conveying systems? Will health insurance not rise excessively in the future? We get lied to. Last year the wheat crop was great despite droughts and flooding. Oh sure. But, in one little corner of a newspaper, this troubled one alert always, read where American wheat farmers were protesting Canada's dumping wheat into wholesalers' bins at unreasonably low prices. Boom crop for Canada and another lousy year for our wheat farmers.

In my day, although I was too young to know what was going on in America during the mid 30's, bamboozling reached an art form. Many newspapers and some magazines came out with some dandy stories. Although I'm fond of the Kennedy's, Joe Kennedy Senior was one tough guy. Maybe he was tough with a vengeance. In his day, and my infant period it was common to see signs over employment office doors stating, "No Irish need apply" or "No Catholics". But the way that one prominent magazine story, several pages of tripe I might add, described how Joe Kennedy was saved from losing his fortune in stocks three days before the 1929 crash was the height of bull excrement. He was sitting outdoors on a shoeshine chair. A young black boy said to him, "Mr. Kennedy, I have a feeling that the stock market is going down, real bad." Reputedly, this was an omen to this man as with all insiders, did not need outside advice but what a story. And of course, the fable went, Joe sold out while the selling was good.

I will again continue describing various experiences of the good old days and the war but leave out what may strike the reader as unnecessary sanguine details. In those days our thinking processes and reactions to unusual incidents, poverty and ambitions molding how we thought of life in general, my having already touched upon so that you can understand why I think like I do and yet not at all interested in cultivating too much pessimism, really sharing rather than teaching. My stories should be somewhat interesting if you appreciate the fact that not knowing history or being bored by it can result in future bungling. Give it a whirl anyway. You've got to give we members of the Greatest Generation a shot. In retrospective, we were tough all right, but not in the leather jacketed macho style. I sense today that the same naivete of yesteryear and to some degree the deceptive emanations that messed up our lives and that of the whole world. Admittedly, coming from the other side of the tracks, not much communication with the rest of America and absolutely none about the world, we were not the brightest specimens. If I had not studied economics in the 60's and stock market peculiarities I would not be so free with advice although all I have expounded so far has to do with public psychology relative to the indecisiveness of their political representatives and an admixture of conflicting views from experts. We therefore have indifference in general, growing hostility by others and the undecided. Some day you may face what some residents of Maine are experiencing. Or you could be wondering who is lying and who is telling the truth. Perhaps a choice of a new location might hinge on weather trends. Who is going to take a beating or who will be spared? Had I known more about Buffalo before I moved there, I would have stayed home and tried somewhere else. The winters were so nasty that my youngest daughter, who despite being a sturdy one, kept coming down with the croup. Furthermore, I lost a lot of equity in a home I had bought in North Tonawanda, (northern suburb of Buffalo). Never buy a home next to any major governmental project. Bell Aircraft shut down and all property took a nosedive from laid off workers. I left there and went to Louisiana. The same thing happened in Slidell, LA. The whole town boomed with 15,000 government financed Boeing and other company employees moving in. New homes, more stores and so on. Poof! The space program some 10 miles north where they tested rockets just ceased. Fifteen thousand employees left. Homes practically abandoned were up for sale but no takers. This happened sometime in the mid 60's. We went to the moon and there was jubilation but not in Slidell a few months later.

If, after reading the article on the next page you are not in agreement with me over how much we have screwed up everything, I suspect that you are one of those sport fanatics who spends hours studying either the size sneakers 35 basketball players use or the batting average of 872 baseball players going back 20 years. I hear them panting over the radio waves sometimes. Oh, I like sports and follow all four varieties and I might drink one cold beer on a hot summer day, but I'm not a double six packer. Here in RI we are told our beautiful but smelly Blackstone River,

Air or water pollution? That's the choice facing Maine's Department of Environmental Protection as it decides what to do about the gasoline additive MTBE, but it's not a difficult choice. ... MTBE is an additive developed in the 1970s that allows gasoline to burn more cleanly. It became controversial after water supplies across Maine were found to be contaminated by MTBE. According to tests carried out last year, between 1,000 and 4,300 private drinking wells could contain unhealthy levels of the suspected carcinogen. Part of the problem is that MTBE is easily absorbed by water, meaning that even small levels of MTBE could cause major contamination if it were to leak into Maine's groundwater supplies.

To reduce that risk, the DEP will recommend to its regulatory board that reformulated gasoline be replaced with a fuel now sold in the South that contains far less MTBE. Trouble is, the southern fuel is not as effective as reformulated gasoline at cutting smog. Therefore, if Maine makes the switch, the threat to groundwater will decrease while levels of benzene in Maine's air will increase.

But the choice is not difficult. As more than 540,000 Mainers get their water from private wells, MTBE pollution of aquifers could affect great numbers of people. Worse yet, consider the time necessary to mitigate an incident of

serious pollution.

Cleanup would be horribly expensive and time-consuming and perhaps not even possible. Beginning in 1993, more than 1,000 Maine citizens took part in a lengthy study to decide what environmental problems posed the greatest threat to the environment, and to determine how they should be ranked.

When the study was completed and submitted to Gov. Angus King in 1996, threats to drinking water supplies topped the list. Major groundwater contamination would be a disaster that would wreak havoc in Maine for years and years. The only real solution is prevention, which is why MTBE must go.

— *The Times Record* (Brunswick, Maine)

winding north to south into Narragansett Bay in a picturesque valley some 50 miles or so, has been treated and is relatively clean. The only trouble, shellfishing has been banned for miles and miles on each side of the bay for years now. Over 100 factories have been dumping their chemicals into the Blackstone River since the 18th century and few have been penalized. Right now a dozen DEM experts are scouring another river, much smaller because it is contaminated with Dioxin, a carcinogenic chemical. Dollars before human welfare – that is the religion with the most powerful following. Sooner or later you'll have to make decisions. Don't think I'm a crank. My father died at the age of 41 from lung contamination, breathing fumes from melting zinc. The Disease Control Center in Atlanta or any decent medical oriented institute can tell you, there are numerous lung ailments caused by insoluble particulates that just collect in your lungs. Overtime I see a big vehicle with an oversized tail pipe whoof by, I wince at it's contribution to future misery. How dense can this American public be, waiting in line for Ford Motors to produce the 2000 year model SUV, with six doors, a V10 engine and probably weighing in at 6500 pounds with a fuel consumption appetite that is unprecedented?

Let me play the proverbial man on the street, the Guru of economics and social justice who berates passing society. We had them in the 30's. I learned the consequences of the wheeling and dealing crowd early. At the ripe old age of 10 I was indoctrinated in management labor relations. Well, it was a start anyway. I witnessed mob violence, all due to gross income disparity – the final product of Free-Wheelers and No Holds Barred economics. Oh sure, some will say there are winners and losers, but have you ever heard of eminent justice? It appears current generations imagine increasingly good times benefit everyone. We have income inequality; I mean poverty versus affluence and wealth. We cannot boost minimum wages to any effective level. It does not bother me if paying an extra dollar an hour would hurt a few businesses – it is our need to stay competitive in world markets that is the stumbling block. Personally, if I were king I would organize covert agitation in slave labor countries and stir up workers. It happened here in America. The king was President Franklin D. Roosevelt who, cruising around, stopped here in Pawtucket, RI and told a protesting crowd, angry at working 6 days for five dollars, "Why don't you unionize?" He was worried about capitalism being in serious jeopardy. Can you imagine him advising a public that way? The sharks were corralling the little fish. Number One hogs were keeping the trough to themselves. The power structure in Wall Street knew how to keep pessimism going and thus facilitate a lengthy period of picking up all the marbles, ten years of wealth, property and stocks siphoned off. I met two young college students studying economics. My chance to test their knowledge of how a quick buck is made "Short Selling". They never heard of it, even though the quantity of stocks involved is listed at the bottom of stock listings. Today, that is not popular, but in the good old days the method garnered fortunes for insiders. W.C. Fields, an ancient comedian use to make what I regarded as a cruel comment: "There's a sucker born every minute". All the insiders had to do in cahoots with the media was to drum up optimism and unload on eager investors. We should be thankful for governmental intervention after that era of immorality. I wonder why colleges don't teach what I just wrote? It's a dirty stain on our history, that's why.

I was bug eyed, what a show! I was initiated in spontaneous combustion, the suddenness of a people having had enough, erupting as they did wondering why our peaceful street for this one day before the noisy bedlam started had so many people all walking in the same direction, some jabbering noisily and flapping their hands around. I followed them from a respectable distance. They wanted unionization and companies called in law enforcement. It was quite an exhibition of grown ups turned wild, shouting challenges and hurling rocks at national guardsmen and state police crouched behind cemetery stones. I could see rifle butts sticking out from various monuments. They had been driven up this square into a burial ground. Somebody chased me away after I had picked up a stone. It was snatched out of my hand and the same man threatened to kick my posterior if I didn't go home. Tear gas finally broke the mass fracas but tension remained*. There was not much communication in those days but it's a different world now. If the cost of living skyrockets from crop shortages in the future it will take more than Alan Greenspan to maintain a degree of law and order. A fiberglass company from Ohio moved in to our area in the late 30's. Their civilized attitude and decent treatment amazed everyone. There were no cops to be seen then, no crime at all. It will be different in the future I'm afraid. Twenty five hundred economists on call, that is more than significant, if you think about it.

I imagine you may think of me as tearing away at everyone involved in environmental issues. I am guilty of finding fault with a political, scientific and social ethos, mixed as it is, that is flawed. If we were to compare it as an army with its leaders, soldiers with good and bad attitudes and an intelligence center composed of quietly bickering experts, it would be classified as ill-prepared. Conscious I have so far nothing good to say about many, such as disagreeing scientists, TV shows, writers, a segment of our population and furthermore preparing to chastise the meteorological infra-structure despite their working hard and providing us with an important service, coping as they do with the elusive. I want to nevertheless make it clear that I have my own strategy. First, a growing awareness some years ago of the subject of changing weather, more tornadoes, floods, droughts, etc… as

Actually 2 people were killed – six wounded.

distinguished from long range climate change, struck me as increasing mayhem in the making, the decent proponents of needed action waylaid and sidelined, bombed from all quarters by the ignorant and narrow minded. The results so far are obvious. A looming crisis is not viewed as such. Weather and it's negative aspects should be the centerpiece and not some side issue to be considered later, along with priority items of any political agenda. For one reason, it doesn't garner votes. For another, if many of our citizens are interested, they are being spoon-fed pieces of facts, don't know who is telling the truth, don't believe because they do not want to believe or don't care, don't understand scientific rigmarole funneling in by shrewd mind manipulators. In all, most feel they want to be left alone. I have piles of newspaper clippings to show what amounts to a continuous show of intelligent reasoning periodically taking a back seat to critics and skeptics, self interest entities for sure, who manage to keep the ultimate deciders, the voting public, partially confused. Did not, on a small scale, Maine residents were called upon to make a decision? I don't want to teach - I want to share. If you are a football coach, do you find fault and concentrate on the stars of the team or go after those players who are not playing well? I am offering a simple logical approach, not as a coach but sharing with you what is truth, avoiding technical jargon which I am quite able to do so that the reader will find himself or herself in the position of being ahead of the game of life and what has a great effect on it, not wondering who is honest and who to trust.

I wanted in these pages to get away from economics, especially that of the past. It might appear as the laments of a loser. But, I do need now and then to mention individuals who write articles and make sense. Frank Levy, an economist at MIT, author of "New Dollars and Dreams" and Iris J. Lav, deputy director of the Center on Budget and Policy Priorities, both consider a tax cut merely aggravating a growing income disparity. They maintain that we will all be losers if we do not compensate through sensible legislation losers during the good times. The consequences otherwise are harmful to free markets. They are not socialists. They are thinkers with foresight, a sorely needed quality. One of the greatest myths ever propagated on the American mind is that we are all born equals. Physiologically, that is impossible, we're not talking about free speech and travel where you may. If everyone had the potential to be brilliant it would be one hell of a world. It was not in the scheme of things, if I may venture a guess as to what the Creator had in mind. But, it's great propaganda by the rich who want you to believe the poor are just plain lazy. Consider, a simple arithmetic computation: A manager, income $100,000 and he deserves it, has under him 100 employees making $7 an hour working for a small factory whose profits are satisfactory for the president and owner. If they, the employees, were to be paid $8 an hour, the manager could not make $100,000 a year unless the owner didn't mind a reduction in profit. My point in this simple and extraneous example, the affluent should not look down at the lower classes but see them as an important supportive base of an economic pyramid where the better educated, the more intelligent and even the heirs of successful parents have, through their own ambitions and drive, the chance to sit higher up the narrower part of the dollar-indexed triangle. If you resent the poor being helped you must remember that they cannot buy the products that your company makes. That is called community spirit. If certain countries south of Mexico or wherever progress very little, it is because there is no community spirit. Immigrants coming here are products of survival without thought to group thinking and it takes time for them to even come out and vote. I see that in our own neighborhood. In their homeland, they think simply of basic needs because the wealthy have seen to it that they remain that way. Because we flaunt wealth, the reaction from too many poorer is resentment and some anger. There is a delicate balance between acceptance and chaos.

I have been recounting memories of long ago just to give myself time these last evenings of February to gather my wits together. Meditating on how to continue examining the thinking processes of certain meteorologists, feeling in a way that they are my peers, without coming across as Mr. Perfect and being too offensive in the process. Perhaps I'm too grim, there being tragedies in my childhood days that have influenced me in a negative way and I beat off by reminding myself that we have come a long way.

Therefore, one little story, a bit of humor even though I wonder if it will be in bad taste. At age 7, or thereabouts, I lived on Hunt Street, Central Falls. Robert Ripley who specialized in a daily illustrated or described "Stranger than Fiction" said Central Falls was the most populated square mile in the world. Half of Hunt Street, one-quarter mile long, consisted of a fairly steep hill and then leveled off. On each side three or four decker apartment houses, some housing eight families, but it was a clean town and no cops were ever seen. One bar but few patrons – there wasn't enough money around for a thriving business. Now, cars in those days, the early 30's, particularly the more popular Mode T's had brakes that were less to be desired. Mechanical brakes were activated by rods beneath the chassis and if you didn't adjust them every month or so, you might end up with one wheel not cooperating with the other three or as it was too often, pathetically inadequate. If I recall Chevrolet came out with hydraulic brakes. In 1946, just to jump ahead a decade and a half, after my wedding I borrowed an uncle's old Buick, which had ancient rod-activated stoppers. The trouble was I couldn't stop in time in Boston and ran into the back of a truck. Not only that, driving from Portland Maine to Boston I experienced two flats and had to buy recap tires to replace the same troublesome kind. Back to the early 30's, I was hit in the rear by a Ford, I guess, while bending over in the middle of Hunt Street picking up a ball. I did a complete somersault but other than a very hurting rump, I wasn't seriously hurt. Run I did, straight through the front door of my apartment home and quickly sat down looking innocent.   Mother came over and wondered what happened, but I couldn't tell her. The driver came in, all wide-eyed and scared and insisted that he take me to a hospital. My mother also wanted me to go. After so much persuasion I adamantly refused with a firm, "NO, I don't want to go to the hospital!" They gave up. I wasn't going to let them find out I was sitting on quite a mess that I had made in my pants.

I deplore the increasing emphasis by meteorologists to allegedly rely on computer modeling. At least limit themselves to use it as a supplementary opinion and not stand there and tell the viewer that the model indicates the storm should move out to sea. Just give us here in Southern New England the edge of a northeaster and perhaps an inch of snow. At the same time, Tuesday February 23, the opinion of the computer, given to this surprised viewer, the Jet Stream believer had to add verification of what the electronic thinker was projecting. A forming storm south of Cape Hatteras would intensify alright, you couldn't miss on that one, a hefty cold mass moving down over the Mid Atlantic States and packing cold air from Canada meeting the usual inert-looking Atlantic High sitting out there pumping the necessary warm moist air. Now, vicariously, I resent without losing sleep over it, the numerous comments that I heard from office personnel Thursday morning, the 25th, fairly heavy snow coming down and winds picking up, their glancing at the clock hoping to get out early, "They missed this one- we were supposed to get only an inch at the most." Blah, blah, all innocent of course but little awareness of what it's about. I just smile and move on. We collected about seven inches here; fairly strong winds while Boston sported a foot and substantially more up in ski country. But, for a weatherman in Taunton, from the very office whose chief seemed to agree with me, laughing at the way I expressed skepticism at the inability of computers to outguess nature's constant shifting values, to be quoted (see excerpt on next page) in Friday's paper

> But what the models were showing as late as Tuesday night weren't much help to Glenn Field and his other meteorologist in Taunton.
>
> "The models were extremely inconsistent," he said, with one model showing the approaching storm taking a track as far as 500 miles from where another model predicted.
>
> Even as late as Wednesday morning – less than 18 hours before the first flakes began falling locally – forecasters were still debating what we would see: a few inches or something much more substantial.

after the storm was over and all was clear, in the manner revealed, that bothers me. It's possible that some hierarchies governing weather-predicting centers want to propagate an image of sophisticates. Am I unfair if I wonder over the pretentiousness of it all or is it just an escape avenue for the uncertain forecasters? Today, for the first time, I heard a gentleman on the Weather Channel say, relevant to a major storm potential out west, "We use computer modeling as guidance while also giving it options." Lately that modeling word has been used often and I wonder if somebody didn't make comments that compelled him to speak out. Guidance can also steer you in the wrong directions. This is a thinking and memory business.

Storm's track proves elusive

■ It's not until late Wednesday night that computer models give forecasters a handle on the northeaster's intensity.

But even then, as the offshore storm moved slowly northward, most computer models used by the National Weather Service were predicting only a slight chance of snow for Southern New England. The better odds were that the jet stream would keep the storm farther out to sea.

Meanwhile at the National Weather Service station in Taunton, forecasters were anxiously studying computer models used to predict the weather over two-, three- and ten-day periods.

Into each model, forecasters plug in various weather facts, such as land observations, offshore buoy and ship reports, and upper atmospheric conditions. That information is then run though computers in Washington, D.C., that can predict if it will snow tomorrow or next week.

Oh sure.

It is not so much what you learned in school; it is remembering similar or reasonable duplicates of system arrangements and weather patterns of times past, even years. It is not a soft touch, definitely. Not wishing to condemn any further nice people for throwing a little variety around, I now would like to explain why both humans and machines, despite computers representing to me the greatest invention of the 20th century, are subject to miscalculations. It is the sudden change in speed of a moving system that trips up the weatherman. Expecting a computer to anticipate the unanticipated has its pitfalls. During spring, the most dangerous situation is a cold front that suddenly changes from a standard west to east (generally) rate of advance to a sweeping 50 to 70 miles per hour rushing across the land. I call your attention back to earlier pages, i.e., a large thunderstorm, associated with a front, killing three in Syracuse, NY, as an example. What did the meteorologists tell me that is different from the above? He said, "It moved across Lake Ontario at 70 mph." He also said nobody was warned. Also, it was not expected – simple as that. In my need to explain everything in detail, particularly the all to frequent surprises or the opposite unexpected delayed frontal or lows delayed advances I have included illustrations to follow in the next few pages. I am not necessarily Mister Grim, but I am guilty of finding fault with a combined political, scientific and unscientific reporting and approaches used and a social ethos that is flawed.

I wish we were blessed with Western Europe's more stable climate. Ours, as weather, since I must differentiate between climate and weather inasmuch as several scientists were quibbling over this stupendous difference in a newspaper account that taught me nothing, ours is far from being idealistic and promises to worsen unless world leaders hurry up and start sorting things out. You can see I am discouraged however. If I wrote that I learned nothing from quibbling climate experts in one particular article, I must remind you I am still the consummate student. My lawyer was right when he humorously called me a fanatic but as I told him that's how you win wars. If a squadron is to bomb a site in some winding valley you don't gather pilots around a map. You painstakingly build a clay model of the hills and mountains and then and only then the pilots see in an operational ready room exactly what they will see from the air.

Mark Twain said, "Everybody talks about the weather but nobody does anything about it." Surely he had humor in mind. I think that we can do something about it. It is not unreal to think so. To be disinterested as so many are, marks those who are naive and very vulnerable. A distraction now, I stop typing. At 4 pm today, March 11, scenes of flooded Fort Worth Texas. Not ankle-deep either! We can be affected in so many ways. Besides death and the mounting cost the possibility of shifting population exists. I saw a young man, disconsolate, his head bowed while a TV man was trying to get his view, following the aftermath of tornadoes killing over 25 in Jarrell, Texas, May 1997. "We're leaving here and never, never coming back." That was all he had to say. It's all forgotten but I have my records and will not forget. Did anyone ever tell you what I have told you? From 600 tornadoes a year we jumped to over 1100 in the

> ■ Oil stocks, recovering from 12-year lows, were sharply higher, following reports that leading exporting nations are close to agreement on rroduction cuts in a bid to shore up prices.
>
> NEW YORK (AP) — Stocks shot to record highs yesterday as oil prices soared, but concerns about technology-company earnings crimped gains in the Nasdaq Composite index.

past decade! But, there is a dilemma. The stock market, it matters most! In today's newspaper, look at the influence of oil company profits on the market. People like Al Gore will be viewed, if not already, as a menace to the economy instead of the real man who simply wants to reduce oil consumption. The propaganda started today on Russ Limbaugh's show – I'll cover that shortly. Worth repeating: Well financed spin-doctors are priming up and after laying down their smokescreens of the past prepare for an assault on America's enemies. So far, good results. An assortment of conflicting opinions and the majority of our ever vigilant public beginning to yawn since winter has dulled their awareness.

As for leadership, worried about votes, ample rhetoric. Some intimidation of course – can't push environmental issues too hard, the enemy is readying to pounce. A substitute radio host for Russ Limbaugh today rattled off at a machine-gun pace just exactly what the Republicans have to promote to win. Cut taxes, get rid of government and take advantage of Al Gore who he laughingly called a walking target. His recruiting base called in, enthusiastically approving. Unsaid, but clearly thought, let Americans do what the hell they want! Then, further pushing the anti-Gore theme, strike fear in the hearts of Americans and their beloved cars, that's it! Governor Bush, the compassionate one has to win! Maybe he truly is compassionate but I also know that wealthy and affluent Texans love cheap labor from across the border, nothing like desperate housemaids to be used. He speaks Spanish and hordes of immigrants are welcomed and get a head start with tax money to boot. According to the radio host who spoke at a gassed-up full-throttle pace, which marks the rather limited thinker, feverishly dreaming of laisser-faire economics, no holds barred and to hell with 30 million working stiffs. Al Gore destroyed his chance when he said that the internal combustion engine was the greatest threat to mankind. Any intelligent person knows what he means. And, reduced government? Already typhoid and E-Coli (animal and human excrement) have contaminated fruit from Central America which reached our shores and infected twelve victims in Florida! Thank God we still have government monitoring! Just think, if our great productive farm lands and livestock take a beating from droughts, heat spells and floods we will have to start importing more foreign meat and vegetables. Nothing like a reduced inspecting staff doing spot checking by necessity instead of total vigilance. Dear Reader, you are surely much younger than I am but have you ever heard the tragic story of Typhoid Mary? Sometime in the 30's if my memory serves me well (probably not) she worked as a waitress in restaurants. People came down with typhoid fever but authorities had great difficulty in tracing the source. Finally, from her working in one place then another and so on over the years the evidence pointed to Mary. Warned over and over again, she defied authorities and kept working in restaurants. In and out of court but finally they locked her up and, according to historians, for life. Warming trend in climates means more disease and they're contagious! To hear some radio host constantly denigrating government as one huge boondoggle, then insisting we can do away with most of it, sets me off, as repugnant an idea as it is. What a world theirs would be! Imagine the FAA not able to effectively safeguard aviation for example. Let the election process decide what individual politicians will serve us best and who should be sent back to the private sector. It is too bad newspapers do not find it worthy of print space to periodically summarize who voted for what in Washington, DC. We came precariously close to seeing our social security fund <u>dumped </u>into the general fund back in March of 1994 when Senator Paul Simon (now departed to private life) drummed up support for his bill, which in essence was a means of vacuuming up your weekly Social Security payment for other purposes. As I already wrote, the bill missed by a mere four votes and a repeat attempt would have probably tipped the scales in favor of such an unsavory resolution. By the way, 41 Republicans voted to approve Simon's bill and only three disapproved. So much for politics on my part.

Several pages back I had in mind a lengthy essay on complications, or should I say stumbling blocks, those forecasters are faced with. I also wanted to include illustrations as a means of broadening one's knowledge, particularly the young and eager whom I hope to reach. Encompassing both the surface weather system affecting New England with our February 25 moderate northeaster and a cross sectional view of what I mean by the dominance of warm high-pressure systems over our end of the Atlantic. But I went off on a tangent. I just had to comment on the daily news, not chancing my forgetting some detail days later and obviously, my reacting to a radio talk show host whose repertoire, repeated every day "Republicans are not mean-spirited." And "How we can win the elections" and of course, "We have to do away with much of the government because we are much smarter, blah, blah, blah…" To hear this radioman rubber-stamping Limbaugh's wildness, constantly denigrating government as one huge boondoggle, then insisting we can do away with laws, sets me off, as repugnant an idea it is. What a world theirs would be! Havoc would reign after vikingry, a sickness peculiar to rapacious types I call larceny of the heart, had torn society apart. Well, back to weather.

I have not intended excess criticism over what many present day meteorologists believe in, that's their prerogative. They do an excellent job of short term forecasting and issuing warnings but I know long range forecasting for seven day periods requires a bit of luck, believe me. While it may have dramatic impact to the unlearned to say that the computers began to spit out heavy snow was due <u>just before</u> it began is rather silly. Two days before, there was the usual jet stream telling the forecaster that the storm would move out further east and little snow could be expected here in southern New England. The point is, there is lack of knowledge as to what controls the movement of systems and if they do not as yet realize how formidable those sluggish looking warm high systems are out over both the Atlantic and the Pacific, they have not been doing their homework. Last but not least, we have a flock of tethered buoys out there in the Pacific, obsessed as some are over El Nino, all instrumental and data transmitter equipped. But, the western Atlantic, which plays an extremely important role in affecting not only routine weather and the occasional storm but also tornadoes and lesser but still violent weather, is a void on any weather map.   By that I mean that there are few data providers out there all the way to Bermuda and south to the Caribbean Islands. I interpret this gap as a problem with top management in NOAA. Well, the jet stream didn't do what it was supposed to do and the computers gave out confusing information. As usual, and I'm afraid it happens too often, the power factor out there over the Atlantic is both misunderstood and misinterpreted. I repeat I could do not better forecasting but I would try a little more logic and be less mesmerized by a computer. Once a high-pressure system moves well out over the

Atlantic it is not finished affecting inland areas that it left behind a day or two before. Somehow the change it may undergo, if any, merits one's full attention. But, there is a problem. Convoluting or not, the illogical arrangement of weather measuring sites, preponderoulsy over the western Pacific and the equatorial zone thereabouts and the inadequacy of information over the western Atlantic, I want to explain that as being symptomatic of flawed reasoning. I'm not alone in understanding this. While talking to the meteorologist I have referred to in previous pages I pointed to a screened display of that day's current map and made a comment, "Things haven't changed in over 50 years have they?" "There's no information out there" pointing to the Atlantic. To which he replied, "Yes, it's a void." I don't believe a reader is going to lap up what is a forecasting problem but at least he can think along with some TV forecaster.

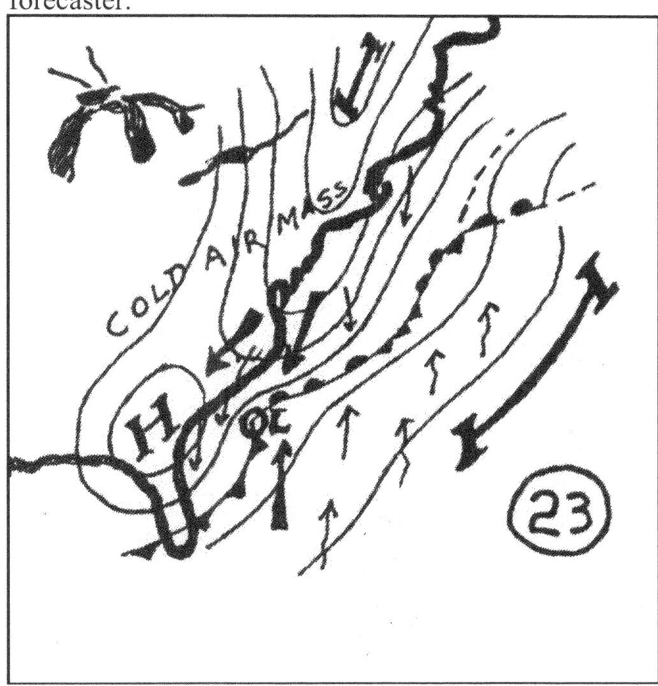

Illustrations (24) and (25) show how warm highs over the Atlantic can either keep storms over our coast as it did the 25th or how it can let them roll out to sea. If by chance your intelligent offspring is impressed by this and wants to be a weatherman, talk him or her out of it. These illustrations, particularly (23) roughly approximate what February 23rd's weather map looked like. The focal point: one forming storm south of Cape Hatteras. Where was it to go? A major southward thrust a cold air mass on the march from Canada, looking tough. They often do. The only certainty, a natural arrangement for a nice storm that the kids are all waiting for. * The elusive factor, a large but spread out warm air mass looking impotent, over the Atlantic. I can't be precise but the warm high, more robust than it

*Not to mention truck owners who have plows.

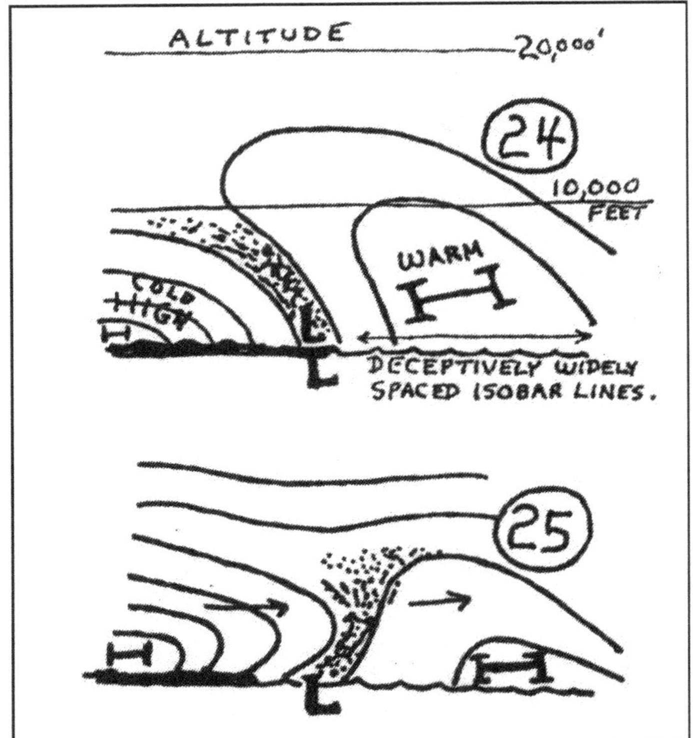

appears, simply blocked and leaned over the cold high domed as it is. Note (24) low-pressure system's clouds and precipitation loaded vortex leaning westward. By February 25th the storm, propelled northward by the unexpected power of the Atlantic high blew over Cape Cod and all the children were happy here in New England because there was no school both Thursday during the storm and some cancellations on Friday. But, as in illustration 25, a warm high is breaking down, contracting in volume or suddenly heads further eastward as it may happen when you least expect it, the leaning action changes. Precipitation will fall at the front and ahead of it while the cold high accelerates it's eastward movement. Such routine storms in this case can't be very violent. Intensity is a function of warm highs bucking pronounced cold air masses. However, it's a different story for tornadoes. Let me get into that later.

From my point of view there is folly in randomness. Without spiritual belief in a created order, believers in haphazard chance rather than a subliminal arranged ubiquitous control over the elements on a worldly scale that truly represents nature's ways, remain but skeptics and retire unproductive. No matter how sensible a contention from the consummate searcher you can count on the usual "We are skeptical." Born with an illogical bent and knowing full well that little can be proven regarding atmospheric vagaries, they endorse nothing and derive satisfaction in confounding issues. They missed their calling. They should have considered accounting, a profession of rigid tabulations whereas the subject at hand, as with physics of decades ago, requires conceptual thinking followed by intense study and hoped for verification.

Interruption: Today, March 14th. The Weather Channel meteorologist is reporting two tornadic cells near Tampa, Florida. It's a little early for that! It takes time for the human race to deal with realities; much better to keep one's head in the sand. It's peaceful there. Shortly I will explain a missive I sent to the Vice President. I even suggested that there were silly experiments going on during the tornado season out West. Foolish experiments made by what I disrespectfully think of degreed comedians should reach a point that someone with a little more acumen in higher circles of management will put a stop to. Explanations to follow of course. There is a new world of enlightenment waiting out there and just perhaps I can contribute something to it – but, unlike scientists, the really dedicated ones, their reports by custom are sent to various societies that they belong to, I'm reaching out to the public. It is they who will, strange as it seems, decide what has to be decided. It is a complex subject and I struggle to express it in understandable terms, but let me assure you also that it is not the great mystery being sold by certain spokesmen. Some months ago, I watched with great interest and then amusement, a documentary on our belated efforts, a la Don Quixote, not the man but the Men of La Mancha, if I can use an analogy to represent a study in futility, to fly into and/or above hurricanes and try to stop them in their tracks. An elderly gentleman, who was a crewmember years ago on such a four-engine aircraft flying over the eyes of hurricanes, per this narration anyway, talking with suppressed laughter, then grinning broadly after describing the consequences of the job he had to do – dump the contents of cement bags, dozens of them, at the advancing eye of the storm. "You should have seen the insides of that plane, what a mess!" "We laughed all the way back looking at each other, like bakers who fell in a vat of flour!" Whoever directed such a mission had no idea of the worldly interplay of systems adjusting. Strictly someone else afflicted with tunnel vision. But I laugh also, it's entertaining, what can I tell you. Another time, what a meteorologist had to say was tantamount to his broadcasting irrational thinking on his part. There he was, my TV screen showing this smiling gentleman (no wonder, was he ready to laugh?) his back to a desk cluttered with envelopes and letters and of course, two lit up computers behind me. After he got through explaining his job, I asked myself if he really gets paid for this. This is what he said: "I am open to suggestions as to ways and means to stop hurricanes." He reviewed asinine ideas and even gave one as worthy of consideration. Everything from exploding bombs and throwing dirt, etc… That

only adds to my disrespect I'm afraid. I was not afraid to tell Al Gore, through the cooperation of a senator who delivered my letter, who also should be a little more cognizant now of what is ridiculous and wasteful. In the following page I've included but one leading paragraph of a five-page letter to Mr. Gore. Of course, lawyers, masters of all that they survey, make sure you understand donators of knowledge are but messengers passing on what they, of course, already knew, after which they find ways and means to adjust to a new approach and take credit for it. The main topic of the letter was tornadoes of course. Al Gore's hometown got hit badly months before my missive and then after this written endeavor on my part to arouse him, Nashville got hit again. Oh, I went on and on with small print from a computer, explaining an aspect of tornado intensification that is suspect but is theoretical. It could be proven but the powers that be are firmly married to ancient beliefs simply because that is what is being taught. I've been that route. In short, traffic jams and I'm talking miles of four lanes jammed with slow-moving bumper to bumper such as was noticed by a weatherman on TV's Weather Channel who remarked, "The thunderstorms are intensifying over busy-busy Route 95 approaching Miami." Such traffic jams can contribute by virtue of enormous heat released by thousands of automobiles at surface levels. Now a tornado, forgetting about it's initiation from extreme cold at high levels to be described later, is, putting it crudely, a vacuum machine. I can write the parlance of weathermen you know. One chap, who now is a forecaster on TV, talking to a radio host said, "After I studied atmospheric physics I gravitated toward this profession, etc…" Sounds like Einstein stuff doesn't it? But all it means and I learned it in about two days: The speed of ascending air is a function of the Moist Adiabatic Lapse Rate relative to the Dry Adiabatic Lapse rate and determination is made by plotting temperature values on a Pseudo-Adiabatic chart. Anyway, you, dear reader, may be heads above me in certain professions but I want to keep everything simple and understandable. Now we should all know that energy cannot be created or destroyed and also that one gallon of gasoline or diesel fuel produces one hell of a lot of BTUs, in fact about 135,000. I wanted to emphasize something I'm convinced of, namely, an incomplete tornado. That is the threatening dark hose hanging down from an ominously rolling cloud, gets a boost from any available ground or surface heat. I think that we all realize that. It moves along, the direction determined by winds aloft, but if I call it incomplete it still is actively drawing subsistence from surface heat but remains nonviolent. Once it's bottom suction end, usually quite wide and limited to whirling winds perhaps one to five miles in diameter, runs into extra surface heat, and it only takes two or three degrees more, that suffices to furnish enough subsistent ingredients to immediately create a violent dark vortex. One such tornado approached

Andre Laferriere
30 Washington St., Apt. 514
Central Falls, RI 02863

April 30, 1998

Al Gore
Vice President of the United States
The White House
Washington, D.C.

Dear Vice President Gore:

In view of your having been witness to the aftermath of a tornado in Nashville would you accept what follows as an interesting twist on cause and effects? Presumptuous of me to volunteer my views in as much as you have access to many experts but there are times in seeing TV shows and reading about violent weather I believe it would not harm to stray from the pack so to speak. I am beginning to wonder going back to basics in meteorology might reveal something more useful than Tornado Chasers wasting their time and TV narrators annoying me with comments such as this one: "If we knew what causes tornadoes we could save lives and property." I don't think the spokesman with his marvelous voice really is quoting experienced weathermen, but one thing for sure, the public is being entertained quite dramatically but remain confused.

The Vice President did answer this letter above although it was a five page report and opinion as to tornadoes will intensify over cities, etc… His reply, of course, I have shown you back on page 4. As you can see I have my convictions and don't hesitate to let certain people know nonsense when I see it.

I didn't want to display such a long letter here inasmuch as I would be repeating information you will already be well familiar with in reading this book.

Miami, dipped down on a crammed expressway, then pulled its tube up, dropped down again on another packed highway and once again, the same action viewed by everyone a third time. After that, the cooler beach area and the ocean where it was headed denied it it's needs. It disappeared. The problem is that people cannot visualize this. One chap I know said, "How can we have pollution, look how the universe in unlimited" pointing to the sky. Oh Lord, I thought, what the hell does he read or what did he learn in school? Anyway, it may have stirred Mr. Gore, since there is evidence that the two tornadoes that struck Nashville arrived unnoticed and the fact I look at also, show me a swath of woods strewn about if the tornado or tornadoes were really full blown specimens before they hit a source of city and traffic heat, preferably between 3 PM and 6 PM. Explain to me, if you will, why a tornado has been observed to rip up sections of highways, usually old cracked hardtop. Well, let me offer the answer to that. Surely you've noticed on a cloudy day, light snow falling and the temperature is say 27 degrees, well below freezing, and yesterday or today's ice and snow on highways or streets, unsalted, melts as it does. The sun's influence still pokes through clouds. Therefore, a highway in Texas, cloudy or not, is always a bit warmer than the adjacent land. Aggravatingly redundant or not, a tornado's business end scoops up heat at surface levels up to probably just fifty feet above. Therefore, it is not preposterous, as surely the habitual skeptic will claim that certain traffic jams can prove very dangerous if weak tornadoes are approaching. I once told a doctor how Vitamin A had healed a lady afflicted with psoriasis. Do you think, as any intellectual with an open mind is inclined to be receptive, that doctor at least nod his head in a pretext at approval? No, he got upset. Do you think some scientist, writing extremely complex suppositions that I have masticated and digested all these years is going to accept the above tornado reality? Four, five years of studying in college and some guy comes along with something logical, something refreshingly new and what do you suppose is the standard reaction from pseudo intellectuals? "Can't be!" "Go away!" and, "We are skeptical." Well, Mr. Gore is now campaigning to relieve traffic congestion between suburbs and cities and the excuse seems to be simply, well, relieving congestion. Wary of upsetting folks hell-bent to do as they please and intolerant of anything that suggests the slightest change in lifestyle, Al Gore is not pushing the air quality advantage of less traffic. Did I influence him? It's possible, but then pushing for group commuting to work is not a new idea.

In regards to stirring tornado letters (blowing my own trumpet) I remained diplomatic with the Vice President, as you can see. But read on. The absurdity of what I have reviewed watching Hollywood oriented shows on TV! Showboating experiments running around yelling, "Look, it's huge! We gotta get out of here!" Oh, the antics are hilarious too, college types scrambling into pickup trucks, laughing and shouting "Lordy Lord!" I laugh along with them. The entertainment value is priceless, I might add. They had, these enthusiastic dreamers, imagining some secret could be found inside a tornado, placed a heavy metal buoy type affair with recording instruments in the path of an approaching tornado. After it was over, nothing could be found. It got sucked up, shredded and distributed to farmlands or woods miles away. Probably some farmer, puzzled over strange pieces of metal that he found the next day, wondered what space aliens had been up to or afraid to fly imagining some parts of an aircraft had fallen from the sky. By the way, there is one of these metal land buoys on display in a museum where it belongs. A concession was made however. Admittedly, it proved useless. Then the happy go lucky weathermen, I guess they were, paid to play, explained how they had sent rockets through tornadoes but nothing came of it and of course where did the rocket end up? Further acting graced my TV screen. Some guy standing in front of a tripod theodolite, that's a telescopic instrument that follows helium filled balloons while also showing angles and azimuth degrees (blah, blah, blah) and after releasing a balloon into a rip roaring black cloud charging toward him says, "We'll follow that balloon and determine the winds aloft all the way to 20,000 feet." It can't be used that way but the theatrics go on. The dandy of all dandies, they describe how a drone aircraft, much money spent on it's iron clad fuselage and wings, controlled by a ground radio transmitter, had been sent into the very vitals of a tornado. Trouble is, they didn't talk about it anymore, the scene shifted to other pursuits. I can guess. All that junk raining down somewhere, 10 miles away is probably the reason some heavy drinkers believe in space aliens and/or flying saucers that conceivably had been hit by a missile. After all, what would you think if a piece of wing dropped into your backyard? How about a propeller blade sticking in your roof? Do you know what we have here, Wizards of Oz, that's what! Perhaps we're all idiots if they believe this is palatable information.

Tornadoes, the isolated major blockbusters or the smaller ones associated with powerful thunderstorms are caused by atmospheric conditions hundred of miles away. But they, if I can call them experts, insist on a laboratory approach, analogous to someone (my imagination at work now) peering through a magnifying glass at the tip of a dinosaur's tail but can't associate what it is part of. Going back to fast moving cold fronts have you ever read of a one half-mile wide tornado, some ten years ago, that swept across two northern states and laid waste a half a dozen small towns (fortunately) that lasted over one hour, unrelenting, and traveled over 100 miles? Two meteorologists, brighter than most, were discussing this nightmare occurring in the early darkness and killed quite a few, although I don't recall how many. They at least thought that such a phenomenon had to be associated with a fast moving cold front, the characteristic of which weather occurs sometimes 100 miles ahead of the surface part of the front. In other words, assuming this is new to the young or formerly disinterested, the cold advancing air is slowed down from surface friction but rushes ahead at high levels where it has the effect of violent lifting of warmer moist air. But, no mention was made as to the major influence of the Atlantic high pressure system sitting out there, fueling such tumultuous systems but also, in a sudden manner, pulling away and thus allowing the cold air mass to rapidly surge forward bent on filling in where the warm high vacated the higher leveled premises, so to speak. Quite a simple explanation for everyone's consumption. If you have ever read the scientific journals I've absorbed you can imagine their authors joking over my approach. I don't even think some of the words they use can be found in dictionaries. You can bet your behind though, they're attentive to anything that will aggrandize their reputation in the close-knit circles they travel in, in terms of the societies that they belong to. However, there are some that don't deserve this condemnation which I intend to inform you of. When I listened to these two sincere weathermen on TV discussing their convictions, if cautiously advanced, they seemed to be inviting the help from anyone. It wasn't my imagination; they were very sincere dedicated people who feel a better job can be done in forewarning the public of grave dangers. Consider the odds of a city getting a piece of it ripped apart by a major tornado. Our nation consists of about 330 thousand square miles of land. Tornado prone areas average anywhere from 6.85 tornadoes per 10,000 square miles to less than 3 per the same area. Some eastern, and in fact one New England state, has a surprising yearly average of three and higher. Rhode Island is lucky (breathe a sigh of relief) 2.2 but it has little area. However, only .23 of one tornado per year after averaging out for a 29 year period. In short, nothing terrifying from these statistics for such a vast land, but, climate change in the direction it is taking means more tornadoes than ever. You cannot discount over 100 in January alone this year.

I think that I was politically correct when I wrote Al Gore suggesting an alternative to what might be called a discreet parting from the ranks by looking over what is not being looked over, or, get away from the pack before they sink us all. The conglomerate, or the bunch, is in the whole running around in circles and when they do finally get their act together it will be too late. My objective is simple. Throw everything I have learned and tie it in to climate change in a thorough explicit manner at whoever is interested and hope this influences enough voters to make a difference. Quite a lofty ambition but I'm giving it a try. This excerpt suffices partially to explain my beliefs but in order to understand thoroughly all that is going on, there being variations at your disposal but causing confusion for some, I have to go beyond just making general statements such as to be found in the world of journalism. I have to make you an impartial weatherman so to speak. I am not throwing theories around. Be ready in the later pages for the vaunted El Ninos and what that's all about. If I seem to lack humility it is because I have my convictions while painstakingly presenting facts and figures that should be believable.

The link between auto Emissions and global warming is strong enough to rethink "corporate responsibility," says a British Petroleum executive.

Los Angeles Times

In this world, my approaching a half of a century of observing, rubbing shoulders with an amazing variety of peoples I make the not too profound opinion – we have the wise, the in-betweens and the unwise. Give me 30 men and I'll show you at least three with an affliction called "Larceny of the Heart." Give me 30 degreed and/or pedigreed men or women for that matter and with sufficient time five will prove to be illogical. Give me 30 managers or bureaucrats and five of them are plagiarists who climb up on everybody's backs. Foot-loose as it may be that's my way of saying that I don't like frauds and don't particularly care for "Winners take all" economics.

I've seen it all. In 1942 while in training at Cape May, New Jersey, I had this powerful urge to be a hero, a fly-boy no less. I wanted excitement. We were patriotic and adventure beckoned. From a dead end environment, exceptionally dull and promising only guiding force, a stepfather having been nurtured in the alienating, stifling Catholic French Canadian philosophy, drummed into heads by archaic priests of the 30's, contrasting from America's general "God helps those who help themselves" populace, inasmuch as ours promoted being humble, turn the cheek nonsense and find your secure niche in one of the many dingy New England factories as being the whole world. I would discover from military life, believe it or not, some new fascinating values new people and opportunities. U.S. N.A.S. Cape May had it's glorious pilots and young men from all walks of life. When I told my stepfather the day after Pearl Harbor that I was going to join the Navy he wasn't about to approve and said, "Why do you want to do that? Are you crazy?" Mother knew better, she said nothing. Boot camp in Newport, RI during the dead of winter wasn't exactly a thrilling change. A typical sea-coast ambiance, dank raw winds of Mid-December while we drilled outside, sniffling noses and an unventilated mess hall with odors that were not exactly inviting, I couldn't wait to get out of there. What few hours of liberty a strict military regime allowed us turned out to be a lonely walk in what seemed to be an abandoned town whose residents we were told didn't particularly care for we uniformed kids. I took advantage of this rationed freedom but once. Patriotism in WWII was glorying in a fat defense plant check on the part of civilians. Everybody waves flags on the 4th of July but out of a population of 15,000 where I spent my teen years, only 10 of us volunteered to join the military after Pearl Harbor. Two died in the war. One, David Davignon, was a prisoner for two years – the Japanese practiced their sadistic habits and repeatedly beat him. Another, Robert Neil, a rousing full-of-fun Irishman, part of our gang that hung around corners once in a while, plopped down in northern France after parachuting out of a plummeting flaming B17 bomber. From village to village, helped by scared French residents, he posed as one of them and bicycled his way to Spain after a harrowing one month of playing hide and seek. It took him months at home in the U.S.A. to shake off the habit of constantly turning his head to look behind him. When we saw him a few years later we laughed at his stories, frantically peddling donated bicycles for all his worth, while wearing misfitted clothing and a French beret. Purely for effects, he described how he always had a stale old half loaf of bread and cheese wrapped in paper in a front basket in full view. One time he thought he saw military trucks parked in the center of a village and grabbing his bike decided to cross a field. There was a group of cows but a bull that he had not reckoned with had decided he would protect his family and came trotting toward Bob. Dropping his bike (all of us roaring in laughter at his description) he leaped over some bushes and landed belly-down in mud. While the bull snorted and sniffed Bob's only means of mobility he watched in horror as the animal butted the bicycle around. More laughter, " I wanted to come back and kill that s.o.b." An hour passed before he could make a mad dash and scoop up his vehicle. The funniest part was this Irishman having learned to speak French in 8 days. One of the guys asked why he didn't stay put and offer that bull some bread instead of running away.

95

The town of Cape May was a dull unfriendly place and since New Jersey laws didn't allow anyone under the age of 21 to drink there wasn't anything to do. An enterprising yeoman in the administration building solved the problem at the risk of getting in serious trouble. For ten dollars he aged me by three years – my new supplementary ID picture card was sufficient proof for any bartender. But that turned out to be wasted money since at that time I was paid the huge sum of 36 dollars a month. Two bars adorned the main street and broke up the monotony of white clapboard houses. Hoping for a sweet one that destiny had arranged for me I walked through the proverbial door "Some enchanted evening, you will meet a stranger across a crowded room". I was doing my nervous best to appear the experienced one, but fearful the grim bartender would spend too long flickering his eyes from the ID to my face and back again, envisioning the unceremonious shove out the door. All I found there were the usual steadies, frumpy-looking clients hunched over glasses of beer. The same for my last shot at romance down the street. Another night I discovered an USO arranged by kind souls inside a cavernous 2nd floor apartment. Two middle-aged ladies did their best to make us feel at home. Harold Gleason, five years older than I, a debonair New Yorker, a flair for impeccable English, who also worked in our weather office, was playing the piano. Two girls stood behind him, quite mesmerized. Older than I, they just glanced at me. No magic spell did I cast. Harold was a night club artist. While his nimble fingers sent sweet vibrations straight at feminine hearts he would turn his head and give his loving admirers an "I'm yours" smile. But, Harold influenced me. We got along beautifully on the job. I began to adopt the King's English, as he called it, and we both would laugh. Society-oriented, which I'm not, Harold married Buster Crabbee's ex-wife. Mr. Crabbee was the first, I believe, Tarzan of Hollywood movies. You know the type, "You Jane, me Tarzan." I saw Harold several years later after returning from overseas. He had succeeded in Cape May to qualify for pilot training but flunked out in Pensacola because he could not judge the distance between the plane and runway on a landing approach. I don't know why we laughed over his misfortune as a married man but we sure did when he explained why he divorced California's best. For all his sophisticated aplomb and knack for penetrating blue-blood circles, he nevertheless was not endowed with sufficient hormone levels. The Tarzan killer had proven too much.

Cape May was half surrounded by ocean waters. We had few planes and in retrospect I believe it was just a training base for various military skills. For one, we would pull down blimps and anchor them. If the winds lifted it up you would release the ropes of course and then try again after the pilot nudged it down. One time at Lakehurst New Jersey, some poor naval youngster didn't let go. It was shown so many times on TV how he hung on for a hundred feet or more and then let go. Of course he didn't survive. This job of pulling a small zeppelin made all of us nervous. We could see them going out to sea on submarine patrols, but I think some German could have shot the darn things down with a 22 rifle. Minesweeping boats, looking like sleek tugs would depart early in the morning and come back late in the day. I never did find out how the hell the Germans found ways to spread mines all over the western Atlantic near our shores where the ship convoys traveled.

Interruption again, Monday March 15. What a fuss over six or seven inches of sticky snow! Everybody in these generations a bit younger than baby boomers are into eye-bulging theatrics. Same old Atlantic high, guiding and fueling another moderate storm, but this time no mention of computers by forecasters.

Ah, but Cape May had a military secret! What excitement it was when it was unveiled. Now the Navy was taking a beating out in the Pacific. The Japanese Zero fighter was slightly superior to ours and so, Admiral Halsey and our banana elevated friend on duty there as his yeoman, spent the first 10 months or whatever after Pearl Harbor bluffing the Japanese by maneuvering all over the Pacific waiting and waiting for better planes to be built. Well, here in our inconspicuous base, hidden in a hangar where none of us were allowed to go, was the fighter plane that would do the trick. When I saw it I knew it was all over for the Zeros. An announcement came over loudspeakers – we all could go up on top of our operations building where our weather instruments were housed, lots of room. A dozen enlisted men and officers watched as the hangar door opened and out rolled a stubby powerful looking plane, all engine, massive and a huge four-bladed prop. Tips of at least six machine guns stuck out of the wings. Pushed out by some men it stopped. Up climbed a Lieutenant Commander pilot. No youngster J.G. for this demonstration. One wonders how a bee can fly. Well, this plane was built along the same lines, short wings but I was certain all that radial engine, huge exhaust outlets meant only one thing, all that power would pull that fighter up into the sky and as is customary, surely the pilot would buzz us at 400 miles per hour judging from all the noise it made when it fired up – the building vibrated. Brakes on, the engine revved up and then the time had come. Down the runway it went. The only trouble was, despite 125 miles per hour and plenty of runway, it hurtled into the drink. Splash, up went a column of white spray 100 feet high! As the plane slowly sank the pilot jumped out, stood on the wing, put his two hands together in a gesture of triumph over his head and allowed himself to sink with his ship. But, when the water had reached his waist the show was over. He swam in after inflating his life jacket. At age 18 I thought that was marvelous, nothing to it. Splash and you climb out. Later, after the war, I saw the same thing in Squantum, Massachusetts, but the water was ice cold. Somebody dived in but couldn't reach the pilot twenty feet down and almost drowned himself. Overseas I wondered about aircraft manufacturers. Was it their philosophy: "Our engineers are the finest, try it out and see if you like it." I witnessed the bizarre and the tragic. Sometimes, perishing in the "Deep Six" never to be found, was a consequence of someone in a hurry to go home and throwing caution to the wind. Customarily, at least in peacetime, naval pilots, before embarking on any flight must make it be known their destination by filing a flight plan, which also calls for the careful to pay a visit to the weather office. If we didn't think it was safe we were not required to sign a release. But, the pilot had the prerogative to go ahead anyway. Reasonably skilled at interpreting weather maps they sometimes played the odds. Frankly, it rarely happened. Overwhelmingly, most pilots live to a ripe old age- I can't speak high enough for their being the greatest. Two Corsair fighter planes, those famous inverted gull-winged planes that proved disastrous to the once superior Japanese Zero, along with our renowned F4U, took off from Norfolk, Virginia. We were never informed that their destination was our Squantum air base, just southeast of Boston. Both pilots were well known to all of us and were intent on coming back home, 340 miles away. Our weather was bad. The ceiling kept dropping all afternoon and by nightfall it was zero. In 1949, NAS Squantum had no radar assisted landing equipment. Why this sad incident happened I would never find out. The powers that be in the confines of the operational officer's room would eventually uncover facts but rarely would they pass the news around – everything is confidential. One pilot bailed out after running out of fuel somewhere near our facility and had pointed his fighter plane seaward before doing so. He had the good fortune of dropping down on firm ground. The other craft with it's pilot disappeared – not a clue ever reported by the Coast Guard. The next day I was told to attend a hearing. No smiling articulate meteorologists on TV were we. The questions: "What went wrong?" and "Who's to blame?" The surviving pilot, looking weary and upset, sat across a table from Captain Smoot and the Operations Officer and I, well, Chief or no Chief, stood a few feet away. Fifteen minutes later I was dismissed – no questions were directed my way. It's not a sensational story. I need to make something very clear. I'm not from the school of theatrics – I wish to be accepted as a disciplined and responsible person.

It seems I was blessed at a very young age with God-sent warnings. I'm not totally convinced however of some guardian angel watching over me or steering our car. Somehow I entertain the notion, beyond what religion may suggest, our fate is predestined and if you are not an air headed youth some wisdom is assimilated along the way. In just one month, the year 1942, I was exposed to three deaths, three harsh lessons, and three military personnel having died unnecessarily. The teenager of today is not likely to be moved into a contemplative state by a newspaper account, quite a few lately of auto accidents, loss of control and sometimes the evidence empty beer cans reveal. Death to immortal youth is remote when it is out of sight, miles away. We saw but a canvas wrapped body in Cape May but that was enough. It jelled. How we managed to borrow someone's car was mentioned of course, in between reminiscing over his personality, even the unkind adding a stoic "He liked to drink too much." The report was right to the point, "Speeding under the influence." The car, an old pre-40 Ford, plunged through a guardrail and sank in a marsh creek, so many coursing to and fro with the tides of southeastern New

Jersey. The wooden steering wheel broke away and he was impaled – that detail set home with us. Then a veteran cook, much older than we, who I would see every day in the mess hall, sallow featured, never smiling, busy floating around vats and mass-production kitchen equipment. But, nevertheless, he was benevolent enough to pass us something extra, when we, my friend Thomas Murphy and I, ignoring closing hours, especially at night when the thought of going to bed without appeasing the usual persistent needs of insatiable youth made us bolder. What a terrible thought, one wooden barracks, rows of bunks, a string of lockers and a large shower and toilet area but absolutely nothing else, no candy machine, and no Coke dispenser. You ate at 5 PM and lights out by 10 PM, by 9 PM distracting stomach noises. To this day that is intolerable for me. The cook was kind but we all knew he had a bottle somewhere, a hard-core alcoholic who managed, because he was needed, staying out of trouble with superiors. He was found dead in a gutter of Cape May's main street, a victim of excess alcohol. I'll finish with this unpleasantness but one final lesson and it had to do with an affliction that other dummies and I knew nothing about, namely depression and how one should respond to someone gravely afflicted. Of course we were dummies! Even radios were a luxury in my younger days. I never saw a newspaper in my home before Pearl Harbor. Hobos would come up and beg for food from my mother while I hid in a garage but couldn't resist peeking from a door window at some stranger devouring a sandwich. She knew, this mother of mine, what hunger was all about. There were a half a dozen marines doing guard and sentry duty at Cape May. They stayed aloof from we white hats. But, this one particular marine, immaculate as they were and still are, again much older, a sergeant, would come in to our barracks in the evening for several days and with a dead-pan expression ask us what was the best way to commit suicide. After he would leave, somebody had to joke about it. He used a rifle to accomplish what he had set out to do but had sought some help and I suppose this is regarded as a subconscious appeal. It impacted on us all.

I mentioned Thomas Murphy. Goodness knows I need to highlight joyous moments also. By nature, I clown a lot and while there were six of us working in the weather office, Tom always responded with genuine laughter at my silly antics. Naturally I took a liking to him. A nice looking lad, freckled Irishman from Lowell that he was, he would overwhelm me in our jousting around in the barracks, my putting up the best of a boxing stance. Our church prior to my entering the Navy had a priest, Father Hebert, a very modern priest; we enjoyed pool and Ping-Pong rooms and a beautiful gymnasium where basketball games were held, thanks to him. On Friday night, a boxing smoker was held. At least 500 parishioners would come in, smoke the place into a white pall while we kids tired ourselves out swinging sixteen-ounce gloves at each other. I remember someone yelling at a Tarzan like opponent "Hit him in the bread basket!" and even though it was a draw after three rounds, I was glad to get the hell out of that ring. Who picked this gorilla to box with me I know not – some old resident decided the match of matches. So I thought I was pretty good until I ran into Murphy. On top of that, I climbed into a ring with some German kid stationed you should know where by now and after seeing stars from a right hook, my aspiring to be noted as a boxer immediately faded. Both these nimble cats, Murphy and that unkind Kraut were former Golden Glove participants. The point I'm getting at is Murph, as I called him, didn't strike me as a man of the world and I was positive if we ever met some girls he would shy away while counting on my setting examples – at least I thought so. Built like a boxer, all shoulders, no waist and spindly legs, his GI issue of loose straight-down pants, lent a floppy look to his crisp short stride when he walked. Of course, I was the elegant one. I had spent money having a tailor peg my pants at the knees and while Murph's allure didn't meet the sought after bell-bottom look, mine certainly did. One weekend and they were not often, we decided to risk hitchhiking anywhere destiny would take us. Hours later, we arrived at Atlantic City. We had about 50 cents in our pockets. In those days there were no casinos that I can recall and the whole boardwalk was a series of stores, honky tonk clubs and bars with the usual beer signs blinking away. We found a joint where for 25 cents we had a mixed drink, quite diluted but it was better than wandering around all night. Whatever young women we saw were not interested in us. They knew exactly what we were worth, economically speaking. The absence of red rating stripes, well known to free-loaders abounding everywhere, meant little available generosity. We trudged along, hanging on to our last quarter. Then, we heard this beautiful music in the distance. This huge pavilion in the distance glowed with promise. I believe it's still there. A summer night, waves lapping at the wooden pier upon which this structure was built, the moon, and the melodious beat of drums we could feel, let alone hear. We hurried toward the heavenly sound. Now, as a side thought, where was this war anyway? Thousands of civilians on the boardwalk, maybe two or three soldiers and a jammed packed dance floor, no GIs to be seen. Well, we stood in the entrance way and scanned the walls looking for some alluring wallflowers. None. I couldn't dance, except jitterbug as any darn fool could do. Dejected we just stood looking on. Two fox trots went on and on. The music stopped. Some man we could see over the crowd, every couple tenderly holding each other standing out there on the floor, announced a waltz contest, warning couples not to be upset as judges would circulate and the less than dexterous would have a shoulder tapped. He mentioned a prize, but I wasn't paying any attention, wondering why Murph didn't want to leave. He disappeared into the waiting crowd just before all these sissies started their Vienna Waltz routine. What was tough-guy Murph doing anyway? The eliminated stood on the sides and finally, who was left out there but Murph, with a huge grin, delicately holding some older woman's hand. His pants seemingly now less frumpy, his catch not being very attractive, but a graceful Cleopatra no less with her whirling dress and a crowd applauding as they slowly spun about in the center, all eyes riveted on this mismatched couple. It was over and I was afraid he would come out with his new partner in life and I would be left alone. Dancing resumed. Murphy burst through the crowd, almost knocking a couple over looking extremely joyous. He had won 25 dollars. That was a lot of money, you might say for two 36 dollar a month seamen. We romped from place to place and by 1 AM we were in a silly state, physiologically speaking, laughing while relish from hot dogs kept spilling about. "Murph", I asked the next day, "Where did you learn how to dance like that?" My best look of dismay of course. As usual he laughed and I joined him. I wondered also, during the day, is everybody a civilian? Later, in December of 1942, my orders to leave came and I would never see my old buddy again. In fact, I have never met a guy like him.

Enlisted men are careful not to show any disrespect toward officers. Nothing new of course, but sometimes it reaches the point of absurdity when two cultures clash. A southern officer didn't like some guy but I didn't know the circumstances. It was some kind of animosity between the two that went on for a few days. Finally, the charge, "Silent contempt." The captain who traditionally holds court, called the Captain's Mast, ruled it frivolous. Speaking of disrespect toward officers, a rarity indeed, I have often told friends that in aviation there was mutual respect. One chap laughed and said, "If I were a pilot I would be nice also." It's not really that, of course. We had this big inter-military branch boxing meet one day involving the Navy, Army, Air Force and even some members of the French Navy. Port Lyautey in early 1943 had taken on the look of Rome during it's gladiator days. What a crowd! We didn't know this Spaniard, Marcel Cerdan, part of the French Navy, was a formidable fighter who several years later became world's champion middleweight. A commander, a three-striped officer, was in charge and stood next to the ring. He had selected a tough street fighter to represent the Navy. Well, each time that chosen one hit the canvas, generally on his back, the officer screamed encouragement. By now, about the third time the canvas thumped, the crowd became silent – quite discouraged. About the fifth time, our middleweight, on his hands and knees, shaking his head and I wondered why that damn fool officer kept yelling, "Get up! Get up!" it happened. Two expletives spat out of that man on the canvas, quite popular in movies today and on the street, followed by "I'm not going to get killed for you, sir!!" Nothing came of it but we thought this assault on the elite was hilarious. The Army won of course – so many of them. I have to find an excuse for our failings, naturally. Then there is the embarrassment felt by some Air Force members who moved in to bolster our submarine warfare using the four engine Liberator bombers. German submarines had moved out of our flying range somewhere nearer to the Azores and we needed long range aircraft. Now, there is no one braver than Air Force personnel, who, during the bombings of Germany with their B17's suffered a near 50 percent fatality rate, such were their being targeted by fast fighter planes and concentrated anti-aircraft fire from the ground. But, we didn't take kindly to being showed up by another branch of the military, coming in to our Port Lyautey base with six ominous looking bombers compared to old faithful PBYs. It only happened once and in retrospect, being more rational in my later years, it had to be a case of the Air Force not having trained their pilots in the art and wily ways of attacking a surfaced submarine or the pilot was more interested in saving his crew. Everything that happens in the military is photographed and there is always a cameraman involved. This

B24 Liberator came upon a crippled (I suppose) German sub and angled down to drop depth charges but the pilot had a change of heart when he saw 40 millimeter tracer bullets coming at him from that sub, some puncturing the fuselage and wings. He turned the plane around and came back to the base. Our officers didn't hesitate to let us hear their comments, reserve set aside and smiling between chuckles. The Colonel in charge of the B24s was furious and grounded the pilot for a month. Of course, this was the talk of the town and I imagine at least over 4000 mouths made disparaging remarks while grinning. At that age there is not much compassion available. Do you know every fourth bullet is a tracer and it's quite a sight when the stream is adjusting your way? In my case, shooting a 50-caliber waist gun one day on my first flight over the Atlantic in a PBY, this being required, I was adjusting a stream of tracers going the other way. A crewmember threw a smoke bomb out and there below, some 1000 feet, a pillar of white smoke poured out of it's floating container and obviously you shoot at it. The only trouble, I got carried away – nobody told me that the tail of the PBY but 15 or 20 feet to my left and perhaps four or five feet above my level, could be hit accidentally. Blazing away, all enthused and thrilled at this power in my hands, it suddenly ended. Some guy grabbed me, shoved me aside and in the din of this sheet metal flying apparatus I heard him loud and clear, "What the hell is the matter with you, can't you see that tail section up there, do you want to get us killed..." A few more words followed. I sat around, almost 11 hours of droning monotony but I was more concerned as to my not having come close to that smoke bomb. Later the yelling one would smile at me. I was sent out frequently for two specific reasons, one, see what the weather is out there 4 or 500 miles out to sea and two, be reminded why you, as a wind-guesser, as we were called, (what was termed "Attitude Conditioning") must be reminded of. The dominant thread running through this blend of anticipated changes in weather and life's experiences contributing to my mind-set, a few stories acting to establish I am not altogether Mr. Gloom and to furnish episodic stages of changing times, is an effort on my part to provide totality in matters of meteorology for your discretion to accept or reject and how socio-economics could be affected. We have to take a long look as to why 2500 economists have joined hands with scientists. I am not at all interested, let alone impressed over long range hypothetical sea-level changes connoted and even insisted upon by others other than the senator. While nature has already given us strong signals over much of the world that normalcy can no longer be expected. I would like to think that I am turning on a light but my expectations to reach acceptance by some hoped for majority is not at all optimistic ones. First of all, there is divisiveness in America. Most affluent who make up the bulk of voters and who work for corporations don't seem to strike me as enthused over government mandates that would prove expensive for their employers or be they self-employed. It is clear also that there is not wholehearted support for EPA for example. While I can't remember every radio guest's name or titles of their books, one author talking to David Brudnoy struck me as being a little more profound than some. More a historian rather than a philosopher, he pointed out some facts that made it easier for me to be more convinced; self-enriching is a preoccupation and so succeeding, it not being evil at all, results in a society not functioning as a whole. The main theme gathered from whatever source the author had access to is very interesting but not at all encouraging. Economic enrichment elevating many creates a class of citizens who gradually isolate themselves from both the government, their allegiance at a minimum, and even the rest of a working society. Pre-occupied with amassing wealth, believing they need not be part of mainstream USA and it's politics, selecting homes in exurbia, they gradually lose interest in national affairs except Clinton and Miss Lewinsky. Therefore, for one day at the voting booths, their choice is predictable. An affable politician who offers reduced government, less taxes and any promise to allow all to do whatever their hearts desire in the name of freedom, gets their votes. That is what Contract with America was about, although I too am annoyed over excess liberalism – but that is a subject that would take these printed opinions out of context with my objectives.

Let me review so far my attempt at sensibly structuring, link by link, connection by connection, what has been strewn* about by different interests during these recent past decades. I started in this book with basics with the sole purpose of connecting one fact with another and to do so without special effects, without sophisticated fanfare. Dirt (all kinds) increases moisture absorption by air and increased moisture content in our atmosphere results in more rain. Excessive rain all over the world at times, for example

1998, is only possible when there is excessive moisture. Ironically, while drowning thousands and ruining rice crops and other agricultural products, precipitation cleans the air – there is, as I must repeat, no other way. Stronger solar radiation then what it was 30 years ago, combined with heat trapping CO2 results in expanding high-pressure systems. Heat air and it expands

Diffusion – that's more Politically Correct

–no one can argue that. That is why summer's Pacific and Atlantic semi-permanent high-pressure systems, more expansive than ever, plague us with abnormal weather, i.e. droughts, heat and floods. Their increasing northward influences, massive in latitude and longitudinally dimensions creates a problem of holding back relieving cooling systems that normally would move down over the warmer latitudes. They are too voluminous and stay around well past their usual autumn contracting cycle. The sky is blue because invisible air molecules <u>refract</u> only the blue of the color spectrum. Dirt in air does not refract – it absorbs solar energy. What we cannot see is there nevertheless. A good example: Leave dust collect inside your vehicle and you will see glittering microscopic specks of silica but only when the sun's rays flashes at your instrument panel, or, stand on any light colored asphalted street and with the help of a sunny day you will see it's surface throw off tiny specks of glittering light. That is because a maximum

Malaysia kills pigs after disease outbreak

Los Angeles Times

KUALA LUMPUR, Malaysia — More than 1,000 soldiers wearing hoods, gloves and surgical masks began killing Malaysia's pig population yesterday to contain an outbreak of Japanese encephalitis that has killed at least 53 people since October.

Nearly all the victims lived on or worked near pig farms in the villages 60 miles south of Kuala Lumpur, the capital, where the soldiers, accompanied by 300 health workers, were slaughtering the pigs. The region is Southeast Asia's largest pig-breeding region.

In response to the outbreak of the deadly disease, Thailand and Singapore banned all imported Malaysian pork. The sale of pork has fallen to near zero in Southeast Asia, even though health officials say encephalitis cannot be acquired from eating the meat.

Japanese encephalitis is passed from pigs to humans by <u>mosquitoes</u>, but cannot be transmitted between humans. The disease attacks the brain, causing high fever, coma and often death. It can be prevented by a vaccination that many travelers get.

Researchers from the Centers for Disease Control and Prevention in Atlanta are expected to arrive in Kuala Lumpur tomorrow to help Malaysia investigate the deadly outbreak.

Health officials estimate that the soldiers will kill more than 300,000 pigs, wiping out the livelihood of hundreds of pig farmers in Neger. Sembilan state.

amount of pebbles were mixed with the asphalt. The constant grinding of auto tires flakes off silica, so light and invisible, but not in frightening quantities that they waft about in our air. Silica is extremely hard of course. I don't point this fact out to you from a hypochondriac bent but as a further emphasis on what is microscopic and remains airborne, that is all. I am not at all interested in spreading fear. I quote sensible and not so sensible comments. I use excerpts for your discretionary consideration. Without resorting to fear mongering the element of expanding diseases was also reviewed. Today, the Providence Journal had an article describing a sad state of affairs in Malaysia affecting hundreds of farmers. The army has been ordered to kill all of that country's 300,000 pigs.

Unrelated to earth warming (I hope) our state of RI has had fatal cases of encephalitis afflicting two children. It caused quite a scare. A small army of DEM employees scooped up mosquito larvae for lab testing but I don't know if anything came of it. March 22, 1999, our Providence Journal also reported mercury dust in the air above Vermont's highest mountain (see clipping below). That is something new and rather disconcerting. Baffling also.I had this pleasant conversation with a professional fisherman, if you recall. My daughter had organized a

VERMONT
Mount Mansfield shrouded in pollution

JEFFERSONVILLE — Vermont's tallest peak is often shrouded in pollution, a new study has found. Recent research shows that clouds hovering around Mount Mansfield bathe the area in acidic moisture, poisonous mercury and other airborne pollution. Researchers have learned that much of the mercury drifting down from the sky stays in the forest, trapped in tree leaves and forest soil. (AP) family Christmas party and he and his wife, being neighbors (when he's home, such is his stay at sea) were invited. Two gentlemen from Hawaii had a lot to say about the fringe effects of tourist-attracting volcanoes that are not beneficial. Of course I didn't leave there despondent over learning that the Atlantic Ocean also has mercury circulating about. Fish caught have minute quantities of this toxic metal in their flesh. After literally squeezing all the information that I could out of this brain, such is my habit, we both went back to the buffet table and steadily scooped up seafood dips. To heck with mercury or no mercury. In a perverse way it pleased me to learn some more bits of information. He did say, while shrugging his shoulders to register uncertainty, that he did not know why but felt that mercury must be part of the ocean in some way. I accumulate all these negative ingredients of our planet as simply knowledge to pass on. It does not portend to the end of the world. But, it is something new for me to think about since mercury dust over Vermont is really both interesting and grim. Can it be some industry somewhere contributing? I recall the news, perhaps 15 years ago, of a large number of Japanese villagers who died an agonizing death after eating fish contaminated with mercury. They resided in a village situated at the mouth of a river flowing into the sea and had eaten fish caught nearby. Investigators discovered a factory had dumped mercury into the river. Consider Brazilian gold accumulating workers, sloshing about along some stream connected to the Amazon River, powerful water hoses slicing out banked escarpments containing specks of gold and how they process what is panned. Mercury aggregates gold specks together; sloshing the two together the gold coagulates into one lump. They then throw the used mercury into the handiest waterway. Wonderful. I don't think dust describes accurately the presence of mercury in air although I'm not qualified to argue. I believe it is in vapor form. It is a strange acting metal if you should care to read about it. Here is something that may surprise you. Some 15 or 20

years ago dentists became familiar with the odd behavior of mercury. In making a silver cap for a tooth using mercury and then by chance it touched an older gold-capped tooth, some of the gold would dissolve. But, that is not the strange part! Somehow a static condition would frequently cause a painful spark between two teeth capped with these two incompatible metals and unattended an ulcer would develop inside one's cheek. Ask any older dentist or maybe your computer will verify it.

What action can we expect to take place in this screwball world if, as explained in a book noted on next page, elaborating on the snail-paced legislation process of Washington, DC, and of course the backward nations of the world, if it takes years to come to any conclusions? Seven years have gone by here in RI and we still do not have an auto emissions test in effect. The latest comment from one of the many in charge, "This is going to be one hell of a tough program to get going." The problem of government extremely slow to react despite disconcerting statistics until there is a crisis is clearly explained in a book titled, "Agenda, Alternatives and Public Policies" by John W. Kingdon. While his examples did not particularly revolve around environmental disasters, since so far other than rhetoric no one is scared, he utilized death figures of truck accidents and other serious problems, placing emphasis on the complexities of bureaucracy, shifting personnel and so one, the end result being prolonged foot-dragging. Leaving PC aside, as it should, he did not hesitate to suggest that politicians need to be smacked over the head before they go into first gear. If, writing of auto accidents or whatever, 50,000 people die, hmmm, they might consider that crisis proportioned. He describes a true account of a young congressman; eager to have his bill reviewed in the hope of having it passed, asking an elderly senator for an opinion. The veteran statesman, seated behind a desk in his office, having patiently looked over a new proposition suggests that the young man should relax and go take a break, because, "It will take 25 years for a bill to pass."

Five years for this, that's not bad but don't count on it. See excerpts on next page.

In recent weeks, as auto executives have shuttled between Detroit and Washington, D.C., for tense meetings with EPA officials, diesels and light trucks have been at the heart of the negotiations.

The regulators wanted to close loopholes in current tailpipe regulations that hold diesel-powered vehicles to lower emissions standards than gasoline-powered passenger vehicles. As more light trucks are being bought and used as passenger cars, they should be held to passenger car standards, the regulators argued.

Auto executives wanted to protect separate standards for diesel engines, especially for sport-utility vehicles.

The winner will be known this month, when the federal Office of Management and Budget, which is reviewing the regulations, releases them to the public with an estimate of their cost.

The draft standards, which will also reduce the amount of sulfur in gasoline, will be the subject of public hearings over the next few months. The draft standards are to be finalized late in the year and won't take effect until 2004.

The federal Environmental Protection Agency is about to unveil proposed air pollution standards for tailpipe emissions. And according to auto executives, the new standards will make it difficult, if not impossible, to produce diesel engines.

Diesel combustion produces a different kind of pollution than gasoline combustion. Standard catalytic converters don't work well on diesel engines, which produce large amounts of the pollutant nitrogen oxide. In addition, diesels produce ultrafine particles that are difficult to filter out of exhaust and

Logic is a commodity sorely in need of restoration. It declines as an expression of conscientious clear thinking into the dregs of unprecedented feverish money-oriented preoccupations. Read those excerpts on diesel-powered SUVs once again. Are they really interested in "smaller is necessary"? That is sick. If Detroit succeeds in palming this off, not only will they make vehicles bigger than ever but will poison affluent suburban homeowners with nitrogen oxide. A diesel produces NOX <u>constantly</u>, even while the engine is idling. Exhaust heat rises but not that gaseous acid! On any given calm damp morning, one idling diesel, the owner waiting for the comfort of a warmed up engine as he or she waits inside having another cup of coffee, that click-clacking engine will spread nitrogen-oxide for yards. The odor is biting, obnoxious and penetrating. Leave your dog outside and watch it shake its head, sneeze and scratch at your door to come back in. Your cat will be inclined to sit on the inside windowsill rather than ask to go out. I have to include some humor here inasmuch as nothing may come of it if Detroit does not get it's way. But it's possible. The Bubonic Plague of the mid 12th century, a consequence of unsanitary conditions was attributed to ignorance. It was death, immediate and far-reaching. Our gradual deterioration of air quality, water purity and ecological disturbances is also akin to unsanitary attitudes but it's insidious and intensification will not kill in a hurry. Therefore it is, in these modern times of rapid communication, a combination of both ignorance and indifference. There are ample clues but each report and there have been many these past years soon fade with time. Of course their validity is ignored once the "anti" spokesmen get through. Only they can lead the way if you take them seriously.

TWO DIFFERENT ARTICLES

WASHINGTON (AP) - A temperature change of just a few degrees may be enough to disrupt a delicate ecological balance in the tidal waters of Oregon, and perhaps elsewhere, according to a study of starfish and mussels.

Eric Sandford of Oregon State University said his study suggests that if a key species is particularly sensitive to temperatures, a slight warming or cooling could start a cascade of rapid changes affecting every animal in an ecosystem...

"The carbon dioxide changes over the last few thousand years have been tiny and slow compared to what humans are doing," said glaciologist Richard Alley of Pennsylvania State University, who did not participate in the study. "We are moving into uncharted waters."

Make hay while the sun shines and to hell with tomorrow, that is the prevailing mood being whipped up. You're doing well if you can find one science magazine in a drugstore. If there is one it probably is buried beneath six magazines on how to get rich. Plenty of clues everywhere but they do not deter the insane amongst us. I am not promoting paranoia nor do I expect spontaneous collapses – just a down to earth collection of technical data for you to kick around. I don't lose sleep over it. Usually, what community spirit remains and to be found mostly in suburbia generally is limited to local problems. Too many demand less government but that is not the way to go. Appoint one individual and have him or her periodically collect the voting record of Congressman and Senators and only then, make rational decisions after all voters discover what is going on. You would be surprised as to what some pet politician has been up to. That is when effective cuts can be made and not some blowhard radio host who keeps shouting "It's unconstitutional!" demanding we all be on our own because we don't need this or that from Washington, DC. If your airline is safe it is because of the FAA, as an example. Cut it's staff? Please. What did our forefathers know about airplanes? If a politician is a nuisance, get rid of him or her! But, at least know what they're about. I'm getting preachy – I'm sorry.

Picture this: You live in a nice ranch type home. It's the year 2001 and your neighbor pulls into his driveway with a huge SUV or a pickup truck. You discover that it's diesel-powered. Put up a For Sale sign and move out! On any calm windless damp morning that monster size engine, which has become an obvious obsession, while idling for five or ten minutes, warming up so as to furnish heater comfort while it's owner has another coffee before going off to work, will spread nitrogen-oxide for yards. Unless some chemistry genius or brilliant engineer finds a way to stop Nitrogen-oxide emissions in diesels you, as a neighbor, next to or across the street from, are doomed to much discomfort. This attempt on the part of Detroit to pursue sinning against nature, very evident in their wish to power up and enlarge their already too massive products marks them as sophisticated madmen and damn tough female CEOs. We worry about asbestos – that's nothing compared to daily doses of nitrogen-oxide, an acid that insidiously gnaws away at a child's lung alveoli while he or she waits outside for a school bus. There are other avenues of poisons to circulate* in one's body all made possible by lymphatic circuitry in your lungs which, up to a time limit, distributes what is alien and destructive to other parts of the body where collection can take place. We're not all born with cast iron lungs. As an aside, don't blame somebody down the street who

*A medical establishment has published a report that industrial toxins have been found in the amniotic fluid of pregnant women.

owns a reasonably sized diesel Mercedes or whatever if your shingles are darkening with the years or vinyl siding has begun to look soiled – that's due to a nearby expressway or a city. If we are the best fed people in the world, why do we rank 17th in health criteria? We breathe smoke that's one reason. The invisible kind. Not to be confused with secondhand smoke.

Summary: Of all the damn developments! Diesel engines to power three to probably three and one half ton SUVs. And Detroit wants to argue over this privilege that guarantees the worst kind of pollution. During the summer we have concentrates of ozone at low levels where we live. Isn't that an irony, increased ozone at the surface and the protective ozone level some 15 to 30 miles above us having decreased in density? Here is a description of ozone that should make you change your mind about moving into the city if a member of your family is asthmatic or if lung cancer has or afflicts someone in your family tree. In the past 15 years the rate of asthma has doubled. <u>Ten percent</u> of all American children are victimized and it is the <u>number one</u> reason for days missed from school. Description of ozone: A blue colored gas having the odor of chlorine. It is a powerful <u>oxidizing</u> agent. It is no wonder that children of poorer families living in cities have a high death rate from asthma. Sulphur dioxide, peculiar to areas of oil refining industries and also from coal burning power plants, has not been reduced whatsoever. It is suspect of damaging the ozone layer. There is no question that our sun's rays are more potent. As early as February and March temperatures in Jamaica have been nudging the 90-degree figure, but not quite above 89 – it's trying. I don't need Nimbus 7 to tell me that.

Everything that I write reflects the obvious. I have no hope of our nation achieving carbon-dioxide reduction goals in the next two decades and judging from the lackadaisical attitude over aging SUVs and pickup trucks being free from having to replace emission devices, I certainly don't see any possible relief from nitrogen-oxide and carbon monoxide for a long time. Cheers. I do not want to be loved. I loathe those who are responsible and make no bones about it.

I am not the all knowing Guru and at age 19 approaching 20, I certainly was not loaded with wisdom or experience in weather. School of hard knocks – that's a pretty good teacher. Trained in eastern USA weather there was enough to be absorbed without intellectually chasing all facts of the world. I knew little about Agadir, Morocco's coastal waters prone to cold water upwelling along the entire African Coast well past north of Casablanca, some 500 miles of it. Sudden fog was to roll in a la San Francisco, blanketing arid sun-baked terrain 25 to 30 miles inland. We had been based there but two weeks. One hell of a start, to be awakened at 4 AM and hauled off to experience the results of a plane crash and the nasty job involved. This fog, it caught me the first time and it was the last time. The day started sunny as usual, after tipping our shoes over, just in case some scorpion had decided to rest there, a quick rendezvous with a lieutenant who wanted to know what the cloud ceiling might be out over our side of the Atlantic and after that, breakfast. Those PBYs flew just under clouds, mist brushing the windshield in the hope that they would not be detected by surfaced submarines while radar scanned ahead for any surface object. At 1 PM it came in, one gray mass and within minutes the ceiling was but 100 feet or so. I wasn't worried about our PBY out there at sea and assumed that the commander would radio to head north to Port Lyautey. I didn't dare go talk to this stoic officer in charge of a squadron since little old me, white hat angled over my left eyebrow, wanting to look the old salt, hadn't mentioned fog during the early morning briefing. To my dismay, at 4 PM, it representing 12 hours of flying time out at sea and knowing little fuel would be left; I heard the familiar rumble of our plane above the fog. Here came the lieutenant commander, all six feet plus of him, and although a true gentleman, he had that look, "Just the facts, Ma'am" you know what I mean. He had a walkie-talkie radio in his right hand but I braced myself wondering what he would say. He wasn't excited or angry and he simply said, "I have to bring him in." If you think this is too melodramatic you must appreciate my anxiety having just lived with the vivid experience of knowing what a major aircraft looks like scattered about. That plane, with seven human beings aboard, couldn't go anywhere else. South, maybe 100 miles away was Spanish Morocco and they customarily kept unwanted visitors around for awhile – besides, it was too far anyway. We had one short runway and on each side rather rough scrubby terrain and worse, at the wrong end of the runway several small buildings. This officer's squadron had experienced the worst in Iceland all during 1942, harassing and keeping German subs submerged before being transferred to Agadir. Just his look was enough for me to realize I had better open my mouth and say something –he wasn't going to ask a kid for advice. "Sir, I'm not familiar with the terrain in this valley" feebly pointing westward where I knew that plane had to go. "Can he come in low above the ground – I'm fairly certain it will be clear 25 miles from here." I wasn't going to tell him what to do but he knew it was the only way. I started to say something else, but he interrupted with a short and sweet "Yep", wheeled around and while walking toward the runway, turned toward me and told me to stay put. Minutes went by. I was very troubled. That officer, transmitter to his face, intently leaning forward, kept talking to that pilot. I could hear the faint sound of that plane approaching and then appearing out of the mist – it was stunning. Try to imagine driving a car at over 100 mph and you can only see 150 feet in front of you. While the pilot could see the ground hurtling by him below he had no idea where the runway was – just a voice, calmly, reassuringly, "You're sounding good." I wasn't the hero, the commander was – he had saved my rear end also.

The debate over global warming – the trend toward higher temperatures across the surface of the earth – has grown in intensity as study is piled upon study the international Intergovernmental Panel on Climate Change concluded in 1995

DAILY NEWS MEANS DISCONTINUITY IN WRITING

To the left, just another bit from the oil company executive in England who, as a breath of much needed fresh air, stated his belief over earth warming and oil consumption being a contributing factor. Study upon study – that gets to me but I can't get nasty and question such a ponderous effort by scores of scientists as being foolish. There are some that I have taken advantage of their reports but truly, I do believe that you and I can handle this by sticking with common denominators. I don't want to lose your interest and by using this brief excerpt I hope to impress you with the concern earth warming is causing. Depending on where you live, you may find it damn unnerving and if not, I guarantee you your pocketbook is going to feel it.

I venture into this minor tale primarily to describe life for men in naval aviation and perhaps an apology for what, in retrospect seemed a cold attitude. Episodes at Cape May – had they tainted me? Maybe it was childhood experiences I never forgot – a forbidden sneak view at night of my father in a casket when I was eight, a three year old girl dragged by a car down Hunt Street, that infamous hill and old cars with bad brakes. Her father ran out, fainted and fell across her sheet-covered body. I was terrified, ran home and hid under my mother's bed, shaking like a leaf. Then, only ten, watching firemen trying to revive a smoke suffocated baby but did not succeed. I still see her tiny arms hanging there. At about age 12, living in North Providence in what was then more rural than suburban, a hobo, probably had been drinking, trying to reach our home with it's lights showing in a snow storm, just got hung up on a backyard fence and froze to death.

PBY crewmembers were a stoic bunch and kept to themselves at Agadir. Not unfriendly but distant in a way, knowing their job was dangerous while that of all of us outsiders were not. Yet, on this trip out to sea one member talked to me – even smiled for a change. He explained why waist guns couldn't be swung too far in the direction of a PBY's tail section. PBYs had some kind of explosive box somewhere aft in the fuselage. If they were forced down in enemy territory the pilot would set it to blow the plane to smithereens after everyone had distanced themselves. Therefore, the guy who had yanked me away on a previous trip, my enthusiastically drifting the gun to the rear as the plane moving away from a floating smoke target required I do just that – try and hit the darn thing – that guy thought me unworthy of learning such a secret. He was afraid of concussion triggering off that powder keg back aft. On this one trip out to sea that started well but ended poorly, I must admit that I was scared. I recall sitting back there with the rest of the crew, they looking more somber than I liked. I remember analyzing of all things, where would my hurtling body hit should we crash. I had this thought, all of a sudden – the upper torso is heaviest, you plunge forward at 150 miles per hour headfirst. No belts. What good would that do? You can't survive. The scene of a previous accident ran through my mind. I thought also I must have gotten hard, unfeeling. I wasn't even shocked at headless corpses, one but a part of his jaw left and I made believe I didn't see him in the first faint light of dawn. I just walked by, stooped over, play acting, looking down. I stared at scorched sheet metal scattered about. This time, knowing we were in serious trouble, I stared ahead at dull sheet metal fuselage sections, supporting harsh looking ribs and then the pilot's partially closed compartment, surely it would come crushing and slicing at us. You sit against cold metal and you hope the noise and vibration don't suddenly stop since the problem of insufficient fuel was on everyone's mind. The crew, the only way I could access conditions, they and not I, familiar with everything, had ceased their usual relaxed activity at the end of a trip. They just sat there, heads slightly down. It was only a few weeks before when they too were part of a body part collecting assignment, traditionally a Navy requirement, placing remains on a tarpaulin. We were lost and could not find Agadir and the pilot and navigator had made a decision to head north where Port Lyautey's radio beacon could assist. Agadir had nothing. On the way, crewmembers had positioned themselves up against the starboard waist gun for a better view, looking at what I didn't know. Came my turn to take a quick peek, wanting to stay out of the way. Nothing but miles of brown cliffs and surf crested in white lapping at their bases. No geographic reference points to go by. No lakes or rivers to seek out since PBYs are also amphibian aircraft. I knew how much fuel these ships had and that was more than disconcerting. Another hour and a half in the air? You can't land in sea swells or coastal surf. It didn't take much intelligence to figure out why the pilot eased the plane down to about 200 feet altitude over water. It meant less impact than dropping down like a rock from much higher elevations should engines quit. To make an approach to Lyautey (now Kenitra) from the sea there was a 150 foot high embankment upon which sat an observation tower.   At it's base a most beautiful but deadly beach for swimmers. I had plenty of time to run a series of movies across my mind and thought how another fellow and myself had laughed moments before fumbling around wreckage that fateful predawn incident several weeks before. In the semidarkness some officer told us all to wait for fires to die off and not to venture too close to the bulk of the wreckage. Off he went somewhere. Two of us, disregarding orders, started easing our way toward the major portion, a few small fires here and there, morbidly curious or to find somebody. Everyone else just stood around at a safe distance.   Quite suddenly something resembling the sparks spurting out of an oversized Fourth of July Roman candle lit up it's source. It hissed and there it was one big depth charge, about two feet in diameter and perhaps three feet long. My friend, grabbing an arm uttered in a low tone, "Let's haul ass outta here!!" We hadn't taken two flying steps when that can blew. Depth charges don't hurl scrapnel but the concussion shoved us down face first into coarse dry grass, my nose hurt. We laughed while scrambling to our feet and put a little more distance between the wreckage and us. We made it to Lyautey alright, but not without a lot of nerve jangling noise, metal scraping, screeching as if the bottom was about to rip open. About 30 feet above the runway, both engines quit and with props rotating idly, no control by the pilot. That's a big plane and it dropped heavily. One wheel collapsed, the left wing tilted down, struck and stuck to the runway, then into dirt and we spun around violently. We scrambled out and dropped down with amazing speed. Lucky me, one sprained ankle. I also experienced a resurgence of religion all the way there.

Have you ever heard of lightning fields before? I know that I never did many years ago.   The Weather Channel, it's speaker rather fascinated (they have more interesting things than I did) was showing one solid mass concentration of red dots in a green field (rain) across the Doppler radar picture, chugging slowly eastward, 400 miles in length from Indiana to Arkansas and 50 miles in depth. You like golf? Better look the Weather Channel over the night before. One lightning victim in South Weymouth, Massachusetts may not have realized the attraction of several miles of water pipes buried just beneath the turf. He wasn't hiding beneath a tree – just running. He had stayed too long. Persevering I grind away at what earth warming promises. I am tiresome I know and that is why Al Gore is now quiet and limits himself to promote less urban sprawl. He even has a Republican governor, Tom ridge of Pennsylvania, writing a long article with a title that declares Gore shouldn't be left alone with his campaign against suburban sprawl. He admits many Republicans have opposed almost any government involvement in environmental management but also, too many Democrats have believed that government is the only player that matters. I have mixed feelings about that but Rhode Island is Rhode Island and it has a long way to go to gain my confidence. It's a tremendous expense and requires an army of experts in many fields and therefore I'm not sure many states are up to it. But, it's gratifying, one major Republican has something good to say about the vice president!

Unless there is a ship out at sea, say 1000 miles east of Africa and somewhere on the Tropic of Cancer latitude predicting when and where a hurricane will develop is not important. I want to interest you as to why they form and what favors more of them one season, less the next season. There is a man, an expert on hurricanes, working out of Colorado University, and because I can't remember his name I do not mean disrespect. He is honest about having missed a long-range prediction of numerous hurricanes for a particular recent year (not important for this time) while being recognized as very worthy of respect for previous attempts.   All meteorologists agree on one point, when El Nino is active there are <u>less</u> hurricanes. The El Ninos in question are not it's <u>one time major impact</u>, when it parked up against California late 1982 and early 1983 but it's peculiar periodic ups in temperatures across the equator between South America and the international date line, or some 5500 miles of abnormally warm water compared with the rest of the ocean surrounding it. Actually, other than 1982-1983 it represents at the most 5% of the total area of the world's seas. Still, it does create not only storms but of most importance, it's warm air works its way heavenward, moves ENE (generally) and adds warmth at high levels above the south Atlantic where hurricanes form. Surely you realize that not only hurricanes but tornadoes also require both a warm area at the surface of either water, or land in the case of twisters, and additionally, <u>cold air at high levels.</u> I was a little put out when some gentleman on TV most assuredly claimed that El

Nino was responsible for a rash of tornadoes, deadly ones, over Alabama and Georgia. I don't know how that can be if El Nino delivers warm air at high levels but it certainly dramatizes the star of all stars in the way of phenomena. I have to extricate myself from this hassle, in fact, the morass of El Nino's fickle ways and meaning no disrespect for a hard working scientist out there in the Pacific, who said it's demise last year caught them by surprise. In short, a great fascinating report by the man, his name mentioned later, but I wasn't out there with him – let it be. The topic now is, why hurricanes? But first, before I'm misunderstood by, how would I say, emulating Einstein in his beliefs anyway, and that doesn't require supreme intellect on my part, we have to separate what religion means to most from haphazardly, as I am prone to do, refer to God as being in control. As a side issue, I listened to a radio poll wherein callers were asked what is the most dangerous, religion or science. By a majority vote it was religion. I have heard callers speaking to Jim Bohanan, who at first were calm, cool and collected, erupt into raving hyperbole because he disagreed with their convictions. Yipes, this is my way to ask that I be exempt from criticism since I don't claim to be blessed with some esoteric purview. I am impressed with what I know to be a very organized systematic control displayed by nature. We have a spinning world. Centrifugally, cold heavy air at the poles is forced outward while gravity keeps it surface bound and the thermodynamics involved between heated equatorial air and modified cold air, creates wind motion, etc., etc., etc., in essence what comes down from the poles has to be replaced. This recycling process goes on and on. And balancing goes on and on between pressure at the poles, pressure in mid latitudes and pressure over tropical zones. To be treated as randomness is foolish thinking. For example, a hurricane is treated as not related to any balancing effort between an over inflated belt of high pressure massively stretching from the equator to sometimes well into Canada and over the Atlantic; well beyond Bermuda; well beyond areas of the Atlantic northeast of Bermuda, and even the Azores for that matter. It is not treated, this hurricane or those hurricanes as an effort on nature's part to excavate as much tropical air and as fast as possible because there is an imbalance beginning sometime in August. Due to the onset of sunless days we have a major build up of cold high pressure over the Arctic area in the process of spilling southward combined with the very voluminous solar energy loaded high pressure belt lying over one half of the Atlantic. Isn't it a fact, (and I'm addressing this to those who treat hurricanes as some entity popping into existence for no other reason than some imagined watery hot spot in the Atlantic about 1000 miles or so west of Africa, the temperature somewhere around the mid 80's and no different than the Caribbean, really) that simultaneously, when a hurricane develops of a sizeable dimension and intensity, there is also a polar air mass rolling southward across Greenland or north or northwest Canada? In short, an excessive volume of warming air such as exist for several months, sometimes longer, particularly of late, has to be reduced or otherwise left untapped the temperatures would continue upward to values we could imagine as being that of the Dinosaur Age. Hurricanes are expedient means of restoring normalcy, the bigger the south Atlantic high and also another bulging up against it, the property of the southern hemisphere as in illustration 26 the stronger easterly gradient between them which in expert circles is conducive to more hurricanes than usual. Summary, the larger the high systems, the more necessary it is to reduce their volume and the more hurricanes to do just that! The mechanism for this move on the part of nature has to be recognized as a high level drawing-effect on the over inflated Atlantic high. Referring to illustration 26 let the upper layer of our atmosphere be represented as supply responding to demand at the North Pole. That is why we have a ceiling over our atmosphere, the tropopause, to insure this conduit (SW to NE) of continuous recycling. I use the word "recycling". You must remember, our atmosphere has a fixed amount of air – after it has been heated it must be returned to the poles for reprocessing so to speak – to be frozen and increase oxygen density. I do not want to distract the reader's attention with various drawn renditions of the not so smoothly curved tropopause as I show it. It has steps within which jet streams barrel along. I am intent only on quick visualization with one reality in mind – the purpose of an atmospheric ceiling under which an expressway allows or forces if you will, warm air to head north and thus plays a major role in balancing what goes on over the Arctic and of course over the south Atlantic. The heavy vertical arrow near the equator represents the effect of a hurricane siphoning heated air where at high levels it is pulled northward. This pulling or drawing force I attribute to a demand for replacement of air at high levels above the North Pole, which I hope to clear up in a few sentences.

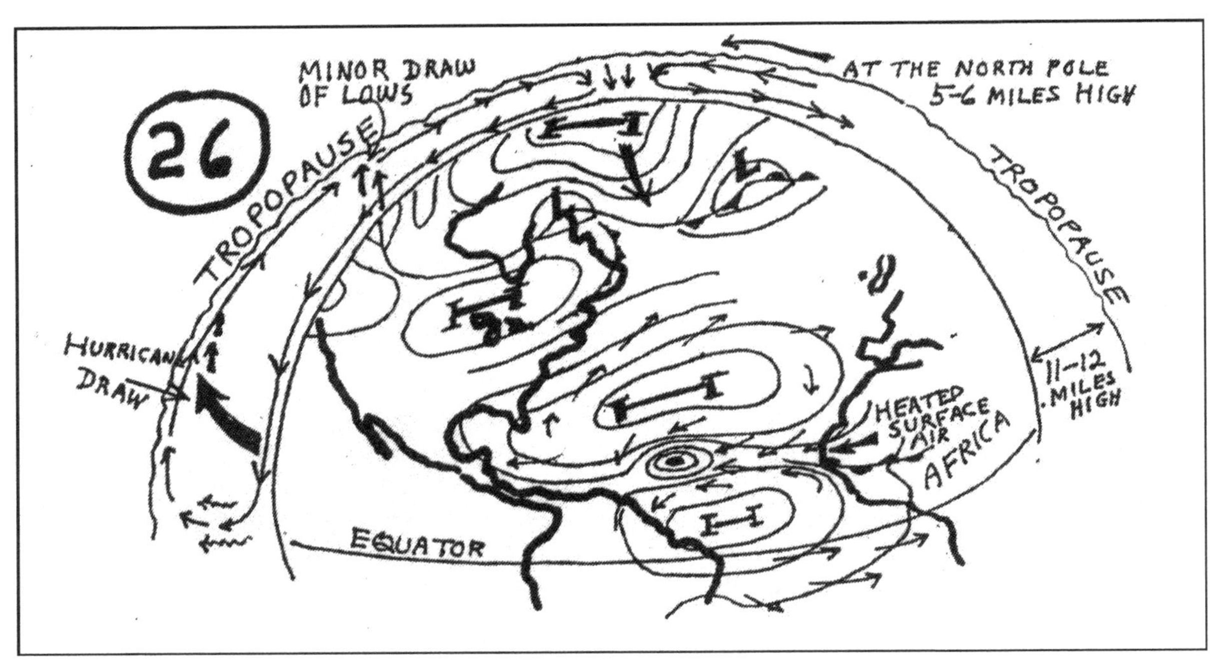

26
MINOR DRAW
OF LOWS
AT THE NORTH POLE
5-6 MILES HIGH
TROPOPAUSE
TROPOPAUSE
HURRICANE
DRAW
HEATED
SURFACE
AIR
AFRICA
11-12
MILES
HIGH
EQUATOR

Illustration 26 offers a two-dimensional view. I am belaboring the topic of hurricanes, covering all facets of causes in their creation, direction they take and most important, their purpose. The destructive power of such phenomena has all too frequently created deadly situations and the extreme expense is a constant factor. Thanks to weathermen, the death toll has been kept at a minimum – at least in the USA. Borrowing from a prediction made by our Colorado University expert who believes that there will be more hurricanes than usual because of warming trends in our climate, really the whole southern half of our hemisphere, I have to go along with his opinion. I am compelled to review every detail. To just read about such storms may be an intellectual exercise to some but the impact of experiencing three as I have leaves a vivid memory. Over 400 people died in 1938 just a few miles from where I lived. Having seen so many film clippings and heard narrations it is hard to believe what danger people expose themselves to.

I am not at all in accord with advocates of hot spots in the eastern Atlantic Ocean near Africa – such a belief does not recognize any particular reason* for nature to create hurricanes inasmuch as water temperatures from the Caribbean to Africa along the hurricane corridor are a fairly constant mid 80's at that time of the year. Surface water temperatures are not at all changed once the hurricane having been formed begins its trek westward. But the hot spot concept is popular – at least it sounds good. If hot spots are all that is required then why no typhoons over El Nino's substantially warmer surfaces? There is a reason for everything – randomness implies there is none. Hurricanes would not eliminate warmer limited areas of the ocean in the first place – therefore accepting this as a cause is to believe that nature has no controls.

Conscious of having to repeatedly use the area-identifying expression South Atlantic High to the point of ad nauseam, let me adopt computer abbreviation style, let's use SAH. Time, if you will, August and September's (usually) (my contention!) the obvious correlation between a surge of polar air breaking loose from the Arctic and very evident on weather maps over Greenland or northern Canada and the birth of a hurricane over the eastern Atlantic. The meteorologist is then witnessing but judging from my experience following their every move, their every word, not aware of an imbalance in pressure systems. In short, too much high pressure over arctic regions, very dense and expansive over the entire polar region, centrifugal force driving it southward and an extremely

*A rule: There is a reason for everything.

dimensionally dominant SAH, or too much high-pressure also. Therefore, nature has organized, or should I say God, a means of adjusting this imbalance. It may require one major hurricane or a series of hurricanes but their assigned purpose is to make room for invading polar air, their function is to literally vacuum up and reduce the over-inflated SAH as it usually is late summer and early fall. When such a mass of cold air surges southward away from the arctic it naturally pulls down air from higher altitudes to replace it. There has to be a continuum at high levels between the SAH and equatorial regions and the North Pole. There has to be a pulling effect over southern regions by northern regions. The North Pole's virtual evacuating process just mentioned demands replacement of air and the main supply comes from semi-tropical regions. The strings of low pressure systems at this time of the year over mid or north USA and the north Atlantic are in a feeble state, rendered so by summer's heat they cannot funnel sufficient air upward and to the poles. The immense swirling power of a hurricane represents an effective vacuuming arrangement, a few feeble lows do not compare.

We now come to the vertical temperature requirements of the potential hurricane – the same as a tornado. A change in temperature at high levels is essential – the air at that level must be subjected to some drop and it is not from some invasive high level advection because major hurricane seasons are characterized by high level easterly winds from Africa and also at surface levels. It has to be this temperature drop at high altitude over a potential hurricane spot, brought about by a pressure drop. Air is both elastic and can be thinned for a brief moment and it also can be compacted. That's what gusts are all about. Give me a series of hills and a wind blowing briskly over them and I'll show you pressure fluctuations, very small ones of course, on the lee side. Low pressure during a lull and higher pressure in the gust. I have reason to believe that the arctic regions demand, described above, pulls at high level southern regions and there has to be a pressure drop, however small over hurricane alley west of Africa. We then need, as in tornadoes, warmer air at low levels. In the case of tornadoes, land surfaces offer that in varying degrees from spot to spot. The steaming heat of West Africa and the recognized by any meteorologist worth his salt of strong easterly winds offers exactly what a hurricane about to form needs. First of all, the African 90-105 degree heated air blown westward by easterly winds doesn't rise when it is over the Atlantic – the water is cooler. Since the easterly winds, markedly stronger during major hurricane seasons are usually in effect some time before the first hurricane develops, it stands to reason that the start of a hurricane needs a drop in temperature at high levels.

Am I alone in some aspects of what I am convinced of? NO. Two gentlemen meteorologists on a TV show revolving around hurricanes seemed to be looking for something new. They want to depart, just a mite, from tradition and made that clear. They were open to suggestions and said so. One readily recognized the coincidental southward charge of cold polar air, in this case, as they were showing, over Greenland and the sudden appearance of a hurricane over the southeast Atlantic. But, that was as far as he went. Busy professionals don't care to venture into the unknown because there are the usual skeptics who have nothing much to offer, ready to pounce, sometimes being generous with their usual, "We are skeptical…" and sometimes saying, "I am amused by so and so's contention…" I have the luxury of plenty of time, unlike paid weathermen and the advantage of age and what memory has to offer. Of course, I'd rather be 20 years old. I wouldn't go back to weather that's for sure. In the first place, you can't prove anything and in the second place, we will never outfox nature. Even Einstein, if he were reincarnated and became a meteorologist would find it impossible to hit 100 percent success in predicting weather one week in advance, from week to week. One high level NOAA official talking on TV said, "We feel it would be a great advantage to predict El Ninos five years in advance." I wish him luck. I think you will appreciate my opinion of this, later.

Please go back to illustration 26 for a moment. I emphasized above how stronger than usual easterly winds from Africa to the Caribbean were recognized as a reason for unusual hurricane activity. There is only one reason – note the SAH and the Southern Hemisphere's equivalent high squeezed up against each other. In weather language, stronger gradient results and therefore a stronger wind action. Summary, A major high over northern cold regions and a blown up summer SAH, a southern hemisphere high which, due to increased solar radiation and Greenhouse Effect, is still intact, both forcing east winds to higher velocities while delivering heated air from western countries of Africa.

Don't sit back and wonder why you need this even though it is rather a lengthy subject. View its implications both here and in the Pacific as potentially altering economies. True, not all of us live in hurricane-risk states. Even now a gradual shift in markets is taking place – we are importing more fruit and vegetables, hopefully not meat, from foreign sources and that has its sanitary risks. Power companies are readjusting. One major plant has drastically reduced its repair crews hoping to reduce costs and not have to raise rates – evidently, there is competition. Yesterday, March 3, a 300-yard wide tornado destroyed a town, killing six residents in the northwest corner of Louisiana. No mercenary view do I have when I write, an emergency like this has to be financed – it's coming – there will be more. How deep is our government's pocket since tax money has to be used excessively this way? What's cheaper, spend money on reducing coal burning power plants or keep on shelling out money time after time to reconstruct towns and salvage lives? What logic, if any? Do you know how much coal we use to fuel power plants? Over 21 percent of energy used is coal, when it should be something cleaner. You cannot imagine how much sulphur dioxide emissions this represents and what, if as it is suspected, the ozone layer is affected? The destructive onslaught of a reactive atmosphere, reacting to pollution, will force all utility plants to increase rates and up will go everything else. It doesn't matter if tumultuous weather resumes this year or the next – overall the next decade promises more. Go back to earlier pages and what do the temperature traces tell you – what is the trend? I can't predict this year or the next, but the whole world and we are moving too slow, beset by military turmoil overshadowing what will prove more costly and enduring. I hope to upset you and then perhaps you will communicate with or evaluate your representatives. Taking me seriously might put you in the position of knowing who makes sense and who doesn't. If this nation shows accomplishment instead of rhetoric and stalling tactics, it can tell international monetary banks we will not contribute anything to bolster nations unwilling to cooperate and then suddenly struck down by the unexpected. Our government has been Santa Claus but the time has come for New Year resolutions by foreign polluters that are serious and not "We can't afford it." What rot! They're always ready at the drop of a hat to blow billions for armament escalation. We certainly don't have a flock of economists buzzing around Washington, DC for nothing, do we? Rising cost of living all over the world promises major wage disparities and the consequences are inevitable. Social turmoil can take on bloody consequences. Several weeks ago what seemed to be more of a military coup and yet it was suspected to be organized by poorer lower-caste Hindus, an upper-caste community was attacked and 35 wealthier residents were killed.

Here we are in the tornado season and this guy comes along with party pooping news! Seriously, I respect this man. For one, he was honest in a previous article last year that he had missed a forecast, a long-range prediction. He's tangling with a difficult problem. Nevertheless, the topic follows the usual Cart Before the Horse appraisal, not that what he has to say is invaluable. Of course ocean currents are warmer! We have a more powerful sun and everything else we are tiring you with, blowing the SAH up,

April 3, 1999
Storm expert predicts 'active' hurricane year

The New York Times

ORLANDO, Fla. — This year's hurricane season in the Atlantic is likely to be "very active" with a chance of damaging storms striking Florida and the East Coast, a leading hurricane forecaster said yesterday.

William Gray, an atmospheric scientist and hurricane expert at Colorado State University, predicts that in the hurricane year beginning June 1 there will be 14 tropical storms, with nine of them hurricanes.

Four of the hurricanes could be major, Category 3 or higher, meaning winds of at least 111 miles per hour, Gray said. Hurricanes have winds of at least 74 mph, and tropical storm warnings are issued when winds reach 55 mph.

Gray discussed his forecast at the annual meeting of the National Hurricane Conference, an organization of hurricane experts.

Based on "climate signals" from stratospheric winds and warmer currents in the north Atlantic Ocean, the "odds are" that the 1999 hurricane season could look like that of last year, when there were 14 storms, 10 of them hurricanes, with 3 of them classified as major, Gray said.

which drives winds and currents further northward. Where does that warm water come from but equatorial regions? Isn't that the source of warm water and its heat being driven into our nation and the once upon a time cooler regions of the Atlantic? Is the languid appearance of the SAH that deceiving? The isobar lines give it an appearance of impotence, not at all dynamic as it truly is, just because they are loose and spread apart in latitudes from Puerto Rico to the equator, as a reference point for readers. Several million cubic miles of stored solar energy sits there but I don't hear them talking about it. That is what I focus on. I'm not looking for some Einstein like formula in heaven or in the seas – it doesn't exist. A thousand thermometers tell the whole story!

Other than several TV weathermen exclaiming wonderment over a massive Pacific High last November (1998) and it having continued its onslaught against our northwest coast, therefore heralded again as mighty impressive a month ago, pumping humidity in at an astonishing buckets-of-rain rate, Cascades buried under snow, it has been only my word so far, insisting a blown up SAH and its Pacific counterpart have prevailed for decades. You might be thinking what proof do I have?

I have saved what follows for last pertaining to hurricanes. Let the little pictures be, let me interest you in what the Big Picture is about and the certainty of it's simplicity. I didn't say I had a sure-fire approach to forecasting. Instead, it is important to think of nature as having a reason for everything that occurs.

That the public will need more education in matters meteorological goes without saying but the way that information is doled out daily during the hurricane season may not allow the TV viewer to appreciate what they are seeing and hearing. We are told, "Hurricane Mimi was here yesterday and today it is here. It remains a force 3 (winds 111 to 130 mph) and moving WNW at 12 knots. We expect it to pass north of Puerto Rico, etc…" They just might add an instructional touch with, "Because of pervasive easterly winds at high levels we expect Mimi to stay on the same course, etc…" But you have never heard them say, "Our massive south Atlantic High has created a steady easterly current at high levels across hurricane alley and it's progeny Mimi (that's really what it is!) has to stay on the same course for the next 24 hours, unless it succeeds to do what nature has set it out to do, which is to siphon off as much warm tropical air and change the SAH from it's excessively voluminous state to a lesser state. Therefore, a change in upper level winds can be expected." It is here, when the SAH weakens, (if it does) that hurricanes start becoming a big problem. They slow down to a snail's pace, generally begin a northward turn and the cause of anxiety for meteorologists who have worried Floridians or whoever to contend with. I experienced the feeling in the 50's when clusters of operational officers and department heads had to decide when and where to evacuate 20 to 30 planes. Of course, these storms approach New England at a good clip but east or west of our station? That was the problem. Hurricane information centers are not in the business of teaching.

I haven't been confined to a weather office for eight hours a day for all these years – I talk to people. There are a lot of people who believe that hurricanes are but independent entities and while they understand how warm evaporating waters fuel them, they do not attach a logical reason for their existence.

Before I present you on the following page with my Coup de Grace, my moment of triumph, reveling over the substantiating material contained that you shall recognize at a glance, which verifies my thesis (I'm being diplomatic calling it a thesis – but not really caring about unproductive skeptics specializing in nothingness) consider the difference sticking out from these two illustrations. The above was not a serious announcement – just had to chuckle a bit. From 1960 to 1990 and still about the same, NOOA a bit slow about issuing more current brochures the majority of hurricanes continued westward into the Gulf of Mexico. Fourteen out of sixteen Force 3 and up hurricanes, unlike those before 1960 when the air was cleaner and our SAH was less dominant, less apt to spread it's western end in ridge form clear past Texas, were directed by the resultant east flow powered by an inflated SAH into the Gulf. Only two hurricanes did not – they, Hugo and Gloria, curved before reaching Florida, as you can see, and took a northerly course, Hugo into the Carolinas and Gloria into New England. From 1940 to 1960, the war and gasoline rationing giving nature a break lasting quite a while, 15 out of 19 Force 3 plus hurricanes were as with those of recent decades, hatched, fed, pushed westward but then directed northward after pummeling Florida and skirted the east coast, some entering it. In short, hurricanes follow the contours of highs because they are but part of them. That's my segment of the Big Picture but it doesn't require genius worthy of Nobel Prizes to recognize it.

Departing from friends at Cape May, November 1942, having received orders to join others at Long Beach, Long Island, scheduled to be shipped somewhere but I knew not where, the call of adventure soon dispelled whatever sadness I felt. "Would I soon be aboard a mighty ship – maybe a carrier?" that

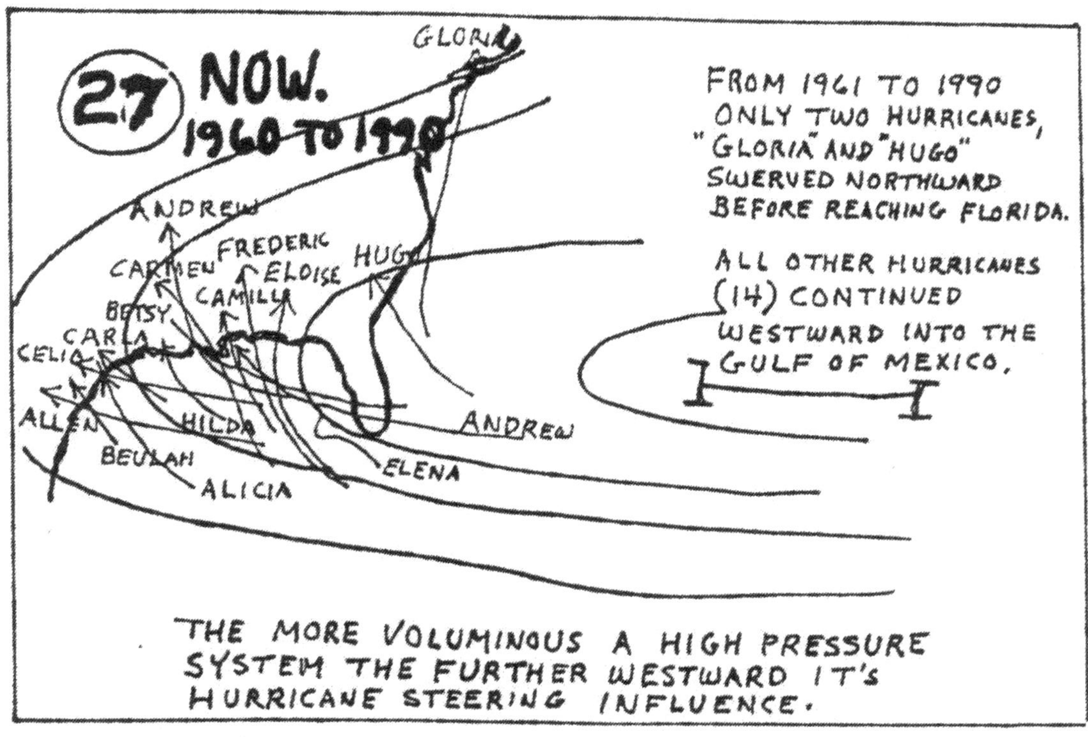

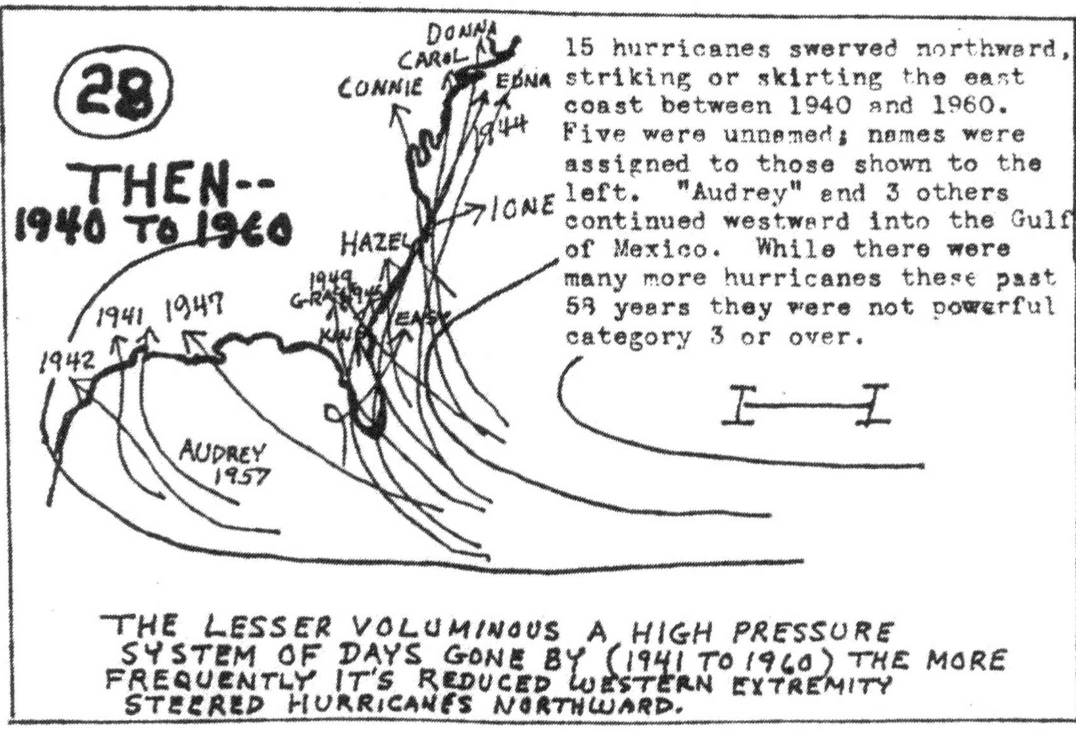

thought flitted through my mind, sitting in a train and looking at neighborhoods fly by. I remember a song, "Laura" some melancholy bit, "She's the face you see in a passing train" some silly thing like that, "She gave her very first kiss to you" hell, I don't remember the words but it ended with "But, she's only a dream." Well, I found no Lauras in Cape May. In those days there was a famous Hollywood comedian, W.C. Fields, quite homely, one large bulbous nose practically glowing red and in a train scene seated next to Mae West, portraying some stranger, busty, broad shouldered and limited to a few sultry lines. Well, Fields turns to her and said, "Ah, my little chickadee." Well, one early evening in Cape May, hanging around the USO I mentioned before that I thought my chickadee had arrived. A very attractive girl walked over to me, 15 minutes later we were walking toward her home. Momma was there. Oh well, I thought myself very fortunate no less. She showed me her bedroom (no, no Momma was nearby!) and lo and behold, at least 100 photographs of sailors and some marines decorated one entire wall. She asked me if I would bring her my photo next time in town. "Sure." Ten minutes later I was gone. There went Laura. She really was a dream but a little flaky.

We knew nothing of the war in far off places. I don't believe anything encouraging was going on – for reasons unknown to me there were no newspapers available at our military base in Cape May. Anything worthy of attention came by word of mouth. Once, we were told to prepare for a possible German commando raid, some officer wanting to unnerve us a bit, describing how a submarine could do just that – drop off assassins who would wreak havoc if we did not prepare. Standing in line, it came my turn to shoot a Thompson sub-machine gun at a nearby target, heck, it was only 30 feet away. I was a total failure although previously at Newport's boot camp I had done well with a rifle. What a weapon this 45-caliber machine gun – looked good in Al Capone movies provided you had Arnold Schwarzenegger muscles. It's a good thing commandos never showed up, they would have had it easy with me. I was pretty tough for a 145 pounder, but this gun was too much – in the first brief bursts the barrel kicked up and I could see why seagulls that normally hung around the premises were not to be seen that day. It's a good thing New Yorkers didn't have to rely on civilian alertness to possible infiltration. If an New Yorker puts on airs over imagined cool acumen because he or she is a resident of that city you ought to tell them this true story that made near headline news. Four characters, from where I don't know, dressed in full Nazi uniforms, shiny boots, the typical short black hat visor, above it, skull and crossbones. On their chests, medals with deliberately enlarged swastika medallions strutted up and down crowded New York City streets. Nary a soul paid any attention to them – they had not the slightest idea of what the enemy looked like, 10 months after our declaring war against Germany. If New Yorkers didn't notice, Long Islanders sure noticed me one night. First of all, I'm not a drinking man but in those days, loneliness and a need to stir up some courage, I would put down a few. After arriving at Long Beach from Cape May, some training followed, which I will describe with a touch of humor, but we were allowed to go either to New York City or Hempstead for several nights. Don't laugh folks, but in those days chaperones were the thing. There was this USO with dozens of young Jewish ladies and if I hadn't imbibed a few Gin Bucks I probably would not have did the usual silly jitter bugging thing while diligent moms looked on or maybe they were just assigned older ladies keeping watch over these pretty girls. One night, prior to going to the USO, the usual stop at a nearby bar. I thought the bartender was Chinese. Some youthful Navy kid sat next to me and we went on and on ripping apart Japanese. We weren't loud but what we were saying didn't please the bartender who I found out was Japanese. On the next drink, what he had put into it, called a Mickey Finn in those days, had immediate effects. My legs turned to cotton and I felt nauseated. I staggered out for fresh air and grabbing a telephone pole hung on feeling ridiculous because despite the rest of my body being of no use to me, my head was very clear. Older couples walking by would slow down and I heard one lady say, "That's awful, disgusting." And others with "Tsk, tsk, look at that." All over New York and Long Island signs "Loose lips sink Ships." I had to learn how to avoid violating that suggestion.

The Navy had leased a large hotel in Long Beach, Long Island. Stripped of any pleasant décor, each room had four army bunks and the general appearance was stark and business like. Hundreds of Navy personnel arrived with us, but there was no conversation. You play the tough loner, you know. Sea bags draped over shoulders, groups were organized by some officer shouting names – then to a room. The next day, we filed into a conference room; some 30 of us and knowing the stage up ahead would soon have a speaker to address us we waited in silence. I still entertained the hope that some warship, docked in nearby New York was waiting my assignment. A middle aged marine popped out from the side of that stage holding a rifle with a fixed bayonet. Tough looking, square jawed, his eyes deliberately scanned every one of us. Then he went through how to charge, thrust a bayonet into the enemy's innards and so on. "Sometimes the bayonet gets jammed in the ribs. At the same time you tug back, pull the trigger, that usually works." The guy next to me, thinking the same thing I was thinking, both of us looking at each other I head him whisper, "Holy s - - t." The next day we were given full packs, strap them on, someone handed a rifle to us and we boarded a landing craft waiting in a nearby lagoon or whatever. It's the first few days of December and it was cold. Off we went looking at murky waters going by. A beach loomed ahead and with a crunch the LSD stopped abruptly, a ramp splashed down and into waist deep icy waters I jumped. The next scenario was getting to me – where the hell are we going from here! On elbows and knees, crawl up the beach, charge some imaginary enemy and finally, we were told to run back to the hotel about a mile away. Well, I tell you, I couldn't play football in LaSalle Academy nor was I any good at baseball or basketball, but I sure could run like hell on the cross country team. Heavy soaked pants or not, about a half a mile left to reach the hotel, the panting behind me started to fade away. Unbelievable, but 1100 ships waited in New York and New Jersey ports or wherever they could anchor and thousands upon thousands of soldiers would wind up in North Africa, many to die. We were a small part of the passenger list; everything was under control by the Army. On this exciting trip I would learn what the Black Market was all about. Our ship, as with most, leased by the US from foreign countries, was a Norwegian freighter, it's interior stripped. There were four layered bunks everywhere. We knew there were German wolf pack submarine fleets out there and while we were not scared, at night laying in the bottom bunk, a steel hull next to me, I could hear now and then water sloshing against it. In the dark I would tap the hull and wonder how thick it was. You were not allowed up on deck at night although I sneaked up there a few times marveling at something so beautiful and so new. This ship had a narrow beam and rolled constantly which made it more interesting in a way. As far as you could see, ships staying in perfect order, none straying from their designated position. Tin Cans, that's what destroyers are called, sped by, cutting through waves and if their sailors have a saying, "Three under and One over" you know what they mean. I asked a friend of mine who spent years on tin cans why they called ships "She". He didn't know, but added, "Maybe we are babies in its womb and dependent on it." We hit the edge of a huge storm two days out of New York and while no waves crashed over the bow the swells were stupendous. I have another reason why they are referred to as "She". They creak, groan and when perched on top of a crest, the screws speed up for a moment, half out of the water and when the ship noses down those props now grab more water and the whole ship shudders a bit. After crossing the Gulf Stream, clear skies prevailed and hundreds of us leaned against railings, no one saying as much, but looking for periscopes in the distance. If some soldier (1500 of them, 100 Navy youngsters) pointed at something in the distance, imaginations soared and those on one side of the ship ran and joined the finger pointers. But, it became unpleasant this 13 day voyage – we were always hungry. Two meals a day and scanty ones at that were insufficient for crisp open sea exposure and sleeping in dank holds. It got so bad that a soldier with a rifle and bayonet guarded crates of oranges that for the lack of space had been stacked in a passageway. We would walk by, hoping to snatch one. Now a few individuals in the Army doesn't mean that service branch has a dishonest side to it; I saw a Navy Chief cook in Agadir get hauled away by agents after a kick back scheme he was involved in. He would go in town, buy food at exorbitant prices from some wholesaler and the two of them were pocketing the surplus. But, we missed the bum – we ate well under his leadership in the kitchen. One day on that voyage I befriended another hungry desperate sailor. We had seen soldiers come out of a forward hatch lugging boxes of canned goods and right there and then we decided to become burglars. Night came. You can't imagine how dark it was below, gingerly feeling for ladder steps, feeling for openings in compartments. Finally, we felt a loosely slatted crate and with nervous strength ripped away. Grabbing at plastic bags we had found a treasure – dried prunes with pits still in them. We spat pits in all directions for at least an hour. I don't recall any dire consequences the next day – my stomach must have appreciated being filled for a change. Fearful our thieving ways would be discovered, we abandoned all thoughts of trying again.

We arrived at our destination but fog made it impossible to see where we were. Quite a scene awaited us as soon as we docked. A short distance away a modern appearing city while below on the pier a dozen ragamuffin Arab boys, shaven heads, their small braided pony tails flopping up and down as they ran to and fro trying to steal from stacks of canned goods and whatever else in the way of food some enterprising officers had collected. A soldier would lunge at these skinny but very energetic kids who managed to laugh even though a bayonet was hot in pursuit. Back would go the soldier, standing in front of meals we had been deprived of, that became a little clearer as I and others leaned over the ship's rail watching this crazy show. I would learn later that the ponytail was a means by which an angel would lift them to heaven when they died. Everything was at a premium in Morocco and these army officers would clean up, selling everything from soap, cigarettes or canned goods. We had missed the North African invasion by several weeks and as reserves we were no longer needed since hostilities lasted but three days. We were in Casablanca. As learned Caucasians from America I heard things to the contrary, "I don't see any lions, this can't be Africa." All 1500 soldiers had departed. We remained on board for one day, exactly December 24, 1942. From there, as I've already written, cattle train shipped us to Port Lyautey. Tents, mud, an interesting fight and a spectacular show of flowers after a Niagara Falls type rainfall except for the eardrum-breaking boom of lightning. Perhaps a week later into barracks with each side of a large compartment, mosquito nets draped down across bunks. We had been warned of malaria transmitting mosquitoes and sure enough, one persisted in trying to find an opening to get at me. Its wings sang a song for a while but it gave up and I fell asleep after tucking loose netting under a mattress. I didn't want to die good grief! A Navy doctor not only showed us hideous photos of venereal disease, syphilis being rampant in this part of the world, fleas carrying Typhus with it's high mortality rate for Arabs who refused inoculation by French authorities but how could you relax at night if some flying insect was out to get you. That was our initial indoctrination of African niceties. And some guy mentioned lions! I wasn't there one week when we heard of six German bombers, knowing there was over 1000 ships in and around Casablanca and making a point of being able to reach us, dropped bombs but no hits. Life Magazine had a cover picture of the sensational concentration of anti-aircraft fire lighting up the sky over Casablanca. It has never been equaled. Every ship had been equipped with five-inch guns, quad-four 40-millemeter anti-aircraft, noisiest things that you will ever hear. The bombers dropped their load from 6000 feet in some fields and while the night was gloriously ablaze, the equivalent of a thousand fourth of Julys rolled into ten minutes, nothing was accomplished. I'm sure the Germans decided to get the hell out of there although our marksmanship revealed the need for some more practice. But, that did it! Here comes the rumor again, commandos were expected to drop down from the sky. The base at Port Lyautey had no fences and at one section of its extreme edge there were dense growths of small trees and brush. Relegated to sentry duty for four hours one night, I found myself very very carefully in my best Indian style walking slowly, noiselessly, my head swiveling left and right. No commando was going to strangle me with piano wire or slice my throat, no sir. I held my rifle at a ready. Then it happened, I heard the rustle of branches, once, then silence and then again. Only a fool would shout the standard, "Who goes there?" I knew where that guy was but he didn't know I was there. Rifle to shoulder, my eye down the barrel, he was a goner. Gosh, he had long ears and huge eyes when his head poked aside some branches. I didn't know Arabs customarily let their burros loose at night. That creature was lucky. It looked at me, wiggled his ears and disappeared. There was a sad aspect also. One young Arab had been shot. We didn't like the guy who killed him feeling he was too anxious to use his rifle. He was a big lummox, assigned permanent duty as sentry but we suspected he took pleasure from the chance to mow down someone. He shot another Arab later but not fatally. The Navy had it's lemons also. As I already wrote we were only allowed in town once every fourth day. The place was too small for so many soldiers and sailors but while initially the first invaders made up of regular Army soldiers, whose white and blue stripped shoulder emblem I recall were gentlemen, having been warned to behave, Europeans cordially accepting them, the replacement soldiers consisted of mostly draftees and some behaved badly. Within two weeks we were not liked at all. There has always been a myth propagated by the misinformed that soldiers and sailors, well, you've heard the one, "Drunk as a sailor" or whatever. That's tripe. Rarely did I hear bad words and if someone was drunk that too was not a frequent sight.

But, draftees…oh, Lord. One wildly driven jeep smashed into an elderly European on a bike. He had a large jug of wine and that had covered the street rendering a grotesque finality, as he lay dead. Another myth: French girls, or even the many Spanish that lived there, they're not out carousing around with the boys. That's nonsense. If you're loose and fancy-free you are stigmatized – might as well have a tattoo on your forehead, so to speak – the whole community knows somehow.

Did the Arabs like us? The merchants certainly did. I don't think the rest did and I wondered how they earned a living, always sitting or standing around. Poverty, stemming from a worldwide depression, had reached abject levels in Morocco. Arab kids would pester you as you walked along, tugging at your sleeve asking for money and you had to finally slap away at them and mean business. I bought an attractive ivory and leather serrated handle to which a curved knife had been artistically attached. It had Arabic inscriptions. Some three years later in the States I made a tracing, sent it to the Smithsonian and asked them what the craftsman had in mind. A month later the answer came, "With my knife I cure sick heads." I made the mistake of wandering into a cluster of vegetable stands, the area jammed with Arabs. Canvas stretched above held by poles served as makeshift shaded stores. Something made me turn to look toward the back of one such rig and there sat an Arab, who, upon seeing me picked up a small rabbit by the ears and slit it's throat. He gave me his best evil grin, which obviously conveyed his dislike for whatever I represented. "This is what I would like to do to you." That's how I interpreted it. Making believe I didn't give a hoot I continued on but once out of sight I got the hell out of there. Here is something ugly for you to ponder extremes in primitive behavior. Postal clerks were upset. Packages sent from Italy by Moroccan soldiers serving in the French army let out an awful stench. If an Arab killed a German or Italian in Italy he would slice off one ear and mail it home. I was taken aback by husbands sitting on top of spindly-legged burros following wives who carried huge bundles of wood on foot. I'm sure that times have changed a lot but there was cruelty toward animals. One European delighted in telling me (I could speak pretty good French) how a camel who had been abused simply got even one day with its master while he slept. It just kneeled down on him and snuffed the breath out of his lungs. That's one dangerous creature, a camel. Perhaps 20 years ago a security guard in our zoo here had fed some carrots to a male camel inside a building during the night. After doing that he walked out a gate but forgot to lock it. He didn't hear the padded feet of that camel. His death made the front page of our newspapers. Previously, jumping from one period to another, I had written how I had been banished from Port Lyautey by a New Jersey banker turned Navy commander overnight. A caravan of trucks headed for Agadir. I was one of 100 passengers. Discarded, unwanted by the big man in Lyautey, just because of a little beer sprayed on his neck…what a terrible sport. A 400-mile trip lay ahead of us, some 10 trucks bouncing over every little bump.

We were run out of town about two hours later – some God-forsaken assembly of three or four garage size buildings and this creature made it clear he wouldn't have anything to do with us. Somewhere south of Casablanca the terrain flattened out and for miles as far as the eye could see, nothing but desolation – a wasteland – not even a tree. Some Arabs could be seen in the distance all headed in the same direction as we were, carrying bundles, dragging possessions on double poles Indian style, a few more fortunate had burros loaded down and from a distance what looked like sheep and goats accompanying others. The sight got eerie after a while. From a few there appeared from horizon to horizon thousands, spread out, perceptively from nearby identifiable rag tags, not an elegant soul amongst them, to a sea of specks in the distance, all converging toward a cluster of buildings we would be passing through. It was market day but we didn't know that. Recalling lessons of killings in Casablanca and obvious dislike on their part while in Port Lyautey for anybody that wasn't an Arab I had an uneasy feeling. It's a good thing there was no appearance of hostility. We were but 100 of us, unarmed, and to see such a mass of humanity all around us was just a bit uncomfortable. We slowed down to about 10 mph as a few Arabs milled about the market center. Two hours of bouncing on hard benches and nothing much in the way of scenery we jumped up, elated, at the sight of a camel who had decided his family was in danger and came at us full bore. From a hundred yards of so his brood looking on, heads held high and evidently confident, his followers seemed satisfied at the fuss that charging camel was making. Deciding trucks were too big an enemy he swerved some ten feet from us and galloped alongside and despite our yelling paid no mind. After a half of a mile of escorting us out of his domain, our trucks began picking up speed. He stopped and from his standing there watching us disappear I'm sure he felt triumphant. As for Arabs and a mild case of trepidation on my part, several years after the war they slaughtered dozens of Europeans. Some years later, their communities having no solid infrastructure they tried to entice European experts back. Morocco is not an oil rich country. Consequently, many immigrated to France. I was there 12 years ago for three summer weeks and some sections of several cities reminded me of Morocco, except that the Arabs had neater arrangements in an open-air market space allotted to them to sell their wares. Perhaps it was symbolic of some change of heart but when my wife and I were walking an Arab woman, leaning out an open sidewalk level window, jumped back and then appeared again holding a baby which she proudly showed to us and from her big smile I don't think she hated us.

What I remember most about Port Lyautey early in 1943 was the arrival of Air Force fighter planes – it seemed like hundreds of them. After two days stay off they went to Tunisia. If the pilots were tired it was understandable and there were frequent accidents on landing. Additionally, we had a slight mirage problem. The runway gave out a double image but with a little instruction the idea was to focus on the bottom one, the real one of course. A few got mixed up and dropped down too soon and too hard. But the fatalities were not from that. Tired? I'll say. From America they flew to some spot in South America, stopped, then to a speck of an island called Ascension for another stay and off to Dakar, West Africa to fuel up again. Finally, they went north to Port Lyautey. Some 6000 miles of holding a stick – no automatic pilot in those crafts. Squadron after squadron flashing around in large circles above waiting their turn to land, Air Force style – charge in at high speed. Just a beautiful streamlined machine in the hands of youth and they knew they had us as an audience. Once they had rested the next day was show time. At least two deadly accidents every week – this went on for about a month. Our weather instrument platform was perched on top of a huge hangar along with offices constituting operations and a control tower. One afternoon I knew this act I was witnessing had not been planned. Two fighter plane pilots had the usual in mind – their favorite game and they knew we loved it. At full speed, flying a hundred feet above the ground you could count on a silent approach and then the sudden roar flashing by our hangar. This time there was something wrong – no communication between two young immortals evidently. I saw one approaching
exceptionally low from the west and another from the south, both at the same level with the same idea – shake the hangar walls, then jerk that fighter straight up, a spurt of white vapor at the wing tips, twist into a tight roll – everyone has seen it. These two had to intersect or cross and at the last split second, the northbound fighter dipped down to avoid the other, but too low. Just a few inches too low. I saw pieces of turf fly off and I knew that the propeller had touched. He tried that pilot who was to die, but other than his plane gaining a little altitude from momentum and a useless bent prop, it plunged down out of sight behind distant rooftops. Surprisingly, despite numerous crashes no houses were ever hit.

One day a new plane arrived. I don't know if there was more than one. It was different- super aerodynamically clean. Instinctively we know it would be the fastest ever. The propeller was there up front of course but no engine where you would expect it. This afforded minimal fuselage diameter from the prop to the cockpit. It had the profile of a missile from the front end on back to the cockpit and wings. The engine, cooled by anti-freeze had been placed behind, between the pilot compartment and tail. There are a lot of engineering degrees around but here and there are the screw-ups. I've seen it in planes and I've seen it in autos. The plane was a death trap and after it crashed there remained the charred remains of what was once a pilot. Surely you have seen how crippled planes spiraling downward fall at a rapid rate. Not this craft though. Most folks up north have seen maple tree seeds* with an attached wing, so to speak, spin rapidly while slowly falling to the ground. That's what we saw a mile away, high in the sky. It baffled us as we followed it's slow descent while spinning violently. No seat ejectors in those days and it didn't take much reasoning on our part to recognize there was no way that pilot could bail out. We speculated for days but it became clear that heavy engine well behind the wings during a tight turn just centrifugally threw the back end around. The prop pulled the ship forward but the tail swung around instead. That was the beginning and the immediate end to a new product. Logic is not handed out on graduation day.

Little can be proven in phenomenal weather – therefore, we are at the mercy of multiple-degreed pundits who can skillfully weave complex suppositions. I read them all and I am fascinated. Logic is not popular. Even non-meteorologists, clever with the pen, are persuasive. The subject of earth warming has been but a Ping-Pong game. Moneyed interests delay enlightenment. I would like to see NOOA try something different with
*"Whirlybirds"
computers but they won't. How about averaging out surface pressures over the Arctic and surface pressures over the south Atlantic in August? If there is too much high pressure prevailing in the northern hemisphere and few low systems then you've got an imbalance and that's when hurricanes form. They develop for a very good reason. Otherwise, you are into randomness and there is no reason for anything – no controls.

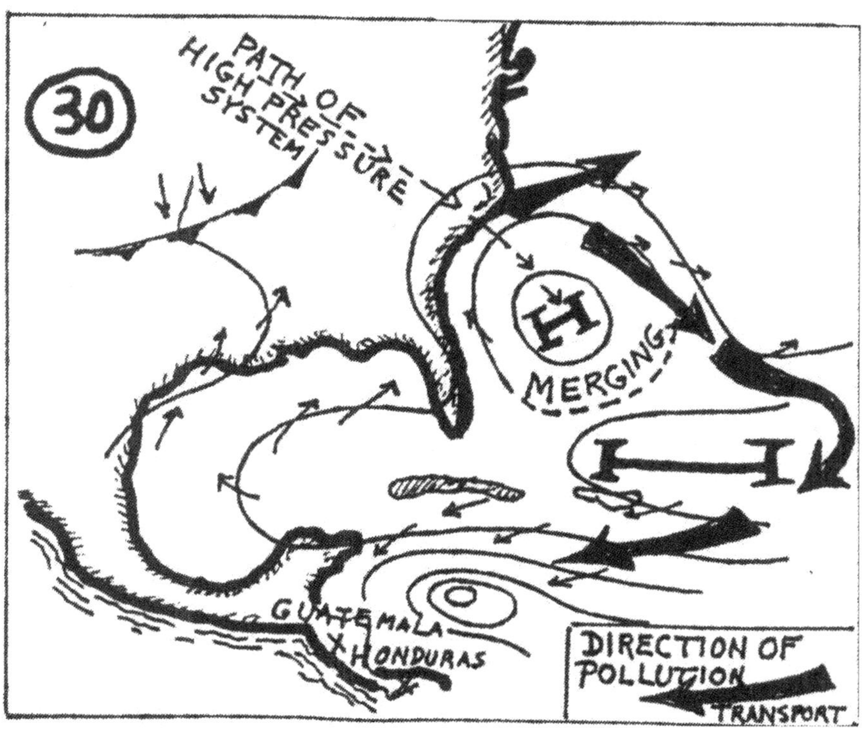

Let's analyze Hurricane Mitch that some gentleman insisted should be called Hurricane Exxon. It wasn't exactly winds that killed over 9000 in Honduras and Guatemala. It was rain, one foot of it every hour – whole villages were buried in mud. Using common sense that took an enormous amount of moisture shipped from the south Atlantic, collecting more over the Caribbean and the Gulf of Mexico where Mitch had slowed down and begun to loiter in a temporary weakened state. There were more hurricanes before Mitch that in the siphoning process had reduced our south Atlantic high into a less potent hurricane delivery and supportive system. Mitch moving in after previous storms from the Caribbean into the Gulf had also helped to shrink the SAH at this stage; therefore easterly winds aloft that drove it along now had lost being boosted along. Remember that it was late in the hurricane season and the sunless Arctic was dispatching its cold air masses southward. One such high pressure system was destined to contribute it's content, better described below and illustrated herein, into a temporary dissipating SAH, thus restoring it's dimensions and once again resuming its more pronounced control over Mitch. That revitalized Mitch's strength and increased its powerful air-excavating function. Tiring of continuously repeating weather terms lacking impact, I depart from convention preferring expressions more dynamic and definitive. Referring to illustration, the main feature of a pending perilous coincidence already highlighted above, a rather large high pressure system moved across the USA, it's path shown with dashed arrows, was losing through land heating, solar radiation and both heat trapping CO2 and accumulated pollutants at a rapid rate. It's uniqueness as a carrier of cooler air and had also collected as all highs do, plenty of particulates from its 1500-mile journey across America. Heated further by the Gulf Stream and warmer Atlantic waters it simply merged into and became part of the SAH. Restored to its previous dimension and influence, Mitch, its creation, now was fueled with more intensity. Within reasonable limits and essentially in principle, the illustration shows what weather maps looked like just before Mitch intensified and moved closer to Honduras and Guatemala. The larger arrows hypothesize the distribution of additional pollutants and consequently the accelerated moisture collecting process adding to the SAH's content, guaranteeing sustained fueling of Mitch. Following Mitch's climax the cold front shown over mid America, it being immaterial as to its exact location or when it was there, drifted down over southern Texas and stopped, not able to continue on. I've insisted and repeat again, cooler air masses do not dominate – warm highs do. The result, one massive one fifth of Texas was inundated. The remnants of Mitch dying off followed by a continuous supply of SAH tropical air pumped in for nearly a week climbed over that boundary line of cooler air stalled across southern Texas and created relentless rainfall. One man, symbolically, insisting that this hurricane should have been named "Exxon", my reacting with one thought, "Finally!" has my gratitude. No one wrote in snide criticism to the editor as they usually do. I believe that most believe him. Well, allow me this much, I give you causes for your consideration while experts give you effects.

While young and foolish the idea of becoming a pilot obsessed me. Chosen Gods, these aviators I looked up to. The uniform, the pay, the glamour of it all, special quarters, periodic dances and the loveliest of ladies in their company, it all seemed irresistible. So very glamorous were these men and their flying machines in my limited world of inferior military status. I wasn't a peon; I wanted to express myself. Having witnessed too much that ambition soon disappeared. I want to tell you I'm not at all the brave and dashing type. From 1946 to 1950 our base at Squantum played host to what we called Weekend Warriors. Similar to the National Guards Navy pilots would leave their civilian life and keep in training Saturday and Sunday along with an assortment of other specialists. We looked forward to weathermen taking our place doing the same old routine while we kidded around, glad to have company. It had to happen. Some weekend pilot crashed into a three decker apartment upon approaching for a landing, but I don't know why it happened. It must have been a very sturdy house as the plane only plunged halfway; it's tail protruding. I went later to see this and that added to my collection of bad memories. Nobody was home fortunately, but witnesses heard the pilot screaming as the plane burned. That did it. It didn't matter to residents, they wanted us out of there, and "Go away, Navy" that was the hue and cry after that. One day I climbed into the front cockpit of an SNJ, a single engine two seater training aircraft, very reliable. The pilot sat in the rear compartment. Earphones attached off we went Boston to the west. His voice came over, "Want to try a little flying? Make a complete circle and if you do it right you'll feel the slipstream we left behind, OK?" Easy, nothing to it. Move the stick a little bit to the left and, oops, pull it back a little. "Great" said the voice. "Want to try a loop?" Pause. Me find myself upside down? "I'd better not." I was a pilot for a whole minute and that was that.. Ah, but the biggest thrill (oh sure) was that given to me by a madman. I didn't know this Lieutenant commander was not only a notorious driver in an auto, having killed some poor soul walking on a New York highway, but a womanizer who this day wanted to scare some lady living in a white house out in the country somewhere in Maine. How could I forget what that house looked like. We were in a two engine SNB and I sat in the front next to the pilot. I didn't expect any conversation – one hell of a grim type. An hour later I thought that he had cracked and believed he was at war in a dive-bomber. Down we went at an acute angle, zoom over this little house on the prairie, up and away. My stomach going back to it's normal position and then two more times for good measure. What some guys won't do over some dame that left them! On the way back he grumbled something about an automatic pilot or some generator not working properly while fiddling with this or that control. I kept looking south hoping to see Boston's buildings coming nearer – trusting in God or whatever. I don't like roller coasters either. The man left for other duties and I didn't miss him. Back to Morocco for more kicks.

After two years overseas a dozen of us boarded a four engined amphibian parked on the Sebu river that courses around Port Lyautey. We were headed home. First to Belfast Ireland and one day free to tour the city. Then we were going to New Foundland and finally on to New York. The pilot decided to play a joke on us. The only trouble the beauty of distant skyscrapers suddenly was switched off, our eyes and ears jumping to a new stimuli. Great! An engine was backfiring; boom and then another boom and puffs of black smoke blew by for all of us on the starboard side to see. It felt like someone in an auto firmly tapping the brakes. Once again, boom, boom. Up front the pilot stuck his head around the open compartment and grinned at us. I have to explain what he was doing. Try shutting the ignition off while going 60 mph in your car and leave it in gear. Turn the key back on and the raw gas in the exhaust system will explode. New York looked so beautiful and this clown had to demoralize us for several minutes. He was piloting a Pan American plane —my gratitude for another thrill. I gave him my best foolish grin.

More doom and gloom. I had good intentions of easing off on what is depressing, even demoralizing despite the pervasiveness of accumulating indicators. Leaving out the morbid side I wanted to technically review El Ninos and tornadoes feeling this was a way out. In ten days six articles by the Providence Journal, the aggregate of which emphasizes normalcy is slipping away. First, an article about mercury

> After years of progress, air pollution from New England power plants has jumped since 1996. This might signify the need for the region's state governments to impose new limits on older plants.
> The increased pollution is reflected in the rise in the region's emissions of nitrogen oxides and sulfur dioxide – which create smog, can make breathing difficult and result in acid rain, which can kill trees and aquatic life. (We note, by the way, that the Brayton Point Power Station, in Somerset, is the region's dirtiest plant in terms of emissions.)

VERMONT

Air quality not up to par

BURLINGTON — A new federal study has found that Vermont has some of the cleanest air in the nation, but the level of seven toxic substances in the air still exceeds the state's health standards. One of the most severe offenders is the cancer-causing chemical benzene, which is 10 to 30 times above the standard. (AP)

vapor permeating the hills of Vermont and then yesterday, April 9th, something relatively new – a powerfully carcinogenic, Benzene, in high quantities over Vermont. New, only because Smog, visible and sometimes odorous, death resulting here and there was long ago accepted as a reality to be addressed. But the invisible? Nothing much has been said. "Out of sight, out of mind." That may account for little publicity. Because Benzene is a combustion product emitted by vehicles using treated gasoline it would be preposterous not to realize such gross quantities over Vermont has to come from heavy populated regions west and southwest of there. But I expect the usual. Bring out the imaging computers, spend five years evaluating a flock of states and keep hundreds of technicians busy. Reduce gasoline consumption! That's all. That's fundamental! I don't think that Scotland Yard will hire me as the ultimate sleuth for that one. Clear blue skies don't mean the air is clean. We will have to wait for lawyers in Congress, who really need large hailstones bouncing off their skulls. "We are not convinced there is earth warming." That's proof of schoolboys rather than sublime authorities. Accept as glib tongued there are nevertheless too many Neros fiddling away. Yesterday another protest launched against the owners of a coal burning power plant in Somerset, Massachusetts which furnishes 17 percent of that state's electricity and recognized as the greatest polluter in all of New England. I went there, curious of course, with the useless idea of photographing its four smokestacks in action. The tell tale evidence of coat after coat of accumulated sulphur dioxide over the years was there. The ochre colored stack tops when breezes play games. It was a clear blue sky but looking in the distance I could see the slightly yellowish haze on the horizon. One lady was angry, "I'm sick and tired of dusting off soot all the time!" Some medical figure was also quoted explaining how particulates imbed in lungs, etc… Today, lo and behold, another article on power plants. While there at Somerset I saw this gigantic pile of crushed coal and a water slip for ships to bring in more. But this article has a shortcoming – it fails to stress what sulphur dioxide does. It's not tough breathing that matters – it is moisture converting it to sulphuric acid. A lot of nonsmokers die from lung cancer but the emphasis is on second hand smoke – not sulphuric acid that I could smell sometimes in Los Angeles, the year 1972. Not only three cheerful articles in the Providence Journal but three more! Am I ready to buy a gas mask? No, but I do have some advice. Do as I do when driving with air conditioning or heater on. Switch to the recirculating position and you won't be inhaling Joe Smuck up ahead with this 2 ½ tail pipe blowing garbage you way or perhaps you are crawling along behind a bus. Every 15 minutes or so (seems silly, but) crack the window open, switch to outside inflow and do that for a minute or so. Got to get rid of stale air. I figure that I'm adding two years to my lifespan. Well, maybe one and a half years. Another article! I'm not exaggerating one bit! I'm not going bugs over bugs they're coming from strange places invited by much longer warm periods of weather than we have ever had. Perhaps you already knew but many northern beekeepers, that is, as a business, have given up. Some mysterious mite likes the taste of bee larvae and has wreaked havoc reducing to a critical point the availability of pollinators <u>needed</u> by fruit and cranberry growers. Then the article goes on. Now Africa, dear old Africa, it has sent us it's latest gift besides killer bees, from it's simmering bowels, the anus end. A beetle now infests beehives in four states, the Carolinas, Georgia and Florida. It harasses the bees to the point where they abandon their hives and disappear. For years northern fruit growers had these southern bee cultivators ship boxes of bees each weighing five pounds with one hell of a busy queen bee being fussed over. Now, northern folks are demanding a quarantine be placed on southern suppliers out of fear that the African beetle will hitch a ride, quite happy stirring up five pounds of bees. Kind of like Kosovo leaders, if you know what I mean. What next. I'm not finished. Again, yesterday the 9th, something very grim. Six American medical experts have been sent to Malaysia intending to find ways and means to curtail the spread of deadly encephalitis prompted by the ominous possibility infected pigs sent to Japan or whatever, the death of a slaughterhouse worker there, many miles from Malaysia, may represents a spread of a deadly disease transmitted by mosquitoes. Included in plastic wrapped cheers from the Journal delivery girl a report on 6 dead and 200 homes destroyed by a tornado in Cincinnati. Plus, a repeat about Mr. Gray and some associated experts on hurricanes predicting more of them this year. I would like to live in the French Riviera but it's too expensive. But, there is something very beautiful after all, Rhode Island is my haven for awhile – it should be exempt from tornadoes and it's nearness to the sea frees us from suffocating heat. Isn't it amazing we have experts in Providence who get paid to attract industry here and yet, I've never heard them mention this advantage.

And now, the quagmire of El Ninos. Only technical types might be interested in this but who knows, maybe I can interest others. When a major article appeared several months ago professing in dramatic terms how the El Nino event beginning in 1997 and ending May 1998, had stunned researchers with its record breaking high temperatures and then unexpectedly snuffed out by upwelling cold waters called La Nina, I wondered if someone should have also explained that this occurred where <u>all</u> El Ninos generally manifest themselves. The 98 special was not in the same position as our destructive 1982-1983 El Nino having been driven up against California, Central America and South America, the latter where it usually has been for a long time, affecting Chile and Peru. Records of El Nino flare-ups have been kept since 1860 and all, except 1982-1983's singular major troublemaker, were quite distant from our country, some 1000 to 1500 miles south of California's southern extremity. Because I have to be fastidious deploring articles that leave a pertinent point hanging in midair, unexplained or out of sight totally, i.e. someone not mentioning the crucial matter of sulphuric acid resulting from excess sulphur dioxide wafting about over southeastern Massachusetts. I must give El Nino's habitual hangout my all. First things first. While El Ninos as photographed through temperature sensing via satellite instruments take on various forms and contours it is safe to say that the average contour looks like that shown in illustration 31. I've given it extra girth so as to appease its enthusiasts or, give them the benefit of the doubt. In dimension it is some 1500 miles north to south and 6000 miles east to west from South America to the International Date Line. At times it has shown more dimensional expanse southward to near Easter Island, shown by dashed lines or stretched westward all the way to Australia.

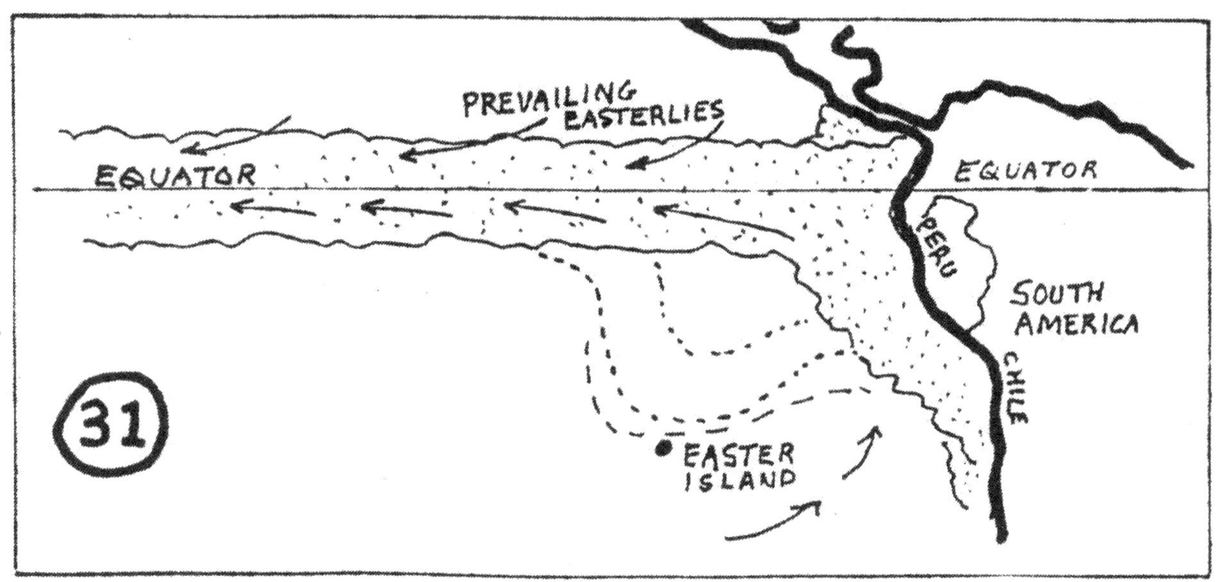

As for El Nino wallowing about near Easter Island as it has before, we have to pay respect to two to four scientists who probed the bottom of the Pacific there with sonar. The results of their study you may find interesting as they discovered a tremendous number of volcanoes at the bottom of the Pacific which will be described in excerpts that follow. You at least can kick it around in your mind, if you will, but I shan't try to unravel the morass of El Nino's idiosyncrasies. Let others out there and a flock of instrumented buoys as collectible information, struggle with its tricky ways. La Nina came along in May 1998 and its cold upwelling waters, and that has been going on for centuries, reduces the equatorial temperatures from their average 85 degrees in between El Ninos raising them above that value, but this time, may 1998, due to some wind forces that experts have found difficult to explain, La Nina knocked El Nino's fever for a loop. I find it difficult to believe El Nino or La Nina both encompassing in their territory but 5 percent of the whole world's oceanic area, have such a profound effect, impacting climate from <u>pole</u> <u>to</u> <u>pole</u>. (A year ago it was claimed one half of the world) and both hemispheres producing crazy weather as a result. A little, yes, but not to the degree that it is claimed.

I have to question El Nino causing tornadoes in Florida, Alabama and thereabouts because its warm air, storm or no storm created, must rise and therefore this warmth traveling at high levels would nullify the <u>conditions</u> <u>necessary</u> for the formation not only of hurricanes but tornadoes also since both <u>require</u> <u>cold</u> at upper levels. Granted there were more tornado frequencies in 1982 and 1983 because El Nino created ferocious storms and as they moved inland tornadoes characteristically were spawned in the thunderstorms that resulted. Therefore, there were more twisters than usual in a few states east of California. Tornadoes at Jarrell, Texas and the southeast were fueled by low level land heated air shipped in from the south Atlantic and not from the Pacific. Weather maps clearly show that! Bear in mind, El Nino disappeared mid 1998 and yet tornado frequencies broke records late in 1998 and early in 1999, did they not? There were over 100 tornadoes in January alone and did La Nina, supposedly flaunted by some sensationalists, who learning this new scourge had reared its head in May 1998, rose to the occasion claiming that record breaking snow would fall on Buffalo (really said something like that!) well, did La Nina having killed El Nino chill off tornadoes? Hell no! The contention pushed, El Nino has lingering effects is not a valid one. We had El Nino until May 1998, and later, Mitch as with all hurricanes had to have frigid air above it and not warmth. That warmth had "Gone with the Wind", that is winds at 20 to perhaps 50,000 feet. The problem is clear in my mind, each expert has a pet project and they want the world to focus on one single factor, theirs. Just what the anti-moguls love! I didn't say I disrespected their efforts, studying technical information they produce, but they should wait until all other causes are put together before pushing for exclusivity. Someone told me during a conversation on the matter that perhaps scientists (not all) are competitors but world climate problems is not an arena for male egos going on exhibition.

Rather than feverishly banging away on my typewriter pumping page after page and reaching the preposterous level of Clinton's impeachment level, of why I don't quite buy the claimed totality of El Nino's influence on our world while its enthusiastic advocates, some having gone Hollywood on TV shows, carefully leave out such factors as increased solar radiation, increased pollution, the ever rising level of carbon dioxide and gigantic high pressure systems rearranging our atmospheric functions, and so, no fighting in mind, since I really don't give a darn, I did the next best thing. If you will, please go back to the beginning of the book to note heavy arrows and small circles decorating the 29-year temperature trends for some American cities. The arrow represents the years El Nino flared up and the circles indicate the years La Nina cooled off the equatorial belt where both phenomena spend most of their time. Do you see anything earth shaking in temperatures? Except western coastal cities of course in 1982 and 1983. Now La Nina (don't get sick of this) supposedly was to cause your nose to be frost bitten if you dared step outside. Do you see anything unusual in the temperature trends? Look again, nothing significant has happened really. Hawaii is the nearest site to El Nino and I assure you, nothing happened. In short, please, warm air doesn't travel along the ground from the West Coast to the East Coast. It is not the reason for prolonged heat spells, for example.

I'm not a crotchety old man picking away at various expressed opinions but I can't tolerate Hollywood types! Environmental problems will escalate and if you think your accrued earnings in investments will go on and on, the cost of destructive weather will cancel much of it unless a lot of ostriches pull their heads out of the sand. Right now, just one of the several, in fact I can think of about six needed items for living well, electricity has a problem developing. Companies are competing and some are now reducing repair crews, one in Massachusetts from 2000 to 1600 so as to cut down expenses prompting several legislators to promote some kind of a bill to investigate whatever, no details given. Quote one: "The next major storm means we will have to wait two weeks to have power restored." "Something has to be done." The next stage nationally, I assure you, up will go utility bills.

I had mentioned dilution of scattered facts that should be kept fresh in mind accumulated over the years is a result of long term memory failure to a degree. Consider in the following pages and illustrations how the works of Ken Macdonald, Tanya Atwater and supportive comments by Herbert R. Shaw, are now nearly forgotten relics despite the intelligence and the logic of their efforts. After you thoroughly read the article, my having eliminated much of it as extraneous, please look over what Mother Earth is up to (see depiction 32 several pages ahead).   She is expanding her girth no less. From California to some 5000 miles south, here seam, at least one major one is separating at the rate of 8 inches every year. That seam, mainly two plates separating, is called the East Pacific Ridge and because the mantle therefore thins out, earth's inner fire punches through. Now, energy cannot be destroyed. I'm sure you know that. Therefore where does this sea bottom heat go?

I believe computers are the greatest invention of our century, really. But, meaning no sarcasm whatsoever I was amused by a lady physicist who writes complicated books on this or that, where we are heading and all that stuff, who, when asked by a radio talk show host what she thought of our developing artificial intelligence some day. Both laughed after she had added a comment to her first statement, "That's a very long way off." Then, "I'm afraid right now computers have no common sense at all." For weathermen to try to elevate their image by referring to a device that they are in command of, connoting it can think and anticipate, is scandalous. Besides, it cannot defend itself when things go wrong. But, computers do draw maps and it has relieved weathermen of the task. Still, there are some that are old fashioned but I know since I have received maps from weather centers, that they do a more accurate job.

That is not important for general wide sweeping weather presentations such as the Weather Channel's fine service, but it will not do for specific requirements that some weatherman has to deal with in a locality that he is responsible for. One grave exception, computer maps are risky when we are dealing with potentially fast moving fronts with ideal tornado conditions. I hope to make this clearer with distinguishing details where such giant twisters occur in advance of a cold front, having astonishing diameters and capable of cutting a swath of destruction anywhere from one quarter to one mile wide. We are so inured to twisters, even though there are many more than in former years. Two hundred homes shredded and blown away along with sixty hospitalized and six deaths, but it makes no lasting impression. But Starr and Paula Jones drag on. Even property insurers say nothing – that would get more attention if they did.

Referring to depiction of the Pacific bottom, Number 32, following marine geophysics article next pages you can see the extent of the East Pacific Ridge, 5000 miles long from southwest of Easter Island all the way into California. And yes, El Nino's red image on computer screens, the visual effect of satellite temperature sensing has been shown to be near Easter Island at times. Despite a variety of depictions as to El Nino's expanse, depending on what motivates who, the majority revealed that there may be a connection between areas near Easter Island and relatively nearby waters whose very deep temperatures are warm and occupy quite a lot of ocean area.

The ridge is but two plates and poetically speaking, Mother Earth is expanding its girth at the rate of eight inches a year. The plates are separating and that leaves a thin mantle – consequently it is the site, if we can apply logic, where volcanic geysers can break through. I have seen a <u>satellite</u> <u>night</u> view of pacific volcanoes called the Ring of Fire and it is awesome. In appearance, flickering candles in the night from the northern extremities of Japan and past New Guinea, many others scattered about the Pacific, including the Hawaiian Islands- my viewing this as if I was sitting up there, the astronaut, 200 miles up, looking down. It's quite a sight. This earth is one supercharged caldron of extreme heat and its lid; the mantle gives way here and there. There has to be sea bottom activity but that detracts from the cause as a contention – El Nino is therefore something researchers can get their hands on and make the most of it. Its near impossible to know for sure the extent of volcanic heat made available deep down in the Pacific but there are skeptics over volcanic influences contributing to El Ninos and they have to make their presence felt. Without realizing it they are suggesting that energy can be destroyed. From my point of view, where does sea bottom heat go and how much of it is there? I'm not a diver inside a bathysphere probing the bottom of the Pacific but after seeing that Ring of Fire, it sure is one hot place! Of course it is important for researchers to chase and try to understand El Nino's idiosyncrasies but it is a cyclical phenomenon and so are many volcanoes! (See next page) Amazing isn't it? Such a revealing find by Ken MacDonald and his cohorts and it has been conveniently shoved aside. By keeping El Nino at the forefront, since it is not at all difficult to track, research funds are being justified and therefore it goes without saying that more money will be doled out. But, trying to establish how many active sea bottom volcanoes or fiery cones, whatever form of acetylene torch heat going on would be too great a project, even futile. Therefore, march on El Nino enthusiasts, they'll stay busy for years, to establish what? We have had quite a change in Pacific air masses and El Nino has shown a predisposition to wander accordingly, but that lacks appeal.

Join my club. No pedigrees nor degrees needed – no fee either. Just bring along common sense and logic and a willingness to review 50 years of observations.

What NOAA should try is to take a few of those deep temperature measuring buoys and anchor them down somewhere southwest of Easter Island but not north. It is fairly well known to which direction deep ocean currents flow. Using your imagination, what would happen if they discovered surprising degrees of warmth down there? That would be very upsetting and the predictability of El Nino's reappearance or departure as a scientific project would, if periodically published be more a study in academic and semantic expression. For the sincere researcher, sheer frustration.

I think that it is very important that we have some advance warning should El Nino, having already created destructive weather on a grand scale which we refer to as the 1982-83 episode, begin a migration toward our west coast. Although it has not duplicated such a massive coast-hugging performance since then, it could of course, happen again. What I find fault with in published reports, separating genuine experts from politically biased journalists, is the absence of very important details revolving around already established abnormalities. In inserting this remarkable article below I am digressing a bit. Published <u>six</u> <u>years</u> <u>ago</u> but it has never been mentioned again!! I would like you to understand why I view recent "El Nino Expansionism" with jaundiced eyes. Please note 3rd paragraph quoting Janet Morton as to density of volcanic activity.

128

In avoiding by intent or by over specialization, other factors in south Pacific weather and subject to variations as it is near the equator one can understand why researches on El Nino make such statements as, "We were surprised by, we did not expect, etc…" "There is disagreement over, etc…We do not know why this happened." But do not misinterpret the above as disrespect. The major published work of a prominent scientist working hard out there in the Pacific proved very instructive-I picked up some valuable information. Insofar as political ramifications, what I have read for quite a few years reflects too much bias, too one sided.

When the first jet plane arrived in Squantum sometime after the war, I rushed out from the weather office and stood up against a fence where I was sure I would get a real close view as it taxied by. Trouble was, the pilot turned the plane around and although I wasn't fried by the blast of exhaust, had I not departed quickly I might have looked like someone with unusually red skin. Gasoline* is an incredible heat producing product of course. Shifting the scene later to the mid 1960's; driving toward New Orleans one day, the thought of that jet engine blowing heat my way was on my mind. Some 35 auto dealer service managers from car dealers in Louisiana and Mississippi had been invited by one of Detroit's auto manufacturers to look over a new jet turbine powered auto. The dealer owners felt this was more suitable for experienced auto people rather than their sales staff. On display in a very large dealership, sporting two huge opened doors to allow cars to be driven in and out it's showroom, a most beautiful streamlined vehicle. We walked around it and noted rather large exhaust outlets under the rear bumper. Fine. When the factory representative said that the turbine would rev up to 25,000 rpm some of us backed away. The whole idea of this presentation was to entice dealers to take a crack at selling the latest from Detroit – didn't matter what brand of vehicle you represented. If you have ever been in New Orleans on a typical summer day, walking along one of its many narrow side streets, the discomfort from 95 to 100 degrees is hard to take. I think that is why air conditioned bars are readily available. The factory man praised the quiet engine and it's power on and on. Somebody got into this pretty creation and we heard a gentle whir build up to a higher pitch. We all ran out of the building, laughing at our unanimous decision to get away from the heat. End of special car. Can you imagine a dozen of these giant laundry dryers in bumper to bumper traffic in the French Quarter?

*Jet Fuel. All variations of petroleum fuel produce 35,000 BTUs per gallon.
I don't believe there is much interest in engines but for the sake of comparisons between various types and potential for eliminating nitrogen oxides, a brief review follows. Later, we were told turbine power, lacking quick acceleration had nevertheless been adopted by tractor-trailer truck companies out west for long range freight hauling where little stop and go conditions exist. I haven't any idea if the economy of such power plants is acceptable nor if they're in use today, but turbine revolutions driven by combustion velocities rather than exploding pressures on pistons causes no NOX. We repeatedly refer to smog because it is visible and threatening obviously, representing density of acidic particles mixed with water vapor but so far we are disinclined to publicize to the point of protest by companies how diesel engines excel in pumping out exhaust laden with smog producing agents. Not much we can do about it anyway, except pursue reducing diesel fuel consumption. As you already know, combustion heat, reaching critical levels of 1500 to 2500 degrees (I'm not sure of exact figures) trigger a chemical change. Piston engines powered by gasoline when stressed and it doesn't take much accelerator pedal urging in autos for example, to reach NOX formation in the combustion chamber, can be a problem under certain circumstances, already explained. Last week a prominent lady radio talk host had to be persuaded by some engine expert calling in that diesel trucks are the greatest polluters. Twice she advanced her belief that autos polluted much more than trucks and finally got it straight. You see what this is all about. One group is pointing the finger at another, back and forth and here is a radio host who in the past has tremendous influence over thousands; invariably the subject of vehicular pollution comes up and ignorance has another heyday.

It may be offensive to use the word ignorance but as a stumbling block along with pessimism it has to be recognized as a deterrent to accepting facts. I just read where Al gore didn't fare too well while holding a political rally with union members in some town in the mid-west. He didn't want to talk about environmental problems but some union honchos had to tell him they didn't like him promoting the Kyoto Treaty, since he allegedly initiated it, declaring this would send jobs to foreign countries. Evidently the idea we may persuade China and other polluters to reduce their trash shipped to us in ever increasing amounts is not important. The article over this meeting inferred a cool reception along with a white veil of smoke filling the hall. I suppose these union members expect our economy to collapse if we reduce an agreed – upon percentage of emissions even though the Senate has yet to ratify the agreement and the timetable is out in left field somewhere. Why should cigar puffing union bosses worry about air pollution anyway? Apparently they feel Al Gore is determined to bring back horse and buggy days for the affluent and bicycles for everybody else. However, you and I can fool them all. We will not have to clear the streets of manure or tinker with finicky bicycle chains and patch kits; we can revert to charcoal as a substitute for gasoline now that we know how it was done during WWII. I have to poke fun at the views of so many. Perhaps union members don't mind tornadoes and hurricanes – they certainly keep tradesmen busy rebuilding devastated communities. Leaving politics aside as too unpredictable I still would like Al Gore to promote visions of the future instead of playing the cool patient one. We do need occasionally, an outspoken visionary.

Here I am espousing a need for someone (anyone for that matter) like the vice president to motivate the public to visualize brand new concepts and stir their imaginations. Instead he was evasive when confronted by union members. Why can't he try tying in curtailing of pollution with ushering in exciting new modes of mobility, futuristic transit systems, harnessing natural power and the evolution of new specialized businesses such as fish and shellfish farms once waters have been cleaned? Fine, but the impression I am making is just another Nostradamus spreading gloom – just the opposite of what I would like to see from a visionary. At this juncture then, frustration over Mr. Gore not saying a hell of a lot which bothers me a bit and my negativity so far which may strike some as counter productive I wish to lay down my objectives. One, know exactly what climate change is about. Two, compile or better still, stack up every little piece of evidence establishing the insidious deleterious developments for a quick, immediate and thorough review. The ups and downs, the ins and outs of information and opinions have been spread out for over a decade and therefore no aggregate, no impact. Therefore, no one is really worried. .Three, have quick access to reports and statements by both pro-environmentalists and those opposed and weigh carefully ulterior motives and the probable eventualities. This is fundamentally, what this book is about. The facts and a fair analysis put together for you to pass judgement on.

Examples of illogical reasoning: We are hearing urban sprawl is becoming a problem and we should seriously work out some plan to stop the present trend which in time may appear as millions of tenements and homes in ghetto like proximity, a few lonely trees here and there. In the meantime businesses want more immigrants, which translates into rapidly growing populations. Therefore, the more successful Americans leave crowded cities and just where are they going to live if not in suburbia and Exurbia? Here in Lincoln, Rhode Island, real estate developers are fighting with residents over the right to chop down trees, extend streets and build homes. We encourage more immigrants by doling out social security, food stamps and other costly help by communities. Once an immigrant couple gets a foothold here, grandparents, sisters, cousins and aunts will follow – they too will need help. The whole state is swarming with social workers. Promiscuous kids, not necessarily all from minority groups are busy making babies. Why not? Cheap rent, free food and the unemployed unproductive boyfriend except for his biological endowments, encouraged at the age of ten by TV shows to view females as interesting creatures, moves in the newly acquired Section 8 apartment with his temporary bedmate. The consequences are obvious. Angry fatherless children plague our society by turning to criminal activity. We have more policemen per capita than any country in the world. That is not cheap, believe me. Who is paying for crowded juvenile reform schools and jails? Don' t blame politicians entirely – the private sector has a lot to do with the problem.

Early April, 1999
No neurosis on my part, in fact I would like you to imagine you and I are sitting around and shooting the breeze, but I see temperatures in the Caribbean climbing up at a pretty good clip, well above their normal of yesteryear. The temperatures are in the 90's no less and now the Weather Channel is pointing out how our southeast is at risk of drought conditions. I see this rise in temps as evidence solar radiation remains abnormal but perhaps some unqualified El Nino enthusiasts with itchy fingers lingering over typewriter keys will have us believe El Nino is squeezing through the Panama Canal or even though it died 11 months ago, it's heated air managed to hang around that long, an impossibility of course. And how about the strong statement made by some writer to the effect that La Nina, the cold one, responsible

for surprising researchers out there in the Pacific by popping up unexpectedly and eliminating El Nino, would surely cause weather to change in the northeast form normal to, well, Buffalo, where my family lived for five years, surely that city would experience record-breaking snow. * It didn't.

One end of our nation broke rainfall records these past months, beginning in November 1998 but Florida has wildfires from excessive heat and little rainfall. We will have to build desalinization facilities, dig up and wall in

Just the opposite was announced by a publicity hound: La Nina causes more tornadoes!

expansive reservoirs, pipe water from one zone to another because the demand for water grows with leaps and

bounds, such is our population growth. For substantial expansive areas water availability is always nip and tuck and yet we go on and on with good old American optimism even though we got burnt quite a few times in the past from this lack of foresight and too much confidence. This figure I'm about to give you is over eight years old but it is astonishing – just one half of our nation consumes close to 5 billion gallons of water every day! What will happen if we are hit again with protracted heat spells and droughts? Confidence we have but memory we don't. History, as one working cohort said to me, is for the antiquated to think about, a typical attitude. The Roaring Twenties, always described by some as boom times was aptly defined by Larry King, who said, "The 20's roared alright but it depended on what side of the tracks you lived on." "Boom to Bust" that's what the "Man with the Cigar" brought on. Don't let the "Down with Government Intrusiveness" sharks ever sell you uninhibited economics again. Do you think we should build a canal from Washington or Oregon all the way to the east somewhere, what with all that water nature dumped on those states? I best back off a little on such an idea by quoting Jim Bohanan when some caller strikes him as being wild, "Let the Prozac kick in and then call me back." What a shame, like the roman Empire we are everywhere. Our soldiers, including my grandson in Germany, are located in over 100 spots on this planet. The horrendous expense of our venture into Kosovo —how can we implement the huge change in our infrastructure that surely is a must?

Everything I write, other than experiences encountered in my life, skipping the unsavory, the ugly I witnessed in Morocco and I might add in America that you will never find in the Internet, intending only to break up the monotony and depressing aspects of earth warming, had but one aim, to connect together political attitudes and impending problems. I had good intentions of putting a stop to my writing about political inclinations or disinclinations, but once more, certain scandalous news, however little newspaper space allocated to it, is too disturbing for me not to jump right back in and continue on. It didn't bother me to hear a certain lady campaign manger for the Republican Party here in RI a few days before two brief upsetting articles caught my eye, resort to express her party's chief punch line, her party's upcoming favorite theme song: "Our party is against governmental intrusion." Remember that I voted for a Republican governor before you misconstrue my attitude to be too biased. Well, intrusive government inspectors found 25 million pounds of meat (that's a lot of money too) unfit for human consumption in the same Arkansas plant that just three months before had intruders from the FDA condemn their supplies. That was one article. Another, E-Coli outbreak in Maine put six people in the hospital. They were lucky. The bacteria from animal excrement (cheery thought) can be fatal. They boast, this political party, exaggerating Ronald Reagan's limited role of having brought down Russia to its economic knees, their leaders holding out tin cups for our help now, but fail to mention one specific bankrupting factor. Russia spent itself out of business by waging a futile war against Afghanistan to a point where for the last months before they withdrew and returned to their steppes, soldiers never received any pay. It made me happy to see their onslaught on robed peasants had turned into total failure. True, we furnished Afghanistanians with shoulder fired missiles with which they wreaked havoc on helicopters and tanks, but I'm certain Reagan did not arrange this war. The Republicans should quantify and qualify what is intrusive and what is not. We had better keep important government agencies fully staffed and ward off conservatives chopping away as they did with the FDA or heaven help us. History may be boring but its details mark the philosophy of some permanently, when they rise again to the occasion, jumping at the lure of more profit I know who they are. And so now, a bit annoyed at this "Intrusive" bit I want to remind you, conservatives (some Democrats included) have also shown a great fascination over the biggest jackpot of all; all that Social Security money rolling in by the truck loads, supposedly billions more than what is being paid out to recipients. They freely dipped into the pool in the past and in 1994, not satisfied they voted to throw all your Social Security money coming in to Washington right smack into the general fund – showing you how little they care about the general welfare of their constituents. They missed by four votes – a close shave. Because they have shown in the past a predisposition to play the parent robbing the children's piggy bank (this time it would not be a loan) it is now suspected by several radio talk show hosts that the government may be using surplus Social Security money to finance our venture in Kosovo or build up our defense.

I really haven't strayed too far from environmental issues but I am losing confidence in tentative politics despite ample rhetoric. There doesn't appear to be leadership. Illogic as an imbedded mental problem apparently is widespread if you consider how various officials argue, given that the Wall Street Journal, acutely tuned in to economics has twice published unusually lengthy articles describing federal, state and town officials quibbling or debating over more Benzene or less, more nitrous-oxide or less while ignoring Detroit's Big Three competing to build the biggest vehicle ever. Ten miles per gallon in the city! That's in the summer! Lord Almighty, we drove two to two and one half ton V8 clunkers in the late 40's and 50's that were more economical! Stop and think! In essence, the quarrelers are telling the public you can choose Benzene and let it infiltrate lymph nodes (it is a carcinogen) and I feel we have to have a reason for a high cancer rate or choose nitrous oxide and its by product ozone and let its acid slowly gnaw away at lung tissue. We need a John Kennedy type – you may be too young to remember but when US Steel had announced it had intentions to raise prices, thus an inflationary threat, Kennedy jumped right to the fore, spoke to US Steel CEOs in no uncertain terms and that was that – no price increase. We need someone to pay a visit to Detroit and suggest that they cool it. If they must sell 40 thousand-dollar vehicles let them be smaller and more economical but add expensive innovations and keep the prices up. Who cares about prices since the affluent blow 50 grand for a vehicle that isn't worth 30 thousand. See what you get when that 3-ton monster is four years old in a trade-in. I wish you luck.

Thirty million Americans are plagued with allergies and many never find relief with medication. I know for a fact, if you live in the city or drive a lot you are reacting to air borne pollution pumped out by vehicles. What is Ring around the Collar? It's not because one doesn't wash up in the morning – it is damp skin to which particulates adhere. The air is filthy, period. Don't let pretty blue skies fool you. The state of Vermont has that

Air quality 4-10-99
not up to par

BURLINGTON — A new federal study has found that Vermont has some of the cleanest air in the nation, but the level of seven toxic substances in the air still exceeds the state's health standards. One of the most severe offenders is the cancer-causing chemical benzene, which is 10 to 30 times above the standard. (AP)

pristine look but it is now heir to Ohio, Pennsylvania and other states' pollution. Somewhere ahead in the following pages I will attach excerpts from Wall Street's best – an article on Benzene or Nitrogen oxide, free of charge for you to breathe in.

It was my intention to bring this book to a close having saved the subject of tornadoes as a fitting end, placing emphasis on understanding the reason why fast moving horizon filling monster sized of the likes we will see more of develop suddenly and then leave half mile or wider paths of total destruction and deaths sometimes without ample warning. But I did not want to finish on such an unpleasant subject before first providing a breather. Digging into my memory bank I was prepared to insert a few stories, although not glorious ones, having a touch of sentiment, perhaps a little humor and it had to happen again – something else that promises trouble on the horizon! A client, a retired oral surgeon who is interested in environmental issues handed me a Wall Street Journal article that I just have to use. For those who might visit a library now and then it is the April 9, 1999 issue. It bears out what I have been pushing all along – authorities, including federal, state or regional and alleged experts, just can't get it together, and again, the public, if not already fed up, discouraging as that is for me anyway, is getting treated to merry-go-round debates over choices to be made while no mention is made of what is really the core of the issue. The subject now being kicked around is gasoline additives. In their adoption a few years ago benefits were anticipated but now they have created problems. It is not, this comprehensive article by New York's best, a twisted version by journalists. There is no question when it comes to future economic ramifications and something poses a threat, the Wall Street Journal has astute sentries always on the lookout and ready with detailed reporting. Fortunately for the more sober of us they refrain from imaginary concepts about weather. In this article's quoted comments by fuel additive manufacturing execs there is the disturbing aspect of accelerating gasoline prices, up and not down! Allow me to end this sentence on a juvenile note, since I have been hopping on disruptive climate trends

that can only lead to <u>adverse</u> <u>social</u> <u>conditions</u>, while I had mentioned previously a very optimistic congressman turned economist who spoke in glorious terms, reassuring Jim Bohanan we could look forward to a bountiful decade to come. Well, in what follows: I TOLD YOU SO!! Bountiful for who? What if rising prices and surely they will rise, hits 30 or 40 million working people really hard? Tolerance has limits!!

Ban on an Additive For Gasoline Debated By States and Cities

Continued From Page A2

upgrade regular gasoline into pricier premium.

Longer term, the outlook for MTBE hinges on whether California's move leads to gasoline shortages or rising prices at the pump. Already, rising oil prices and refinery problems have pushed gasoline prices in California near $2 a gallon in some places, limiting consumer tolerance for further increases.

California officials hope to avoid that by convincing federal officials to waive the requirement for oxygenated fuels, which make up almost all of the market for MTBE. If no additives are needed, then shortages and price increases are unlikely. But since MTBE makes up 40% of the demand for methanol in the U.S., methanol makers are bracing for, at best, 1% global growth over the next five years.

First the Wall Street Journal's last paragraph to the left, since the possibility of two dollars for a gallon of gasoline is a sure fire way to jolt the indifferent citizen – especially in California where big engine aficionados excel. Trouble is that state can also set the lead. Can you imagine filling a 30-gallon tank featured on massive SUVs? Maybe $50 or $60 a pop – not bad.

On a very serious note, higher gasoline prices affects everything – everything that requires mechanized cultivation and shipped to cities it is far more reaching than imagined. The argument spreading over the United States is whether or not to use MTBE in gasoline. We covered the problem

One MTBE maker and refiner, Valero Energy Corp. in San Antonio, says the result of a widespread MTBE ban could be catastrophic for the gasoline industry, and for consumers. "Anybody can look at the numbers and see that this is a train wreck waiting to happen," says Valero spokeswoman Mary Rose Brown. "We're really heading for some serious gasoline shortages and some serious increases in prices. "The consumers could get killed."

when Maine residents were asked to vote yes or no over its use. MTBE is Methyl Tertiary Butyl Ether, quite a mouthful. In a more focusing perspective, Governor Davis of California has ordered banning MTBE by the year 2003 because it's combustion produces a carcinogen, Benzene. Without MTBE we're back to square one – smog, visible or not visible and respiratory problems – that's putting it mildly. Depending on where you live your late age existence may depend on an oxygen respirator or if you do not have a cast iron constitution, something else will lop off a few years of your potential lifespan. *   The reality of more and more gasoline consumed, exacerbating pollution, shortages now threatening and escalating costs of living – clearly, our approach for now is a study in illogic. The height and folly of skirting the truth defies common sense. I draw from this an analogy, the Roaring Twenties such is blatant disregard evident when rapid monetary gains have priority over everything.

In adopting to some extent a diary mode of writing simply because day by day occurrences, as probably universally posted for everyone to see, provides a link between the reader and I, there is also the advantage, not from any condemning interest, but calmly seeing for ourselves through step by step evaluation what makes sense and what does not, given that we also have to put up with specious material. I've collected all information and all opinions putting it all in one package. Otherwise, what was said, showed or printed over five to ten years ago may not be clearly remembered. When a lawyer goes to court to present his case he or she does not rely solely on memory. There is also a briefcase by their side. This book could be more than temporary paper sheets.

I'm afraid that the modern climatologist or expert is striving for formulations out of his or her reach, happy in the rarified atmosphere of complicated studies, showing a talent for describing in highly sophisticated terms and

Hear me out all you big city super folks!

invariably projecting the idea a great secret or development guaranteeing extremes in long range forecasting awaits as a goal soon to be achieved, when in reality, there is none. There are however in fact, the simplest basic fundamentals, which can offer answers. For example, today April 19, a most disturbing report of droughts and wildfires and two governors of Texas and Florida declaring a state of emergency. This is grave because we are only halfway through April. And yet, a climatologist had this to say, "We can't predict when and where droughts will occur if we don't know why it happens." Now I'm not malicious but there is apparently a disassociation of facts. I find this admission rather strange! It is not predictability that interests me, believing that there is always the unexpected, but causes! How can not one see the consequences of unrelenting sunshine in the tropics this winter (early 1999) scientists having already established that UV (ultra violet) rays are more intense than they were 30 or 40 years ago? That is not a difficult thing to do. How can they not recognize the evidence, clearly for everyone to judge by, southern Caribbean sites reporting temperatures in the upper 80's and exceeding 90 when they should be four to six degrees lower? And past statistical averages? Doesn't it mean something? Can't they recognize this influence on our southern states, an influence not only permeating northward but also from Mexico to Africa? Do they believe this immense quantity of heated air is a separate portion of our atmosphere, just a Mecca for sun worshippers? Even though it briefly contracts its extent in winter, it is always there waiting for early Spring's intensifying effects of a returning sun. We are told 20 species of Costa Rican frogs have disappeared but aside from the usual obvious attention paid to temperature and humidity changes as the probable cause undecidedly attributed to earth warming, that fiery ball in the sky remains the object of little analytical reporting. On one hand, in the same depressing article on droughts, in addition, wildfires in Nebraska killing hundreds of cattle, someone states it can't be El Nino since it died an abrupt death over 11 months ago but then another exuberant contributor* to the article adds a touch of drama. "Recorded El Ninos going back 10 centuries (a thousand years no less!) show cataclysmic droughts always follow each El Nino episode." Unless he overvalues tree ring studies showing evidence of rain or dry years, that in itself is inconclusive as to whether or not El Ninos played a role, there is in that statement the usual need to dramatize. We have had but two major droughts of a catastrophic nature. One in the mid 30's and a lesser one in 1998 but other than 1982-1983's coast hugging El Nino, there are 15 recorded flare ups of this Christ Child phenomenon in the equatorial Pacific since 1930! Furthermore, we have only been measuring such flare up activity since 1860. I'm not sure how reliable

Endless publicity seeking.

such information is as to its equatorial extent in the late 1800's since I have to suppose whalers and adventurous seafarers merely dropped a thermometer, recorded the reading in the ship's diary and continued on. As for the article, sad as it is, if one likes a potpourri this one excels. In all fairness however, it is after all, the work of a journalist who may have added his own touch here and there.

Here we are embroiled in a war after a lengthy period of distracting impeachment hearings. Now, we have droughts and wildfires. I am certainly not an advocate of law after law, dear to the hearts of aspiring politicians anxious to make their mark on history, but there is a crucial need for restructuring so much of our nation's infrastructure but I don't know if we can come up with a Franklin D. Roosevelt or John F. Kennedy who were men endowed with fire and brimstone unafraid to dramatically stir up emotions and take action. Al Gore fades in polls while Governor Bush gains approval and, for reasons of my own, that is disheartening. Time will tell. Furthermore, nature knows no political boundaries.

RELIEF IS CALLED FOR

Conscious my outlook is a dismal one and perhaps even demoralizing, although I would like to think something clearer can emerge from my efforts since there is no end to journalistic rigmarole over who, what and where this and that environmental factor will take us. I would like you to also know, at age 19 believe it or not, while warrior inclined as most of us were (not draftees!) I had my moments of contemplative sentiment – easily moved by what I experienced. Can you imagine, feeling sad over one lonely stork, for example? That certainly wouldn't bother most folks I have run into! And so, time for a little relief from problems peculiar to America which for now stirs few. For example, every day I hear meteorologists say the same thing over and over again, "Strong winds have knocked down power lines." How long will this go on before utility companies are faced with the inevitable? I have already went into the economics of this and what has to be, unless this turtle paced government decides they start to use some money and help companies put up something more storm proof than flimsy shaky wires strung from poles ready to topple over whenever winds blow over 50 miles per hour. Well, I got one gripe out of my system. I'm really weary of hearing the same old, or in effect, "Obsolete electric transmission systems have again failed and 200,000 homes are out of power." But, and this was one of my favorite statements years ago when boasting of our military might, "We can hit a fly in the ass 17 miles away." That has priority over everything and that is the price of being the world's sheriff, requiring a huge defense budget which I hope some day takes a backseat to more important needs-yet, I suppose this is not being too realistic.   Even though I have delayed providing relief, prompted by current events to make a supplementary point of my profound distaste for progress delaying quagmires that only please defense plants and fist shaking leaders safe in their offices, I'm afraid we lack the necessary leadership to push other nations to do more. Harassed as the whole world is by insane tyrants, you must allow, mercifully, understandably, some time to consider I just may have learned something in 75 years.

I have to give Moroccan Arabs credit for something despite my disgust over their chauvinistic treatment of wives, who seemed to be treated on the same level as beasts of burden while they, the turbaned robed ones sat English lady style on undersized creatures, two legs dangling to one side, throwing their midget steed off balance although its four thin legs compensated by taking very short steps, urged along by a steady tapping of haunches with a branch. And yes, there was the veiled wife, a huge load of wood on her back, walking in front of her lazy husband sitting on a disproportionately small burro if you consider the loafer was almost of the same weight. I felt bad for both the lady Arab and the creature that trotted behind her. It revealed to me clearly how some who are not evolved sense no guilt over imposing involuntary servility. Before I describe one virtue demonstrated by a breed that I have no fond memories of, I was not alone in feeling disdain and wanting very much to knock some hawk-eyed tyrant heels over head off pitifully overloaded creatures. An army truck slowly rolled by bringing a few Navy personnel back to the base and I watched a well aimed cantaloupe sail through the air and nearly giving an irate sailor the satisfaction of bowling over his intended target, missing the turbaned head and bouncing off a shoulder instead. Another melon flew also but splattered on the street and with that, fists raised and a few unflattering yells, the assailants moved out of my sight. While the intended victim seemed unaffected I reversed course not wishing to be viewed as belonging to a gang in need of reprisals since there were a lot more natives around and we were few in number. Well, the Animal Rights chaps had their day.

I must continue on but with a kinder view of Moroccans; that is the civilian segment. Many had been inducted into the French Army and did quite well in the Italian campaign, but I'm afraid there was one who was trigger-happy. During the invasion of Port Lyautey, it's role as a strategic site never having been mentioned in documentaries that I sometimes watch on TV, the Vichy French, an embarrassment to most, resisted our Army invaders but quickly gave up. A colonel by the name of Shaw, sitting in a jeep, holding a white flag on his way up toward a group of waiting French soldiers was shot and killed by a Moroccan. There were no reprisals. Apparently he had not understood instructions. The airbase was named after the colonel but since life in those days had expendable value I don't recall his name remaining fixed in our minds. Remunerating my intent to be more merciful I first must admit that I have been depressing, consistently painting a dark cloud and for the love of me I can't come up with any sunshine to clear it off. Not even a few brief rays, such is the unclear hesitant attitude of our government and strange doltish thinking on the part of too many citizens. I'm sorry, another damn interruption: Today's small time newspaper had a full page article describing in precise figures (which is impossible) trucks only produce exactly (oh boy!) 8.8 percent of nitrogen oxides and volatile organic compounds but autos produce 39.6 percent. Three dignitaries from unscientific organizations followed through with a slew of impossible to calculate emissions from everything that uses gasoline. Cars produce tons of carbon dioxide but no how, nowhere, given devices that function to cut out a lot of NOX, do they produce that much. Besides, how do you determine a constantly shifting atmosphere's content of this or that toxin and a million vehicles on the move, varying in speed from near zero to 70 mph? Sorry for the discontinuity but I'm really fed up with all this continuous tripe! But, truck companies can rejoice and sing, "Release me, let me go." No "Please" needed.

I have briefly described Morocco as being inhabited by destitute and to a degree rather primitive peoples who, except for merchants showed not the slightest interest in us – not that we cared for them either. Veiled women shuffled along, never a glancing eye – besides we knew better to even attempt nodding a silent hello. Small groups of ragamuffin boys, some even wearing an army hat, each with their own territory and range of operation would either pester us or follow a soldier or sailor who they thought was not too alert to their wily ways. An erratic gait sufficing to send a signal of possible generosity but if the opportunity arose, one might find they were not just satisfied with a dollar. Off with a snap a wallet would be separated from its owner and after these kids broke the 100-yard dash record, sling it aside.

Considering my untraveled ways before Pearl Harbor, one trip to New York city in some friends' "Shake, rattle and poop along" Model A Ford which took us an eternity on old Highway One and if I-95 had existed then I'm sure a trooper would have pulled us over. The mountains of Morocco, the first time I had ever experienced such beauty, revealed the other half of Moroccan Arabs, the Berbers. Once, having valiantly fought the French unsuccessfully in the 1900's, they returned to their pristine way of life, such was the impression I had of seeing a few of them on high grassy plateaus tending to sheep or small farm lands. Our ten truck caravan which manned by some 100 of us on the way to Agadir, had to be the first time these very different people, very unlike others of a related breed living in squalor along the coast, got their first look at what must have seemed an intrusive army moving in. At about 6000 feet above sea level, two trucks decided to quit running. We all hopped out while somebody fiddled under lifted hoods. Some 50 yards away a lovely little Berber girl, standing in front of a flock of sheep, stake in hand, showed no fear at all, entranced or perhaps determined not to abandon her responsibility stared at us until we rumbled off again. The trouble having been vapor locked fuel pumps. One has to see these peoples were a blend of Arab and Roman blood since many look more like Europeans than their counterpart occupying lowlands. Evidence of Roman influences is everywhere and of course, L'Amour Toujour changes things and I doubt very much if it was the rape and pillage of Hollywood renditions. Consequently there are thousands of very unique, neatly attired and very independent Berbers. If it's truth one wishes the English are the best authorities on Roman history – one book, Hadrian the 2nd tells it like it was sometime in the early to mid AD's. He was the first to build orphan asylums, feeling sorry for children, court systems were introduced and the first London was to his credit. True the Roman Empire collapsed but many of it's soldiers never returned home and found it much more to their liking to settle down. Getting back a bit to weather, it is clear the Sahara has slowly expanded its influence since there are many sites in Algeria and Tunisia which were once upon a time cultivated but now are but parched remains of farms. Unless you can classify scorpions as evidence of life there is not much else.

Finally, something virtuous to write about for lowland Arab habitants, trusting the Sultan of Morocco will not read of my contempt for ill treatment by his peoples even though wives and husband could resort to the quick expediency of divorcing without lawyers by simply tearing up a marriage card of sorts. I had photos of the sultan who died a long time ago, his son now in Rabat, on a supremely decorated white horse, dozens of mounted guards on lovely Arabian steeds but my children have done a number on us and have managed to take all my treasures somehow. Scattered amidst rows of ramshackle native homes were a few roof tops that stood out from the rest proving its residents had a soft spot in their hearts for a winged creature nearly extinct. Undisturbed, perhaps for reasons I would guess had something to do with good luck, perhaps reverence, maybe a token of prestige, since such a huge impressive bird as a stork had carefully selected ideal peaked roofs to build the gaudiest of nests that you can imagine. I was told there were very few of these birds in the 30's and there was but one remaining when I was there in 1943. If diminutive birds display engineering talent for weaving together intricate homes than the ungainly appearing mythical baby carrier wasn't fussy at all. A stack some two feet high and somewhat wider made up of coarse branches and whatever else kept it together was deemed, apparently, despite what we here would not tolerate since property values would collapse from such an eye sore, a sacred move on the part of Allah – not to be disturbed at any cost. That creature had no fear of humans and once it had selected a place to stay it knew it was secure. For storks to migrate some 1500 miles from Morocco to what I believe is somewhere around Austria is quite a feat for the impression of awkward confirmation it gives when perched. But in flight it is a beautiful sight – in retrospect I now feel it bore a vague resemblance to the Concorde, one long slightly bent neck, with it's pointed beak straight out, cutting the air and effortlessly moving at a surprising speed. I had seen this one stork gracefully motoring along while making weather observations on top of our hangar and if I had not been told by a young Frenchman who served as an all around helper and liaison, acquainted with everything new to us in this acquired aviation facility, that this white bird and it's graceful movement of wings was the last of the breed, I would not have felt what I hope you don't attribute as mushy sentiment on my part. It actually saddened me. The lonely one – it is no more. Why? It troubled me. But don't forget, if I describe it's graceful means of propulsion as impressively large and powerful, you must realize carrying six to ten pound babies requires quite a wingspan – human babies that is.

Today, my friend, the oral surgeon, asked me why Florida had gone without rain for quite some time. He was a bit troubled by this since a member of his family living there registered concern over the phone over wildfires a long way off and yet the odor of smoke had become stronger. Flattered at his "You should know – tell me why?" quite naturally, this would make part of my day a more pleasant one. A chance for a brief lecture and an attentive educated man – what could be better. It's redundant of me, since by now you are approaching the level of a full fledged weather expert (really, it's that simple) but Florida's condition is a matter of stability as opposed to the desired state of instability (rising air, clouds and then rain). Consequently the air is descending, gently of course, simply because for reasons already belabored, high pressure dominates and is not either moving off or shrinking. The smoke spreads but cannot rise. Having did my thing and mind you, despite gloom and doom I persevere over, I try to find something more cheerful in conversing with others. I offered the doctor a simple lesson of up or down movement of air since ideal conditions existed above us. Humor would have to wait. The sky was blue but many puffs of white clouds were forming and so I explained whatever invisible pollution we have in the air was being lifted up but, on a clear blue sky day and no clouds whatsoever, then the air is gently descending. Deceptively, that is when pollutants remain at lower levels. Appropriately, it was the surgeon's turn to make a contribution. I asked him how his "Moocher" friends were doing. That worked before for a cheerful end to our daily brief talks. I was quite amused over crows who arrive every day, right on time, about ten minutes after "Doc", as I call him, is finished eating. He leaves a rectangular aluminum container containing some tidbits in the backyard. One crow now has the habit of picking up the empty container, once his buddies are through, and with it held by a beak, trots over and drops it upside down against the back door. The doctor interprets this, delighted at his idea, the crow is saying, "Don't forget, we'll be back." I told him that the crow might have other ideas, like, "How about some seconds."

Having previously indicated day by day media articles or television presentations particularly those manifesting something to do with unusual weather, serve me but waylay continuity in my writing, hence Florida fires and the surgeon as a side item. Furthermore, using stories of ancient experiences not only help to add variety to a subject quite morose but perhaps I can encourage involvement and although I would like to find episodes of the past offering an inextricable connection with the main topic of the book, I am hard pressed to produce one after the other. But, if a reader cannot connect with a writer and appreciate some degree of compatibility then fair or superior literary efforts have no lasting effect. Too many think of youth as a passing stage and not much credit is given to many that I believe are quite profound and sensitive for their age. We were hard nosed from two influences, a terrible depression and indoctrinated into a warrior mentality – quite easy to do – not much urging needed if a worthy cause dominates all thought. I've seen soldiers feeding swans overseas and talking ever so gently or passing candy and cigarettes around while others showed their feeling about going to war and vented what was on their minds by behaving violently wrecking bars and trying to start fights with anyone handy. My moods swung from wanting to be a hero and that meant nothing very peaceful to also feelings of sadness for lost souls. Poverty in all it's ugliness but here and there beauty to behold. I am leading into something less than fascinating – I do like to insert some lighthearted views I felt as a youth. I wasn't finished with Morocco's strangeness and it's peoples – the surgeon having interrupted what I had started with his genuine concern for relatives in Florida.

Remnants of ancient structures, burros and desert attire gave the feeling of being back in Biblical times minus the balm it offers. Abject poverty brought about by the Great Depression and arid lands bordering the Sahara offering little, I would nevertheless see, if there had been wealth before, it found it's way to a palace. The French as with all colonialists had done well but there were no chateaus, just neat homes and cars were a rarity, and food was rationed. But, if there was exploitation I would later theorize this had to be perhaps, as I believe, a godly arrangement and once a colony has risen from the dust to modern metropolises, revolution hands back everything to its original owners. But, instead of negotiated transfer of power as in Hong Kong the Arabs, disenfranchised by extremes in poverty resorted to violence. The first move occurred in 1945. Five Europeans sitting in a restaurant did not get out of the way of a bomb thrown. Several years later, Arab women stood on rooftops screaming encouragement while their man hacked away on the streets. Mercilessly, women, children and anyone European were killed. Why would female hearts change from what they were designed for to deriving satisfaction from murder? Simple, that is what money in the hands of a few and nothing for anyone else creates. As a child I saw mobs and if it wasn't for overwhelming numbers of armed soldiers and police I don't know what anger and frustration would unleash on factory owners. Do you connect the above with growing wage disparities?

The truth is some adult told me to get my ass
out of there before I got killed.

Before I get back to what I hope is more pleasant having mentioned previously, lightheartedness but Morocco's initial impression on me necessitating first the ugly side. Casablanca in 1942 was quite modern, so was the capital of Morocco, Rabat and in all fairness some of the more ambitious Arabs held important jobs. But Humphrey Bogart's Casablanca in the early 1900's, just 25 years or so before World War II, wasn't that at all. It was a mass conglomeration of flea ridden ramshackle shacks and huts. Disease was rampant. Along came the Europeans, up went progress at a dramatic rate and then, as with all such arrangements, at least those created by the French, English and to a lesser degree other Europeans, the cosmic programmed change takes place. Patience, I'm leading to a brief adventure that fiction writers would envy. Don't believe that, of course. Casablanca was a remarkably modern looking city. Because there was no air conditioning in those days a major movie house solved the problem by opening the roof at night. Even though, after we had docked there at the end of a trip spoiled by hunger, except for one night of a prune festival, we were confined to the ship for an unbearable 24 hours, we managed to beg and receive permission to spend a few hours in the city. We were forewarned of course not to venture into native sections. In my anxiety to test what little French I knew, some lad discovering this tremendous advantage, we decided the two of us had an Ace-in-the-hole and surely "Chercher la femme" would prove rewarding. Lo and behold, there stood a blond young lady, who having emerged from a store, stopped and seeing the proverbial "Ze French Lover" as if the breed had invented sex, my buddy staying in a neutral zone waiting to see a Tour de Force take place, just put on her haughtiest look. Unfazed, my credibility at stake what with a partner looking on, I proceeded to ask directions but never got a chance to complete one sentence. She snapped something that smacked of German, implicitly guttural as that language can be which I interpreted to mean "Get Lost!" Oh well, unlike the sad Bogart asking Sam to play it again, no agony of defeat in mind, we trotted off.

While Morocco was under Germany's thumb there were none of its soldiers present – just officials in several major cities. It's possible my being undiplomatically rejected by that girl was prompted by her possibly being a family member of occupational representatives prior to our rude intrusion – like 500 ships, 20,000 soldiers and a bunch of noisy airplanes. Gosh, couldn't she have been more polite! All our armed forces had done during an invasion six weeks before, that I had been scheduled to participate, in was for the USS Massachusetts to punch a hole in the Jean Bart's bow. It was a beauty of a battleship docked there. It had a hole big enough for a jeep to drive through. And then of course, they set off maybe 50 more booming rounds, sinking four or five small gun boats plus aircraft carrier planes dropping a few bombs here and there while others knocked down several obsolete fighter planes, then turn 8 to 10,000 soldiers loose in the city. Did she have to take it out on me just because of this! Meaning no disrespect for the captain of our battleship whose 16-inch guns steadily poured it on, the man had gotten carried away to a point where one gun burnt out. A chief yeoman by the name of Shields, also stationed years later at Squantum Massachusetts who served alongside the illustrious captain, just thought the man, evidently frustrated over months of no fireworks was hell bent to blow everything up. Having told me that Shields burst out laughing.

Our brief venture in Casablanca ended on a pleasant note and of course if I had any skill at writing I could make a lot out of nothing. Never cared for fiction anyway. After bumping into Germany's finest my friend who thought the broken French that I spoke would lead to something more interesting, although we had no lecherous intent (really) when we spotted a large department store. Our paychecks hadn't caught up with us, but penniless or not some semblance to America awaited us. There were counters, the usual goods and of course, a girl behind each one. I talked to one and two more moved in as we were a curiosity, really. Sometimes they giggled when I couldn't put a sentence together while my friend asked, "What did she say?" It was time to go but I had a warmer feeling and decided I had to improve on my French. The next day, sea bags slung over shoulders, we embarked on to a sorry looking open train car, which had to be designed for cattle and although clean it suggested something less than exotic awaited us.

If some of a hardier nature in the world of men think all male office workers are more on the delicate side they would be mistaken. Inevitably there is always amongst large assemblies of youth in uniform a combative edge, bound to pop up – some individual ready to test you – a rooster type. I got along but spent spare hours outdoors in a makeshift exercising area, a gym if you will, trying for the body beautiful – parallel bars and whatever else promised muscles. Even though we were not allowed to stay in town after 5 PM, limited to liberty but once every fourth day, somebody in charge relaxed restrictions if you could establish there was someone's home you could stay at until 10 PM. An elderly gentleman who I had befriended invited me and for once I had the same privilege as officers. It was the last time after that – walking back to the base in darkened streets, a wartime requirement, and not a single light to show the way proved to be too harrowing an experience. Let me first give a touch of history since the old gentleman had quite a bit to tell me. If Hollywood movies of today provide exciting versions of heroic impossibilities they also may dull interest in side episodes of World War II, but I'll give it a shot anyway. You can always tell movie buffs when calling in and insisting over radio waves that we should go in and take out Saddam Hussein or Kosovo's Milosevic. There is nothing to it of course. I could relate the sanguine too, like a friend of yesteryear, Raymond Lee who charged a machine gun nest and was cut in half, but temperance is my choice right now. For the sake of history I must tell you the lot of dedicated Frenchmen under the Vichy regime was a dangerous one. This elderly gentleman, his wife and son, also in the army, narrated how his group of clandestine patriots helped sneak pilots to England but their plan to cut communication wires to help American invaders was foiled. Some traitor turned the old boy in and he was packed off to a concentration camp nearby. Had it not been for the speedy invasion and rapid consolidation of everything useful he would not have been around this particular night. What follows is also funny. I would not have engaged in war that pitch black night on the way back home. My weapon consisted of a long cloth bag containing one liter of red wine, a treasure that I knew the boys in the barracks would appreciate, six oranges and a pair of Arabian sandals – weight about eight pounds. It was bad enough that I was nervous, but when a voice out of the dark barked a firm, "Halt!" I froze right then and there. Immediately such an order means a rifle might be pointing at you but I managed to ask what did he want. No answer. There was no "Advance and be recognized" furthermore this halt business didn't quite sound American. I saw him moving toward me, no rifle and, well, let me dress this up a bit. If you have ever seen disc throwers readying to go into a wild spin, that's what I did. I wasn't superman but the adrenaline out of fear sure can make a difference. Left foot forward, slowly, unseen, my right arm and the bag at the end of it went straight behind me, calculating the distance where I wanted his head to be, about five feet, six inches – had to be exact you know. Hope to get a smile out of this, by the way. An overhead shot wouldn't do thick skull and lots of hair – a shot to the side of the head a must. If I missed, well, no telling and truthfully I thought as a scared rabbit might – bound away. Down he went after a thudding smash to the side of the head and away I went. I wasn't going to ask him his state of impairment. Good thing the orange side of the bag hit him instead of the bottle, which fortunately stayed intact although the sandals soaked up some orange juice. Did the guys in the barracks believe my having come through enemy lines to deliver their Vino? All I got was "Awe come on." No oranges to salvage – that was sad.

But there is a moral to this true story however commonplace. I have a point to make for the 30's in America. Poverty during an era of little communication didn't manifest itself on a wholesale level other than a few skirmishes between union-oriented groups and factory owners. Very few were killed in massive confrontations. If, and I believe it will, wage disparity takes on an immoral and dangerous disproportionality my "noblesse-oblige" compelling me not to offer an anemic prediction for the sake of avoiding being tabbed a gloom-spreader and writing as would some wary Casper Mittletoast, I want to list exactly how the environmental problem can create an unforeseen economic disruption. Oh, the affluent won't care but I'm concentrating on the bottom of humanity – not the top. We are in an age of ultra communication and wealth is perversely and continuously being flaunted for everyone to see. An example in yesterday's paper, two executives relieved of their positions following the merger of electricity providing companies each are to receive over six million dollars, aptly called "Parachute Compensation" (poor devils) but in providing readers with all the juicy details of sufficient stock options to fill up a pickup truck they were the talk of the town, people were quite amused at photos of these two who sported grins from ear to ear. In 1930, who, amongst workers and unemployed, knew (early historians were great liars) alleged disappeared money really had not totally melted away. Therefore no real hate on the part of the public. Earth warming can destroy croplands and livestock and you don't want to read anymore about devastated property. Consider the great possibility of Canada inheriting a large piece of our agricultural capacity since warmer temperatures would benefit that country. Property insurance has to go up and so tenement owners and privately owned high rises, also eventually affected by the cost of electricity and maintenance will demand higher rents. Already one elderly high rise arrangement in North Providence has notified some 300 elderly that they are requesting from HUD permission to jack up rent – and right here in a state more or less free from violent weather. Rent, food, gasoline, heating fuel, electricity and medical insurance will be more expensive. Many employers will not be able to give inflation compensatory raises due to their own increasing expenses. You are told crime rates have gone down and your taxes shouldn't bound upward but you have to consider how many more policemen will you need. Drug dependency and the criminal element also chasing wealth means addicts must steal. That is what I mean by pollution in an age where nothing else but money matters. If the Great Depression spawned John Dillinger and Pretty Boy Floyd we will not have that problem – instead daytime burglaries will increase when you are at work. Costly alarm systems, some specializing in rather expensive pay-as-you-go constant monitoring will be needed unless you want to feed two

Rotweillers for about 35 dollars a week. I suggest male and female. You can breed them and sell puppies – it will help defray the cost. Pawtucket and Central Falls have hundreds of Pit Bulls now – even the poor are worried about burglars. There is a price to pay for abusing nature!

I'm no socialist, but I was surprised to read a group of organized bank employees based in Boston had demanded a president's income should not exceed seven times what the average bank help made. That was quite unreasonable and futile but a major leading bank chairman of the board collecting eight million dollars last year motivated it. Of course their demands were voted down but it smacks of unrest and discontentment in what seems rather serene surroundings. The mayor of Pawtucket just announced an increase in city taxes and two weeks ago made a public appeal to help stop increasing graffiti and vandalism. This city is not the Bronx – other than a few areas of poorer immigrants and some lost souls it's quite respectable. I am expanding on the effects of climate changes and not what is at hand politically, socially and economically. We have a worsening immigration problem and to a major degree the private sector is responsible. They want cheap labor, period. I talked to one chap from a South American country and he told me stealing is a religion there! And so, more overtime pay for cops and jammed courtrooms plus loaded jails. Another gross headache: Across America probably 15,000 well paid experts poke around contaminated waters, measure toxins in the air, which has to get worse after I reveal the latest put out by the Wall Street Journal, April 24, 1999. That really bothers me and it is not a knee-jerk reaction coming up somewhere in the next pages – what a letdown on the part of the President, Al Gore and whoever is meant by White House decisions! A real lulu of an agreement made with power plants and industrial polluters. In the meantime oblivious or trapped state and city politicians fuss over details that don't promise common sense approaches since the truth of the matter is still under the rug, swept there by the likes of spin doctors. Each week, each month some analyzing chemist or whatever discovers something worse than ever in the air, therefore you can expect more and more experts put to work. One chap in Maine, who discovered mercury over Vermont hills, is quoted in the Wall Street Journal as saying, "We tell them it is raining mercury and they think we're nuts!" I don't know why Russ Limbaugh remains popular since he has said that the American populace is ignorant. I don't think they are ignorant – I believe they're too darn lazy to keep informed and react misinformed at the voting polls.

Fear can be very damaging. All I hope to accomplish is recognition of the facts of weather and then just maybe developing a consensus leading to persuading Congress and the Senate to take that extra step – nothing drastic while promoting the brighter side to what change can bring about. Right now there is anxiety and a tendency to frenetically pursue wealth; it permeates all psychological persuasions in advertising, even rapid-fire auctioneer sales pitch to save a few seconds and make more money, from electronic broadcasting to magazines, but it also pollutes. Today, April 28, 1999, FBI agents hauled off two figures working in the tax division of the city of Providence, evidently they want a Lexus and a cottage by the sea. Tapped telephones and tape recordings now weigh heavily as evidence of bribery going back a few years. Little people want to be big people overnight and simpletons turn to a fast buck. Politicians need millions to be re-elected and so on. The same all-pervasive high prevailed in the Roaring Twenties but thank God we have today intrusive government as epitomized by Alan Greenspan and other imposing departments ready to lasso rambunctious loose bulls. The condition of slaughterhouses before the government intervened

> BOSTON (AP) – An estimated 4,500 people now living in New England will develop some form of cancer from poisons in the air, according to a new report.
>
> Nearly half those cases of cancer, 2,200, can be expected in Massachusetts, and 1,400 are estimated for Connecticut, according to the Environmental Defense Fund report.
>
> The report, issued Tuesday, estimated that 360 of every 1 million people in the United States will develop cancer because of toxic chemicals in the air.

(the S.O.B.s!) sometime in the 40's was appallingly filthy. As a teenager I would strive to make a quarter by cleaning one Ted Gillen's gas station on Saturday. His father was wealthy and owned a slaughterhouse. Ted use to grin at me and tell how his Dad absolutely would not let anyone in the family eat hot dogs – after which he would laugh. I didn't know any better, in fact, I thought a wiener would be nice to have. But lately I can't put the president in the intrusive category, especially after eloquently stirring Congress and Senate representatives to rise up and applaud during his stirring state of the union address a few months back when he promised to help industry switch to more environmentally friendly fuels. Then the Wall Street Journal reveals nothing of the sort if I am to interpret an agreement by the white House (whatever group that means) most gratifying to Big Business. Tiring as it may be I have that information on the next page.

A TEMPORARY BOUT OF BITTERNESS – "C'EST LA VIE".

"The reason is because of the complexity of the task," explains John Stanton of the Environmental Action Network, which has been involved in the debates over the regulations. Western governors and lobbyists for oil, coal and electricity industries have all played a part, he says.

Most of the players realized that getting the air in parks back to pristine conditions would take a long time, Mr. Stanton notes. Environmental groups had argued for a 50-year deadline; industry wanted it to be much longer. The Clinton administration finally settled for 60 years.

I had good intentions of putting a stop to inserting any more substantiating excerpts, feeling it had reached in the aggregate effect a stage of fatigue for the reader rather than my intended purpose of continuous validation. But, sure enough, two articles are too significant and I must lay it on once more. I make no bones over what it signifies – dollars before deaths! The first item, already an old story, did not stir me so much even though a flicker of contempt for ignorance and lack of resolve crossed my mind. To publicize estimated deaths from toxins reaching 90,000 for America, if one resorts to a little arithmetic without the least inference (everybody yawning) a crisis looms ahead (at least a hint, please) is blunderous callous thinking. We can also count on the ever present philosopher, who most likely in retirement years contributes most to escalating Medicare expenditures has to say, "We gotta die of something, right?" A serious thought: What good is building up your nest egg if at retirement age you become grievously ill? Big city pundits should take heed!

There has to be the future of economics in mind or otherwise America's second leading newspaper wouldn't be so dutiful as they have been lately in this matter of air quality. In the astute effort by staff writers, surprisingly showing interest over various opinions, one, a Maine based scientist quoted, "We'll tell them it's raining mercury and they think we're nuts!"* First, the article is about Al Gore wanting to manage cleaner air in national parks and so on. If you're interested it is the April 24, 1999 issue. Evidently, the Vice President is cautiously skirting the edges of a national problem and temporarily concentrating on auto infesting public tourist attractions – probably scared of panicking 30 or 40 million American car owners as he once did. Then, details the same EPA mandates resistance and so on fill in quite an extensive report by the Wall Street Journal. Relatively speaking, here is a snippet of the final collapse of resolve. That really got to me. Sixty years and I thought, just maybe, Clinton and whoever had plans to perhaps assist one lousy coal burning plant at the rate of one a year. Therefore twenty lung-poisoning coal burners, now spewing mercury (I just discovered that!) would in 20 years mark a turning point in our industrial history, but no.

Since this is a turnabout, and suspiciously hypocritical in view of a rip-roaring presidential address, I'm not sure the agreement is etched in stone. But, maybe when hospital wards can't handle coughing crowds holding

Had to repeat that one! That's how I feel also.

handkerchiefs over their faces as scenes of smoky Hong Kong show, some new and most certainly, fearless politician decides to do something. Indisputably, a 60-year agreement is preposterous! This regrettably will brings back the acrimony I once felt for tragedies that could have been avoided except that we had three listless presidents, the Do Nothing Presidents, Coolidge, Harding and Hoover. What a beaut that last one – an exploding stock market and he listened to Andrew Mellon* who did quite well before, during and after 1929. The money in America disappeared all right – a chunk of it in his pocket.

Multiply this story below by several million why I feel contempt and disrespect now but had hopes some months ago when I began to write this book. We have buffoons both in charge and buffoon voters who haven't the slightest idea of what slow lingering death is about!!! The implication of increasing amounts of benzene, mercury and sulphur dioxides in the air and we do not have as yet all the testers needed to prove the extent of this grave problem, although, as I've repeated too many times already, the solution is not a complicated one and at least some relief could be had.

Having experienced living with a father that I never knew, during an age where practically every child had a father, taking five slow years to die, the final 12 months in a charity ward, my mother telling me a decade later, a human skeleton in a crowded ward, his body spotted with bed sores, for a period from 1926 to 1931, spitting pieces of his lungs out, four years previously to hospitalization, hiding in a bedroom after being told by flimflam doctors of those days he had tuberculosis and thus spare us from it's danger. An autopsy proved different. He had collected zinc particulates at work and like all insolubles (think asbestos is bad?) plugging up to a point of necrotic decay the only means by which the rest of the body can function normally. I do not hesitate to forward ungraciously the opinion that there are a lot of imbeciles in this world, if it's just a matter of "We have to die of something." No wonder Einstein drew humorous comments from writers for his occasional giving the Bird to collegiates walking by staring at him. That April 24 Wall Street Journal article gave me the Red Ass** as we

Southerners are prone to say. Might as well show my coarser side, that's a good way to show contemptnevertheless. I didn't know my father had died when I was eight. Mother shuffled me off to an aunt's apartment

*Another repeat: When Hoover expressed concern over a boiling stock market Mellon told him it was evil to tamper with capitalism – per John Gailbraight. Mellon and other insiders cleaned up with excessively slack "short selling" rules!

**Because of steaming heat, one working outdoors can develop chafing.

several houses away from where a casket filled too much of our small parlor. But I knew something was going on at home. Curious, I waited till dark and very carefully opened a side door while a small group of family members were busy in the kitchen. Oh yes, I had disobeyed and walked quickly – had to find out why all the people that day and I hadn't been allowed around. Have you ever seen pictures of King Tut as a mummy? The face shrinks to bone but the nose, a Roman one no less, doesn't. An outdoor street lamp sent a beam right through a window and I saw a glimmer outline the ridge of that nose I inherited but the rest of that face was but a dark unclear silhouette. I quietly shut the door, having recognized who he was but truthfully I don't remember my feelings. Perhaps it might explain why, other than sincerely grieved by suffering, the numerous deaths I have seen don't affect me as it should. Then some months later, my younger brother and I in a combination reform and orphan asylum, the only service available, both suffering from malnutrition. Boils* all over my body and my 6 year old brother worse than that have to see some man punch a 10 year old around to nearly a pulp because he had run away. We both managed to move away from this frightening scene and sat at the far edge of an athletic field, holding each other and crying. But, brave one here, looking at Massachusetts' Merrimac River a half-mile away said, "We'll steal a boat and go back home." We didn't of course. Brother did much better after one year in a rehabilitation facility located in Wallum Lake, RI and during WWII did something that few now would dare to do – his job, a volunteer no less, neutralize bombs and shells. He was involved in our Army crossing the Rhine and served in the Korean War. Retired after 20 years but now ravaged by Parkinson's disease, which I attribute to a bad physiological start in life. I didn't want to display such a sad aspect of life in the 30's but those articles remind me of something.

Heroes are labeled
show valor in combat,
enemy soldiers and/or
process but there are
My brother was one
soldiers waited at a safe
calmly walk toward an
or shell, kneel down
and defuse the damn
snapshot was taken in
his tour of duty during
As teenagers we
tough guys, but both of
into boxing rings
flail away for three
opponents or us none
for an occasional red
gloves weighed 16

as such if they
maybe capture
are wounded in the
also the unsung.
such person. While
distance he would
unexploded bomb
with his toolbox
thing. The
Kobe Japan after
the Korean War.
weren't the local
us would climb
Friday night and
rounds- neither the
the worse, except
nose since the
ounces. But I'm

tough, believe me…well, mentally anyway. I wish that song "Yesterday" would not have been written; you know what I mean.

1896-1998

1920-1925

At age 93 Mother was clear headed, even on the same day she died. A failing heart did her in – no wonder. She

Was a talented piano player and performed with Hoagie Carmichael who in New England was a well-known musician. She tried to make me into a violin player but squeaking and squawking away for two weeks, whatever inspiration I had faded away.

After my mother married again life was much better of course but hard times affected so many that even stray cats weren't safe. One old duffer, always leaning on a cane, waving a hand at anyone glancing his way, dressed in rumple old clothes with what was once a hat pulled down to his ears, living in a shack that town officials tolerated despite his home being an eyesore, was exposed as a cat killer. He would invite occasional hobos passing through this piece of RI called Marieville being neat and respectable because, as he confessed, "I feel sorry for them." He would ask them to join him in a rabbit dinner. Gradually now I'll leave what is grim behind (just want to chop away at laisser-faire economics and what it did) but not until another bit of incidents I remember so well. A big gasoline truck rolling down a hill totally lost it's brakes but our town's corner street approaches were flat – therefore the tanker was probably going less than five mph when it plowed into a building with a barber shop at it's front end. The driver didn't want to run a red light. It went right through the house from the backside and came to a full stop with the engine compartment and cab protruding onto the main street. In front of the truck's bumper one perfectly erect barbershop chair. Sitting in it, a white sheet still wrapped around it's occupant a 12 year old boy looking quite shocked but unhurt whatsoever.

On a more humorous vein (we need that don't we) my stepfather would have fitted in very nicely in beer commercials although he was strictly a one night embiber. One of his drinking buddies owned a car and this foggy night decided to try another place some 8 miles away, a city called Woonsocket. Lo and behold, Dad's barroom friend hit the brakes and both of them thought they had drank too much at the apparition looming out of the mist. A herd of elephants (they thought it was a herd) slowly crossed a dozen yards ahead of them. In those days circuses traveled on foot between towns but only at night.

I never went to a circus and despite movies for 10 cents I only had the privilege twice. The second one, "The Mummy" back in 1933 frightened me – I just ran out of the theatre – couldn't take seeing the mummy removing it's wrappings and what it's face would look like.

Back in 1932 a wealthy man wanted to marry my mother but, as she said years later, "I didn't love him." Perhaps when I was in my late twenties I teased her about that. During our trying days in the Home of the Guardian Angels, Massachusetts she got a ride to visit us. I saw her in a new Chevrolet pulling up where we waited outside the front door. To me, sitting in the back seat, soaking in all the luxury, trying to see through the back of a neatly attired man driving, analyzing glimpses of his face, out for a short ride, I thought couples were automatically to wed. When they left I couldn't contain my excitement and said to my brother, "Mom's going to marry a millionaire!" He smiled which was a rarity – six more months there I'm quite convinced he would have died. In such tight money times no system of help would have paid for one year in a semi-hospital to take care of my brother unless it was a desperate situation. No tears, no sympathy do I solicit, but 67 years later I still see that Merrimac River in the movies flitting across my mind.

Having mentioned being tough or just maybe survivors here is a suggestion. In fact I'm going to go overboard now with advice for the ladies revolving around that true story of my giving some bum a headache in Morocco. Might as well let you know I learned a few things from Marines and including some French Legionnaires. Try to find some humor in what follows even though it's a bit rough. Ladies, as I stand here on this stage lecturing to you fine people as to ways and means of defending yourself against some demented drug addict or someone who should be castrated, don't get foolish and allow some creep to get close with the idea you can simply kick your way out of trouble. If you have to work nights or maybe walk some distance from your job to a car then let me introduce "More bang for the bar."

However, it's not recommended for men as it can be lethal. Get yourself a large bar of laundry soap and a long man's stocking – you can guess the rest. It will nicely fit in your purse. Man, I mean Woman, what a zonk that will give and doesn't even leave a mark and if an officer looks at some figure sitting there in a daze you can always claim you didn't really do anything other than scream and he fainted or whatever. Remember that five feet, six inches advantage. If you're right handed, left foot forward, please.

I have been less than poised in the beginning expressing contempt for certain writers, a few displaying abbreviated emulations of William Buckley Jr.'s skillful compositions and also degreed, in what I don't know, whose ulterior motives are obvious. I was also less than generous with my opinions of certain personalities, usually titled, functioning on TV documentaries of violent weather as trusted authorities, never backing up their contentions with the evidence that <u>only</u> a weather map can provide. I know showboating when I see it. Now, a brief additional comment on another group, the climatologists but with a kinder outlook. I have collected their articles – they strike me as good sports as they bandy about between each other the difference between weather and climate, for example. Sometimes one will step out of courteous bounds and head the first paragraph with "I am amused by so and so's contention." But, mercifully let me add this, long range forecasting is one tough nut to crack. I'll let them keep trying and hope they might consider different approaches.

More Evidence But You Will Not find It In the Internet

Speaking of climate and change as it is, let me introduce to you other signals given to us by Nature. This is for the sport's fishermen of which there are thousands. And something for commercial fishermen also. I did not need a scientist to suggest, and that's all he did, that warm ocean currents have worked their way further north. In fact there was timidity to his announcement. Of course warm currents have pushed further into colder latitudes!! Some thermometer readings on record and very significant, what is happening to fish for example, supports the undeniable prevalence of abnormally sustained wind forces previously described and which only 50 years of weather maps can substantiate. <u>We</u> <u>now</u> <u>have</u> semi-tropical type fish being caught right here of RI and clear up to the tip of Cape Cod that 25 years ago was <u>unheard</u> of! Their home territory for most of the year is adjacent waters to Florida and in the Gulf of Mexico itself. They are Bonita, False Albacore, Trigger Fish, Dolphin (it has a head like a porpoise) and Jacks. These fish are leaving the edge of the Gulf Stream in summer while near our New England shores and are quite comfortable in ambient warm waters which were substantially cooler several decades ago. Striped Bass, a favorite, once upon a time would arrive early May – now it's early April that has fishermen getting their poles ready.   The bass may also be confused since some do not return to Chesapeake Bay at the end of summer and can be caught all winter long right here in little old Rhody. As for commercial fishing, flounder is a big time business but something is going wrong. According to an oceanographer they are not multiplying as they should because they need much colder waters.

I have to go off on another tangent again. Blame the newspapers. India told Kyoto Treaty organizers that they could not afford emission curtailments, which they excel in after the USA and China. But, this past week they have become very busy blowing million by lobbing shells at Pakistanis. Always, huge sums of money are available to make war.

Again, some humor. My oldest son, who was president of the Sports Fishermen Society for a year, was seated beneath the Jamestown Bridge doing some night fishing with a friend. On one side of his buddy a lunch bag which unnoticed began to move away. However, it was too large and too heavy for a four-legged thief to carry. The sound of paper dragged across rocks upon which they were seated aroused them and my son's friend jumped up and made threatening gestures, swinging his hand at an enterprising raccoon. It stood up on its hind legs, showed him it's front paws which meant (Let me get a hold of you just once) and hissed fiercely revealing an array of long white teeth, sharp as needles. Two more tries from a safer distance but to no avail. While looking for a stone the opportunist thief scurried off, the bag bouncing next to it's head. My son couldn't stop laughing and consoled the distraught victim by offering to share his food.

How would you react to this one? You have rented a nice cabin next to a lake in New Hampshire. It's 11 PM and your two children are sound asleep from a busy day boating and climbing hills. My son was watching TV with his wife and suddenly detected something in the window and, as it would you, it scared him. Jumping back about four feet he said something to the effect "What the hell is that!!" There looking in the window, both black eyes shining from the cabin lights with one large round face. After collecting himself he realized it was a raccoon – went outside to chase it off. But no, this mother with three youngsters did not want to leave – again, the stance and hissing. It finally left and my son rationalized it all as a Mom checking out the cabin. Maybe some goodies for her family.

In my need to relate to you and also to let you know I've experienced about everything there is to experience, one more simple story. I just love animals and they do talk to humans in their own way. After my youngest son was killed in 1971 and perhaps eight months later I found escape from reality and sadness by spending a few hours each morning with thoroughbred horses. I bought a young one from a farmer for a thousand dollars and his name was Fid's Star. Managed to win two races, which is about average for cheapos – it paid the expenses but more fun than a barrel of monkeys. Now, I never rode a horse in my life and consider it dangerous, particularly nervous racehorses. This one morning I got disgusted waiting for an exercise boy to take Sid as I called him out to the track for some exercise. I weighed about 170 pounds, which is too much, not harmful to a horse but a maximum of 140 pounds is the practical limit. Very nervous I climbed aboard Sid but after three steps he kept turning his head and with one eyeball twisted to the left, I knew he was saying, "What are you doing up there?" I gave him loose reins all the way to the track and normally thoroughbreds are obedient and would not crane their necks around to look at what they classify as the boss sitting up there. But Fid kept giving me that look all the way around the track, even in a slow canter – he didn't like it and told me so. It was the first and last time I rode a horse.

I have to break up depressing subjects with some humor. We don't have a wild environment here but talk about

Laughter when some 85 year old lady, enough money to go to Dunkin Donuts in her purse was confronted by a teenager threatening her and reaching for her remaining social security money. Discouraged after getting swatted four or five times the youth took off. What a demonstration!! Must have known all about calculating how close the target has to be.

We forget many experiences as children but traumatic events stay fixed in our minds. Mine is more symbolically oriented, chafing at schools of thought reminiscent of days of old. I did not seek a tear-jerking story but certain political inclinations trouble me. Today Republicans want to increase funding the war in Kosovo from Clinton's requested six billion to thirteen billion – something like that. You can bet on a repeat performance of previous years. Your social security fund is going to be dipped into, as if they give a care about your interest. I feel mean, take a look at your chosen leaders and manipulators on the next page, although some are probably back in business of lawyering since their unsavory wish almost reached fruition back in 1994.

...of the prevailing winds that carry urban pollution up from the big cities of the East Coast and east from the smokestacks of the Ohio Valley. Sticking up in this foul

After reading the article two pages back or whatever I got a little bit riled up and discovered the excerpt snippet I wanted to add along with another section had been overlooked. Here it is. Maine and Vermont, two beautiful states and let me tell you, anytime of the year, whether you ski, like autumn foliage or indulge in lobster in some village at the edge of the Atlantic, that's where it's all at. Now, scientists discover benzene and mercury permeates the air. Did you know how concentrated are the "Foul Toxin Producers" in the Ohio Valley and western Pennsylvania and why prevailing southwest winds deliver it here in the northeast quadrant, the most populated part of America? I'll complete the last

Robert F. Bennett, R-Utah
Jeff Bingaman, D-N.M.
Christopher Bond, R-Mo.
David L. Boren, D-Okla.
John Breaux, D-La. ◆
Hank Brown, R-Colo.
Richard Bryan, D-Nev. ◆
Conrad Burns, R-Mont.
Ben N. Campbell, D-Colo.
John Chafee, R-R.I. ◆
Dan Coats, R-Ind.
Thad Cochran, R-Miss.
William Cohen, R-Maine
Paul Coverdell, R-Ga.
Larry Craig, R-Idaho
Alfonse D'Amato, R-N.Y.
John Danforth, R-Mo.
Tom Daschle, D-S.D. ◆
Dennis DeConcini, D-Ariz.
Bob Dole, R-Kan.
Pete Domenici, R-N.M.
Byron Dorgan, D-N.D. ◆
Dave Durenberger, R-Minn.
James Exon, D-Neb.
Lauch Faircloth, R-N.C.
Dianne Feinstein, D-Calif. ◆
Wendell Ford, D-Ky. ◆
Slade Gorton, R-Wash. ◆

Bob Graham, D-Fla.
Phil Gramm, R-Texas
Charles Grassley, R-Iowa
Judd Gregg, R-N.H.
Orrin Hatch, R-Utah
Howell Heflin, D-Ala.
Jesse Helms, R-N.C.
Ernest Hollings, D-S.C.
Kay Hutchison, R-Texas
James Jeffords, R-Vt. ◆
Dirk Kempthorne, R-Idaho
Herb Kohl, D-Wis. ◆
Trent Lott, R-Miss.
Richard Lugar, R-Ind.
Connie Mack, R-Fla.
John McCain, R-Ariz.
Mitch McConnell, R-Ky.
Carol Moseley-Braun, D-Ill. ◆
Frank Murkowski, R-Alaska
Don Nickles, R-Okla.
Sam Nunn, D-Ga.
Bob Packwood, R-Ore.
Larry Pressler, R-S.D.
Charles Robb, D-Va.
William Roth, R-Del. ◆
Jim Sasser, D-Tenn. ◆

Daniel K. Akaka, D-Hawaii
Max Baucus, D-Mont.
Joseph R. Biden, D-Del.
Barbara Boxer, D-Calif.
Bill Bradley, D-N.J.
Dale Bumpers, D-Ark.
Robert Byrd, D-W.Va.
Kent Conrad, D-N.D.
Christopher Dodd, D-Conn.
Russell Feingold, D-Wis.
John Glenn, D-Ohio
Tom Harkin, D-Iowa
Mark Hatfield, R-Ore.
Daniel Inouye, D-Hawaii
Bennett Johnston, D-La.
Nancy Kassebaum, R-Kan.
Edward Kennedy, D-Mass.
Robert Kerrey, D-Neb.
John Kerry, D-Mass.
Frank Lautenberg, D-N.J.
Patrick Leahy, D-Vt.
Carl Levin, D-Mich.
Joseph Lieberman, D-Conn.
Harlan Mathews, D-Tenn.
H. Metzenbaum, D-Ohio
Barbara Mikulski, D-Md.
George Mitchell, D-Maine
Daniel P. Moynihan, D-N.Y.
Patty Murray, D-Wash.
Claiborne Pell, D-R.I.
David Pryor, D-Ark.
Harry Reid, D-Nev.
Donald Riegle, D-Mich.
J. D. Rockefeller IV, D-W.Va.
Paul Sarbanes, D-Md.
Ted Stevens, R-Alaska
Paul Wellstone, D-Minn.

sentence after the word "foul" since I unintentionally chopped it off. "Sticking up in this foul stream, the park's (Arcadia of Maine) 1800 foot Cadillac Mountain acts as a kind of smog magnet." The writer meant well but it's not a magnet affair but rather a spot representing a vast area of something airborne.

Below on your left, senators who voted for Paul Simon's Balance Budget Act pushed in March 1994. Design: Scoop up all your Social Security money, truck loads of it and dump it into the General Fund. Out of those 54 senators approving 13 with the emblem after their names <u>did</u> <u>want</u> <u>to</u> <u>exclude</u> Social Security money and other trust fund programs from being sucked up. On your right the 37 senators who voted against Paul Simon's pride and joy. It would have been joyful! If you don't think I'm a pretty good artist here is a picture of Simon.

Despite what may appear as one wholesale effort on my part to comment on everything I still am maintaining some written contiguous arrangement between personal experiences, feelings and attitudes and the principal theme of this book. I have good reason to feel a bit of apprehension over future economics and if I may, let me draw an analogy as to how the incomplete man, brilliant on one hand but oblivious to nature's presence as an unbridled force can result in, well, what better word than crash. Mr. Zeppelin upon having constructed the first gas-inflated ship sometime before World War I, did not have, apparently, any knowledge of what goes on in the sky above. Spectators, their necks craning and mouths open in wonderment looking up at one glorious zeppelin move higher and higher were in for quite a show. Somewhere around 1500 feet altitude (didn't get far) that wonderful silver banana in the sky suddenly lurched sideways, blown off course and noiselessly piled into a hill. Rapidly disinflating, taking but a few seconds, there remained but a sorry looking mess of fabric and protruding metal ribs. I suppose that there were some crunching sounds. From "Oooh" to "Oh No" as in 1929, except that year it took one day while insiders, justifying everything they did as an amoral right, rejoiced over their immediate successful short-selling ventures.

What do I think of hybrid gasoline-electric vehicles? Wonderful except you are swapping sources of emissions from tail pipes to smokestacks. Unless we rid ourselves of coal-fired power plants then the problem remains unsolved. Evidently a fundamental rule of physics "ENERGY CANNOT BE CREATED OR DESTROYED" is not perceived exactly as it should be. For every X amount of energy that plant has to provide for that vehicle to motor along then the power plant must correspondingly consume more coal or the equivalent X amount of energy needed by an electric motor under a hood. On the assumption that it will take some doing to persuade the American public to consider divorcing themselves from attachments to monster vehicles, we can therefore, if hybrids come along, count or look forward to more mercury and sulphur dioxide in the air but much less benzene and nitrogen oxides. The irony of this, the illusion of less smog and therefore clearer skies will delight spin-doctors since very poisonous substances are invisible. But, fixed permanently unless we go to nuclear power or natural substitutes, carbon dioxide as an irreversible factor will continue registering higher density values in the atmosphere year by year from data gathering Nimbus 7, no matter what. But Doctor Gloom here may have his predictions fall apart, oh say 15 years from now, if the much talked about fuel-cell engine ever reaches mass production. But, an enormous problem alas – what will the rest of the world be doing?

LATE MARCH TO APRIL 1999

If sulphur dioxide affects the ozone layer negatively I don't believe the sun's rays will be altered as they should and the baking of tropical zones will continue. My youngest daughter and her family just returned from an eight day cruise from Florida to making the rounds of the Bahamas and some points south of there. Visiting us as she frequently does, her sporting a red nose and cheek bones, arms and legs the same, despite her not being a sun bather, quite naturally prompted me to ask, "How was it down there, lots of sunshine?" "Like Godzilla!" she replied, "No let-up the whole eight days!" Did Godzilla in Japanese movies belch fire or was it his arch enemy the dinosaur imitation? Heck, I don't know. If it's already 90 degrees in southern Texas what is it going to be in July? No wonder Governor Bush is looking for a job in Washington, DC!

April 29, a late news bulletin: Honda has abandoned making electric cars and angry environmentalists express disappointment. For once this may have been a God-sent development for the better and regrettably, clear air advocates so upset were not thinking analytically if they thought electric cars would alleviate the problem of certain air borne toxins – not as long as electricity is provided by coal burning plants. In regards to Maine and Vermont's problem, something to ponder, the reality now of existing benzene and mercury will not hit home with indolently inclined citizenry who concentrate more on baseball statistics, "What's for supper?" and TV tripe. Furthermore, regional faithful allegiance to one political party typical of northern New England while disdaining an aspiring vote-seeker of the opposite party who may offer something good, guarantees that "Long walk on a short Pier" will go on. Toxins are both insidious in their effects and invisible. What unconscious types need is to see tottering living skeletons with decaying skin as I experience, whose gaunt faces, sunken eyes and oxygen tubes in noses, portend death is just around the corner. Eventually with my "Why" mentality cultivated at an early age after tragedy affected my family in the 30's, motivating me to first study the history of Wall Street shenanigans. I invariably, but gracefully please, find out if the human wreck before my eyes smoked and "Do you take vitamins?" or such other questions to which they respond well, sometimes smiling, knowing from frequent contact that I am sincere. Predominantly they have always lived in the city and yes, not all were smokers. We fuss over a few dead in international incidents which is the moral thing to do but leadership count dollar bills first while turning their backs at a growing scourge that will slowly kill thousands. More population growth means more air conditioning, bigger TV sets and whatever else. Of course bigger cars also. And so, more poisons drifting over the most populated section of America, the northeast quadrant. Medical associations and their tepid statements "Respiratory problems are on the rise" tire me. What does that imply? Coughing, sneezing, drippy noses, watery eyes? Couldn't they really lay it on? Asthma deaths? A little stronger, true, but the average person probably thinks of this as a genetic predisposition when in many cases it was the healthy child but not gifted with an iron constitution who ran about in a crowded neighborhood and passing vehicle after vehicle added to what is being shipped in from regions not too far west.

My opinion only: A good movie is enhanced with touches of the outdoors. I've not taken literary license for the sake of dramatic effects. How can anyone be infatuated over movies specializing in some California beach boy crashing through glass windows, shooting down at least four enemies while dodging a barrage of bullets and then rescuing some terrified girlfriend, directors making sure her silicone breasts are amply showing accentuated by exaggerated breathing. I like an occasion movie, sure. Two of my favorites, quite old now, were "The African Queen" and "The Black Stallion", the latter praised for it's outstanding photography. How many people know how that horse obeyed over 100 commands, which he heard through a radio, imbedded in his mane, transmitted from a trainer standing on a nearby hill? Overseas we had movies practically every night – I saw two. I'm not a dullard however. If folks want to sit on their cans trying to pluck substantive living out of a cold computer screen that's their prerogative. I preferred doing what a song "A horse with no name" has to say and I still lift every stone in life just to see what is underneath it. Just for taking a peek back in time as young teenagers we managed to make a working radio out of a piece of crystal. I don't remember how it was done but it awakened us to something mysterious since how did a voice come out of what amounts to a rock. But to take everything for granted and not take advantage of a hands-on feel and experience wonderment is not for me.

If I write about myself it is not an ego-trip whatsoever. I want to interject light-heartedness, sell the nature-boy and sentimental side of me also. Can't leave the reader feeling some crotchety old duffer is prone to bitching and putting a fly in one's soup out of some need to antagonize. Aha! You thought I was going off on another unrelated bit but flies, Oh Lord, you couldn't sit in any restaurant in Morocco without cupping your hand over a glass, otherwise the damn things would hurtle themselves by the dozens and parade inside dampened glasses frantically seeking moisture. One guy on the next table, probably dulled by cognac, lifted his glass of something but stopped just short of swallowing these pests. He got up after slamming his glass down and walked out. But Aha! This has to do with heated climate and the "Aha" inspired by America's chief advertising agent assuring all radio listeners conservatives are Godly and everyone else is stupid and even evil. A classical example of limited thinking, probably the genetic types who would condemn suspected witches in Salem, my friend Russ Limbaugh leaped to the fore with the following, April 30th: "Whacko environmentalists thought deformed frogs were affected by climate change but scientists have found the problem is parasitic worms…Aha!" Well, here's my Aha! The more persistent a warming trend the more <u>proliferation</u> of strange new microbes and insects and whatever else needs higher temperatures and humidity. Historians, just last week, elaborated on how the Bubonic Plague had it's origin in the Gobi Desert even though native populations nearby were immune but the rats were not. Fleas, just as head lice (Oh boy) thrive in sand. What four-legged long-tailed creatures survived spread southward to seaports and you know the rest. Now, no alarm meant but we know Gobi Desert dust particles are over our western states, maybe further eastward than we know but, another "Aha" for me and you, the west to east air aloft blowing stronger from the Orient. Winds aloft are also blowing stronger from Africa (remember coral reef dying from airborne bacteria) toward our southeast. I have already stressed how blown up high-pressure systems from heat means nearly year round unusually powerful high altitude express shipping – anything can be transported! Thank goodness tsetse (palpalis) blood sucking flies causing fatal sleeping sickness need year round steaming equatorial African climate. Also, if they could hitch a ride at 10,000 feet, destination Florida, they would die from the cold. Did you know there are poisonous weeds in that state fatal to horses and cattle? Good thing we have intrusive government. I constantly hear bean-brained radio speakers say, "We can take care of ourselves – we don't need government." Lord almighty I can't imagine relying on self- destruct prone citizens for my safe existence! God No!!!

Here is something bizarre that I was witness to in Miami, March 1972. If you don't believe me try contacting libraries there or maybe the internet can help. It was a cloudy day and hundreds of small parrot like birds from South America plummeted straight down to their deaths in the streets of that city. Were they disoriented, lost and then too exhausted to go on? I really don't know. Probably got a good dose of toxins. Why did they fall like hailstones if not unconscious? I left there a week later and never found out. I'm a pain but four months of a pleasant winter results in the usual frame of mind "Everything is fine" and all is forgotten from last year. I again have to quote a Maine scientist: "They think we're nuts!"

You can't discount the ominous if you add everything together! Yet, I didn't say the end of the world is bound to happen – maybe some unintentional self-inflicted means of population control is gathering momentum. Gosh, when I was a child, polio crippled children I knew. Doctors then had limits. One classmate of mine in grammar school, Lucille Cardin, had an adjustable chrome-plated turnbuckle integral with two long threaded shafts, one on each side of her shinbone. I hated to see it. Two pins, part of that rig went through the bone, just below the knee and the other above the ankle – the idea being a doctor periodically would adjust tension so as to stretch the bone, since polio had shortened it by two inches. But Lucille was always a very cheerful girl – amazing when I think about it. One day she turned around and gave me a resounding slap since she sat in front of me. It was an accident – my pen and ink, typical of those days, had jabbed her in the back but it resulted in a loud female "Ouch!" followed by a noisy "Whap." Truly embarrassing particularly when the nun teacher briskly moving toward me had to say for everyone to hear, "Well! What is this about?" Good thing she didn't have a ruler in her hand – that would have hurt. Lucille would laugh with me later over this embarrassing show, especially with 15 other kids looking on. Pathetically, nothing worked and it troubled me to see her peculiar gait. I don't know what life had to offer her. I left home after Pearl Harbor and did not return there except for a few days visiting my mother. "Ah those were the days my friend, we thought they would never end." Like hell. I'm glad they did.

I have to go back to weather. It's a lot more weighty a dilemma than some speculative estimate of rising ocean levels. Yet, I didn't go overboard with some wild forecast. The end of the world is not around the corner. There is no need to foster anxiety but a lot more concern certainly would help. Increasing disease rates in various parts of the world doesn't mean we are exempt. But, we're smarter scientifically and that's comforting. Right now I'm trying something new. By keeping a trace, day by day, of high and low (24 hour) temperatures of Caribbean sites for months I can estimate increasing carbon-dioxide or it remaining fixed in density – perhaps even dropping. Minimum temps occur before dawn or thereabouts and if they get higher than the norm for awhile and max temps stay roughly the same it means that there is something preventing cooling.   I wish the Nobel Prize committee would give me a break. Nothing sensational in that approach but it doesn't require 500 scientists and satellite zooming around. Good thing all of them barreling around up there don't need gasoline.

Do you know what the "Old Man and the Sea" movie means to me? Those sharks ripping off pieces of a giant tuna the old boy wanted so badly to bring home are, figuratively speaking, plagiarist fakes thriving on someone else's hard earned knowledge. It happened to me in Morocco although it wasn't anything of scientific importance. I can draw well and thought something new in a land we were not that familiar with needed a descriptive touch. Of course I was looking for Brownie points. This unusual low-pressure system, the same one that lingers over the Sahara continuously, drifted westward and into Atlantic waters where it built up in intensity, then came back. The staff wasn't exactly baffled but it did start a little more conversation than usual. And so, while our commander, formerly a banker in New Jersey, much to my distaste, strutted about I found time to describe what had transpired and drew, with great care, the weather pattern involved. I handed it to Lt. Jorgensen and practically forgot about it. About six weeks later the lieutenant with an amused smile, allowed me a sneak peek at a letter from Washington, DC, something to the effect, "Thank you Commander ___, etc..." But I want to be fair, they did push for my taking an examination and I obtained another stripe on my sleeve and about 15 more dollars a month, what the hell. Just one time though, the commander, happy at his having penetrated the inner sanctum of operations officers, holding on to their military secrets, trotted in to our office with a stack of photos. Even I and other lowly ones got treated as if we were worthy of sharing the ultimate with him. A trapped surfaced submarine, fighting off several small Canadian warships called corvettes while our plane, having accomplished something with it's depth charges, now flew round and round with one photographer clicking away. Time for photos – traditionally practiced by all branches of the military. Then a photo revealed the bow of one corvette elevated much higher; a mass of white wake behind the stern and obviously under full throttle just rammed and tipped that submarine over – its crew jumping into the water. The banker commander was exhilarated and kept saying, "Look at this one!" He even smiled at me. Did I like him then? Of course. But the next day, same old Lord of the Flies. I did write of German prisoners before – these were the same Krauts who had to take a dunking. Insofar as the commander, certainly I am condescending now. Nobody but nobody in the Navy and I'm proud of it's military elite and traditions, ever looked down upon me, except this one faker.

No Viking am I. I can mull over in my mind scenes of the past reliving in my private reactions over something that is beautiful, sentimental or perhaps ugly. One day, while driving between Buffalo and Rochester, New York, part of my territory for five years, I saw a doe running with an arrow deep in her rear. I have read where archers claim deer die of shock when hit with an arrow. That's crap. They die either slowly from bleeding internally or days later from infection. I can't print my views of these louts and their idea of sport. But, when you feel sorry for fish that's too much. If my oldest son is an ardent sports fisherman and belongs to two ecology-oriented societies it is because of my showing him how to catch speckled trout in Louisiana. Sticking a barbed hook through a live shrimp for bait bothered me. I found out fish have a sense of pain – sounds crazy doesn't it. My fishing interest started to fizzle one day. Silly story but that's me. There is a salt water fish in Louisiana called a Croaker. When I started to tear the hook out of one I had caught it's tail thrashed while making the damnedest sounds – that turned me off and I threw it back in. Perhaps a month later I took the family to Biloxi, Mississippi and paid some craggy featured diesel skiff owner forty dollars to take us out trolling in the Gulf. For one hour I sat holding on to a sturdy rod while longingly looking some distance behind the stern for some action. Nothing. I called my youngest son, then aged nine and told him to take over. Sure enough five minutes later I heard the whirring noise of a spinning reel – something big out there was running away with yards of line. The captain of our ship (he didn't look like one) jumped forward and growled, "You've got to set the drag!" and did so, obviously annoyed. I tapped my son on the shoulder and told him I had to take over. Another male for the world to put up with – "No, it's my fish" and valiantly struggled with the crank. Fifteen minutes later there lay a gorgeously colored King Mackerel, all 2 ½ feet of shimmering cobalt. I wanted to throw it back in – couldn't keep something like that in a motel and then drive several hundred miles back home but the skipper wanted it. Woe is me, the beautiful thing had to show me it's final moments – from sparkling cobalt it gradually turned a dull gray. Couldn't it have stayed the same color, good grief? On the subject of color, did you know the Sargasso Sea is the only large area having waters of that color and no artist can duplicate it?

If I write these petty nothings I want you to know I'm very human, very curious but critical too. And a plus, the escape from gloom it offers. Also, without the usual smugness accompanying such a boast, I've been around.

How would you like to have 600 dollars in your pocket, now worth about 2500 and there I was free as a bird for three days in New York City, August 1944? Imagine devouring wieners, you know the ones with lots of grated onions – that's class, isn't it. When two other Navy guys left Port Lyautey in a big four engined amphibian plane, we looked kind of tense, sliding down the Sebu River and wondering if the hull would unglue itself from the water – engines roaring for all their worth and a rather short availability at that. It had a bend. What a feeling this ship gave us. Suddenly, it squirts free and you feel a slight surge forward and a nice sense of relief. My first experience and no doubt my last. Anyway, this working mate of mine, a new Yorker by the name of Mazzarella, talked all the way about a restaurant he was going to take us to – steak was his obsession and mine was quite different. So we ate steak and the darn fool warned me not to use ketchup, even though I had no intentions to, "It will upset the cook" he said. Sorry, I'm not a gourmet and I was glad to go off on my own after that. It's amazing how so many New Yorkers in those days thought that anyone who didn't live there was a lost soul. Nevertheless that city was then a marvel to me. What follows has something to do with 5th Avenue even though I have to again mention a song.

"L'AMOUR EST UN OISEAU REBELLE QUE NUL PEUT APPRIVOISER – PRET GUARDE A TOI!"

I would balk if my wife played the radio for hours listening to Oldies. I think she allots herself but two hours but suspiciously, it's usually when I'm typing. However, while not being drawn endlessly to melodies some songs inspire a few more pages of "Yesterday". At age 16 I became very fond of a girl who had graduated with me from elementary school. Attractive but not exceptionally pretty she drew me to her like a magnet – her personality appealed to me. What are you supposed to say about puppy love? This is another simple story and while most certainly not the equal of "From Here to Eternity", that being fiction anyway and besides I didn't do the Burt Lancaster thing on the beach, it is mine nevertheless. What you must understand first, it's amazing how this world can be small. I want to drop a little impact over the subject of tornadoes to end this book but the radio derails my thinking processes – some guy was singing, "She's sweet sixteen, she's beautiful and she's mine" and that started me with the above. I thought my school girlfriend was more than a pal. Four of us, 20 cents to my name, holding hands would walk a mile and buy sandwiches in a Chinese restaurant. Even though girls were supervised like Mona Lisa's painting in a museum we managed to play Spin the Bottle in the other girl's home garage. Things were so disciplined in those days – only one girl I was familiar with got pregnant and she hid for two years. My chickadee really sizzled the wrong brain cells in that garage. But boom, some guy rode off into the sunset with her. He had a car and in 1939 that was the ultimate achievement. I got over it. What a beautiful city New York was to me in 1944. Crowds, everything open, walk all night, music, bright lights – I loved it. Population? What? Eight million? How could this be, walking toward me, all dressed up in a blue Wave uniform, looking a spruced up maturity, very handsome, my lost Laura of 1939 – five years had gone by. I mean the Laura of a song. Her name was Reina. I think I looked a lot better than five years before. All that money in my pocket and so off we went that night to Jim Brady's Diamond Horseshoe. If you didn't throw a think stack of dollar bills down on the table the waiters never came back and drink or eat you must or otherwise some Gorilla with a black bow tie would be lurking nearby and discreetly ask for your table. We danced my Mona Lisa and I – pretty words flowed along with more mixed drinks. The dance floor was very crowded and while rear ends bumped graciously we were so glued together there was no roguish glances at other passing eyes. I wasn't use to booze after two years overseas. After untwining ourselves it was time to go. She had to get back to some military housing of sorts and so I walked her to the subway. The conversation got serious. We stopped. I got callous. I told her I was on the verge of becoming engaged and she started crying. Maybe it was the booze or maybe a subconscious wish for revenge, I don't really know. Well, if you listen to Oldies as I have to listen to, this song will be repeated a few times – such a sweet voice too! I lie not, it's how I felt and the radio just had to, "I'm sorry, I'm sorry, I didn't mean to - - -!" But three days in that city started fine and the first night was kind of noisy. I first went to a big USO complex, attracted there by posters of Xavier Cougat whose band was scheduled that night. He had this South American band, you know, rhumbas and all that and of course his leading lady was his wife. I don't know how many times he married but they were the epitome of what you expect of every Mardi Gras of the world – Ole, Ole!! It's old I know, but if you have ever seen the Conchita Banana Girl making commercials you would know what I mean by gyrating over-decorated show girls. Outlandish head gear, huge double ring earrings, tight dresses and hips that you would not see on lady marathoners, constantly swiveling them around whether they were standing, waiting to be introduced or singing, or whatever. Xavier Cougat the orchestra man looked like the proverbial Latin from Manhattan except he probably was from Argentina. Sporting a long thin mustache, modest sideburns and slick black hair he grinned constantly probably very satisfied or perhaps knowing he had more to look at than the audience did. His only love, luxuriously attired, a dozen bracelets tinkling away, preposterously long earrings, inch long eyelashes, bounced around singing Ha-cha-cha numbers or whatever. She suddenly announced that we would have to form a Conga Line. That was that. It seems I was sticking out as a loner. There were quite a few soldiers and most were accompanied by girls – I suppose most couples there were New Yorkers. "Hyper Hellion" just didn't invite we GI's – she darted toward us, grabbed arms and hauled us out one by one onto the dance floor, told me to get behind her, hold on to her waist which proved hazardous later, such what was more animated a number than I imagined, then pointed to some girl, "You Rosita! You hold heem like dees." And so on until a dozen couples were hooked together. "How do you Conga?" That worried me. Dancing without liquid false courage and an audience to boot – I felt what is meant by stage fright. Bongo, bongo went the beat, trumpets blared while her hips went zing to the left and zing to the right. Basically, this silly dance should bear some semblance to a sidewinder snake moving along – there was first, hesitation, then stamp a foot down but the whole line fumbled along awkwardly, much to everyone's amusement. We got the zigs crossed up with the zags. I didn't meditate on swivel hips – it was imperative to get the hang of it all and not look the bumpkin. If excited gals were Cougat's cup of tea they never were mine. She was pretty no doubt but I wouldn't at that time bring a picture of her home to mother. But I can always tell folks that I danced with Xavier Cougat's wife and really had quite a hold on her!

My first night in New York, so much glittering, beckoning and memories of desert heat, flies, poverty behind me I quickly tired of noisy brassy Latin music, always the same tempo with emphasis on bodies doing gymnastics. I left, walked and walked, gathering in what seemed so beautiful. It was August and everyone seemed so carefree. My eyes and ears took it all in – I didn't feel the occasional sense of loneliness of previous years. "I like New York in June, how about you." I always did. Had to throw that very oldie in there!

Years later I took my family to the Bronx Zoo and a museum with huge dinosaur skeletons that my children gaped at – the vastness of the zoo requires all day to see. Took in the Triple Crown also at Belmont. Talking my frugal wife (thank God she's that way) to bet 30 dollars on Affirm and I splurged a lot more. Always wager on trainers in major races. She's not a gambler and my expedition into such halls was spread out over the years. When Allydar put his neck ahead of our horse at the head of the stretch my wife practically collapsed, sat down and said "Oh no" while 50,000 people including myself were standing beckoning the jockeys to do more. We won but I never went again – it took an eternity to get out of the parking lot. One day, at Saratoga, she ripped a ticket in half while giving me a hurt look, you know "I thought you picked the right horse!" The horse had come in second, 30 dollars from her, and 60 from a son with his wife and my credibility shot to hell. Hotter than blazes at that ancient track stuck in an airless valley. On top of that try to get a seat in the grandstand. So we had to walk through hundreds of bodies and ended up somewhere in the woods with a TV screen hanging from a tree – no way could we see the race. Well, my son ripped his ticket and just dropped it in the grass. I had told my wife not to throw her two pieces away. The place was steaming hot and worse, four idiotic young folks sitting in front of us, giggling like loonies, were smoking pot – the odor was stifling. On the way to Saratoga, bucking traffic I kept boasting of this horse, Great Contractor and how, "Don't worry, he'll be last most of the race but he'll win" and "You'll see!" The TV screen could only show the first four or five horses and as the end of the race approached my son kept raising his voice, "Where is he, where is he?" Here he comes, second. Doom. Broken hearts and people throwing tickets around. So darkened spirit-wise, we started to push our way out and the most beautiful sound came over the loudspeakers, "There is a steward's inquiry." Not a jockey objection – big difference. My son, Bob, jumped two feet in the air when it was announce dour horse had won. But, what a sight to see, on his knees looking for his ticket pieces. Others were humped over barrels fishing around to salvage what they had thrown away. Well Bob and my wife had to stand in a long special line where the tellers tape your ripped tickets and that was cause for much laughter on the way to our car. If I tell you these pleasant but silly incidents it's to advertise myself as not being some rigid non-life type, with four power glasses, studying all the time.

What do I recall about different cities in WWII where I was able to surmise civilian attitudes toward the military? Including America? Indifference mainly, no real friendly hellos anywhere except Belfast, Ireland. I boarded a bus and some 25 Irish folks, each and everyone greeted me – quite disconcerting nevertheless wondering how to respond to some many strangers, all beaming at you. Did they like Americans or was this some sort of custom – I really don't know.

In the past year I have accumulated a two inch thick stack of newspaper clippings but for the past three months the quantity has been surprisingly high but very unpleasant – not an iota of good news. It has emboldened me to continue on, hammering away at what is gross negligence by so many. I wanted to end this book on a technical note for you to totally understand what tornadoes are made of and why some have reached enormous proportions. Just think, some guy on TV some months ago said, "If we knew what causes tornadoes we could save lives and property" remember? Well, let them, should experts read this, charge me with heresy or even call me some fly by night maverick. I'm going to put you, the reader in the position of having an insight you never thought of. But judge the material, not me – I did ask that in the beginning. If you like frills in English composition I'm not your boy. "Just the facts, Ma'am" is more substantive in the long run. Why waste the reader's time.

First, the last few news items. The six medical experts in Malaysia are troubled. Ten Singapore slaughterhouse workers have died from encephalitis after handling pigs from Malaysia. The problem: How to stop it's spread. How can they actually, warmer climate and more humidity means more mosquitoes. It's one of the multitudes of side effects from earth warming. Now, a client tells me he heard some person on the radio say that the earth had changed it's axis tilt by one degree, therefore more sun beaming down and so forth. That's a beauty of a new spin! Naturally it tosses earth warming by pollution, NO2 and a wrecked ozone layer into the wastebasket. The hand of God has grasped the globe – we're doomed! The worse the weather becomes destructive the worse kind of showboating is to be bestowed upon us.

May 3, 1999: A shattering experience – tornadoes, one or two reputedly a mile wide, 43 dead so far according to CNN and 3000 homes estimated destroyed. For years and years we have had tornadoes but most passed by cities except for the last few years. What with rapid expansion of urbanization the odds are changing. Oklahoma City all the way to Wichita one twister after another. Where are the El Nino buffs? A moderate typhoon skirted Japan last week and some guy is blaming La Nina, the cold one. El Nino has been gone almost a year. Supposedly the cold Pacific phenomenon would chill America. Did it cool off warm soggy air fueling all those tornadoes? Give me a break. Texas' tail end a scorching 90 degrees and some points just south of there in Mexico close to 110 degrees. I'm a thermometer man, not sticking my head up in the stratosphere looking for the impossible E=MC2 answer.

Here we go again – time for school. Tough for any Rhode Islanders to get interested in this but your food, as one example, comes from areas that play an important role in your life, and they, the providers should begin to get more interested – in fact, better they start doing some serious inquiring. I see insurance adjusters in Oklahoma on TV. Can you imagine what will happen to insurance rates in the near future? And the cost to the government?

We have been lucky so far, this country plagued by tornadoes as it is. Of course, I'm referring to periods before May 3, 1999. I find it incomprehensible the lack of preparedness out west or even closer east. Why all those homes in Oklahoma just tacked down on a slab of cement and no cellars? Cost no doubt. Couldn't they at least dig a hole in the backyard and lid it down? That's not a crazy idea. Never mind running to a shelter, if there is one – some monster twister will make short work of that. Something coming at you somewhere between 40 to 70 miles per hour doesn't give you much time.

When are meteorologists to take a good look at what has been pumped into their heads in college that a cold high is master and that it pushes such and such a low and front along when it is always the warm high that it is bucking that controls that. I'm leading up to big fast moving tornadoes lasting very long as compared with the smaller ones, although vicious nevertheless, usually spawned by frontal thunderstorms or even isolated ones. That's a situation where a tremendous amount, a very dense amount of ice crystals are thrown out of the tops of thunderheads and chill the air, so much so that as you already understand, creates extremes in instability. But why the funnel? Mind if I play the teacher now? First, it is a myth that tornadoes dip down but for brevity's sake I've always referred to that dramatic action as doing that. Follow me please. First, one more jab at the errant ones. I heard this weatherman say, referring to a low pressure system off the coast of New Jersey, which has kept our weather pleasantly drippy (I dislike too much sun) as, well, here is what he said: "There's nothing west of that low to boot it out to sea." And out there in the Atlantic one fat Bermuda high, blocking any such move on the part of that cooler high from northeast Canada to roughly the Appalachians. First, understand the process from tornadic development at high levels to the illusionary stage of a dark funnel dropping down takes but a second, more or less but my explanation is one rather lengthy detail. It has to be.

Before you think I'm a maverick on the loose let me treat the subject systematically – step by step. It's a habit I have, to go for precision but it annoys people. I'm afraid the Navy insisted on it and at an age where good or bad habits are formed. But, don't expect anything of the sort from me in a multitude of other subjects – I'm a big zero, a goose egg in math, geometry, calculus and so on.

Air is very elastic – compress it and it's temperature rises – molecular proximity excitation they call it. Reduce it's density by one means or another and it's temperature falls – less molecules and further apart and of course, less excitation.

The most violent and massive tornadoes occur when a warm fueling high begins to change from a slow moving resisting force up against an advancing cold front to a much faster east or northeast movement. Then, the cold front at high levels moves faster in unison but has limits. Remember, a string of high pressure systems from the west coast all the way east to well past Bermuda are but inverted bowls nudging each other eastward. Sometimes one, for example, it's center over Tennessee, is subject to changes that weaken it and contracts it. Worse, being then smaller, may scoot eastward over the Atlantic seaboard. The Jarrell, Texas tornado or tornadoes, killing over 25, was typical of a cold front, <u>suddenly</u> advancing, but not so much on the ground because of friction but because it's upper cool content although moving faster could not temporarily catch up with a warm high, deteriorating and

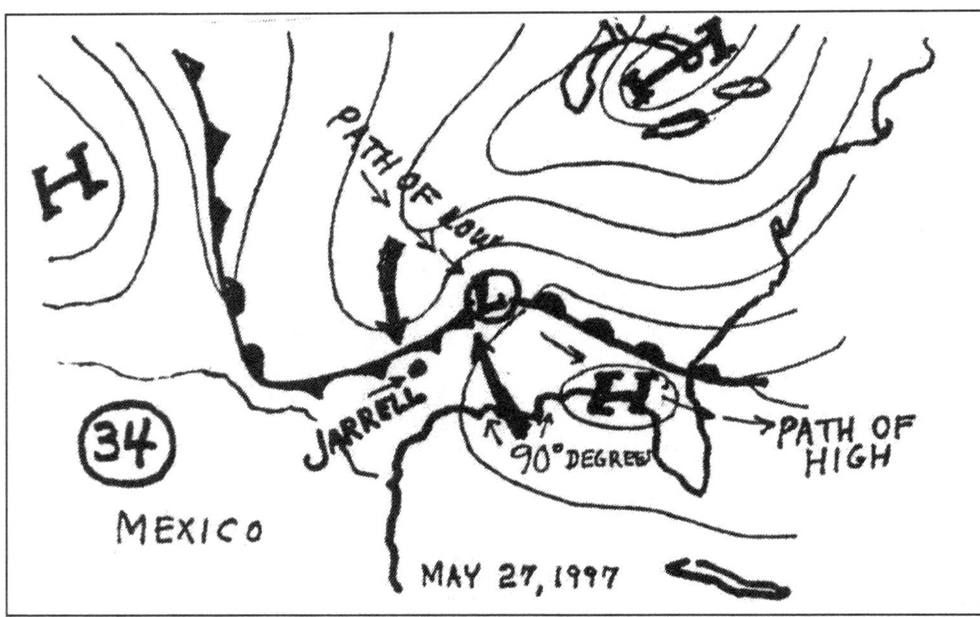

running away, shown in below illustration (34) positioned over Mississippi and points east. This is what the weather map looked like. Just a brief peek at a situation where the high over the southeast began a rapid move shown by arrows. The low followed suit. The surface cold front shown did not but the upper part of it jumped southeastward. But not fast enough.

Just one illustration makes all the inner workings of conditions leading to violent tornadoes hard to follow. For now, just visualize in essence a collapsing system over the southeast and everything northwest of there following suit. The cold high was well up into Canada but furnished only a ridge of it's cold air approaching Jarrell. Since it was moving eastward and not southward the ridge of cold and it's upper level strength was at a minimum. Despite it's contrasting coolness there was not enough muscle behind it higher up and therefore a warm high moving off at an accelerated rate means air pressures at anywhere form 3000 to 20,000 feet (nobody is sure) has a drop – it's momentarily stretched. Let that be until I can describe this terrible change from acceptable stormy weather into a nightmare in more details as I go along. Any New Englander who is indifferent or perhaps about ready to yawn let me remind, much of this section of the USA can also feel and experience such calamities at the rate of weather changes now occurring. I saw and heard a TV tornado show personality, a meteorologist say, "If we knew what causes tornadoes, etc…" no need for me to repeat what I previously wrote. Well, I hope to put the reader in the position of knowing a lot more about them than this gentleman. Naturally, I agree with the age-old explanation frequently handed out by weathermen to the effect that excesses in instability is behind tornadoes and of course hurricanes but from where I stand, some 55 years of doing a little thinking and the time to do it, it takes a hell of a lot more than that to create huge tornadoes more than a half mile wide with the unusual characteristic of racing along at some 40 to 70 mph some lasting one and two hours. They move in the direction forced upon them by winds aloft and if they suddenly form and race off then it stands to reason something up there in the sky above is also undergoing drastic changes, first in pressure and instantaneously in temperature drops. It takes a lot more than ice crystals scattered widely above thunderheads, a lot more! Temporarily, leave the subject of 90 degree temperatures at the surface such as they were at Jarrell, at rest. That's getting old anyway. For the young and/or novice, an example: Those vapor trails behind airline engines means just that, instantaneous change in pressure, poof, temperature drops to sub-zeros and ice crystals immediately form. Same thing when you release a can of Coke – pssst, a little vapor for a second. Illustration (35) coming up to show three high level zonal changes that can occur very rapidly and/or continue on when a warm high either dissipates and/or runs away. Remember that we are talking about speed of movement and not average moving cold fronts with thunderstorms spawning the usual smaller but just as vicious tornadoes. Speed at high altitudes! That is where all tornadoes start, not on the surface. Zone 3, the warm high is pulling

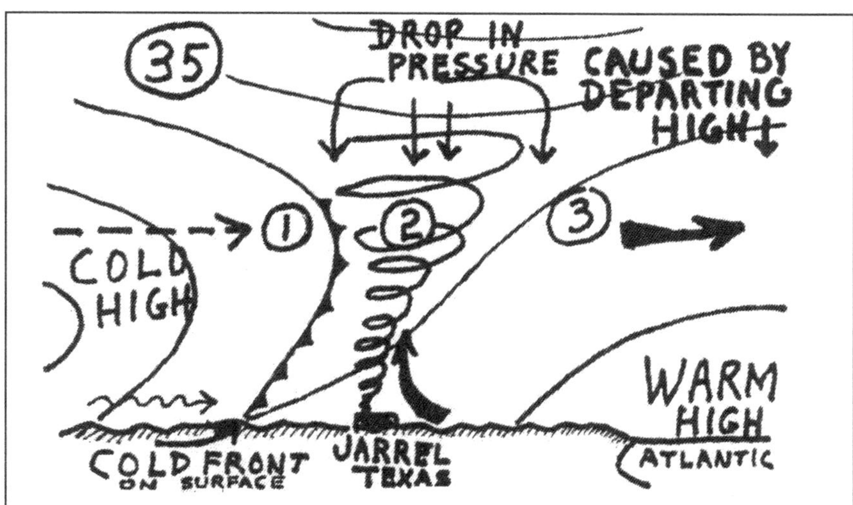

away and zone1, the cooler air quite removed from it's source of power, the high center, doesn't quite have the mass and inertia nor the push to catch up to zone 3, therefore, if I can resort to plain English, a feature not popular in higher circles, the air in zone 2 is thinned out, stretched if you will and therefore a lowering of pressure and of course, a drop in temperature. Looking at a surface map, this drop in pressure at high levels in fast moving systems is also evident in the way isobars show what they call a trough above. So be it, troughs or what have you, that is immaterial to me. The pressure drop is up there and if it's great enough and expansive enough you cannot imagine how wide the tornado cell (as they call it – I prefer Big Vac machine – more punch) truly is, or otherwise the immense width seen of funnels delivering so much suction would not be possible. In short, a funnel, say one half mile wide must have at least an upper level powerhouse (Big Vac) having a diameter at least two miles in diameter – it has to be! I'm sorry if I don't use the word "cell" – it's just too feeble to describe what is so awesome. I think I'll write to the International Meteorological Society and request that they not use that expression – not enough oomph. Trusting this is not viewed as callous, not with 45 deaths this Mary 3rd.

In order to dupe the public and for that matter quite a few elected officials open to anti-environmental propaganda, Carbon Club missionaries have been capitalizing on El Nino's alleged influence, in this case tornadoes over Alabama April 8, 1998. This required a Coup de Force. Call in the elite and infallible – show time! On TV, one of those wild and wooly depictions of violent weather – youthful tornado chasers doing their Yahoo bit, here comes a dapper looking gentleman, smiling most assuredly and of course, he's no ordinary meteorologist – a TV host of course addresses him as Doctor So and So, where from I don't know as there seems to be hundreds of titled weathermen who know secrets of the universe and their word is final, even, like all of us, have their share of missed forecasts. "Doctor, do you feel that El Nino caused those vicious tornadoes in Alabama last month?" "Most assuredly," replies the astute one, adjusting his tie, "We're positive of that." Exit one agent – the scene shifts. In the minds of millions, exit earth warming, exit carbon-dioxide permeation of our atmosphere, exit pollution and exit common sense. No weather map was shown – that would never do and furthermore no one else ever does! Results: El Nino is responsible and there is nothing we can do about it. First of all, there has always been a steady northward flow saturating our southland with heated moisture-laden air after it traverses the Gulf of Mexico. It originates from a vast area of sun-drenched semi-tropical seas since time immemorial reaching maximum temperatures beginning in spring and lasting deep into the fall season. We all know how this hot soggy air fuels tornadoes. When I left New York state in 1960, traveling alone heading for Slidell, Louisiana looking for a home while my family waited for Doctor Adventurer here I had to stop somewhere in the middle of that state. I jumped out of the car feeling I had made a terrible mistake – I couldn't handle such heat – no air conditioning in my fairly new 1958 Plymouth Wagon. What a dog by the way. Even in New York it kept overheating – brand new at that and dealers did not want to do anything about it. Getting back to tornadoes they also are triggered off by invasive cool air from northern regions* such as Canada, naturally – sometimes colder air slipping down across our Northwest after leaving the Aleutian area. The actual map for April 8, 1998 is shown on the next page —clearly it shows exactly what the circulation was doing – quite a common arrangement by the way. The air over our Southwest and California was being pushed away, not bringing in El Nino warmth. The Alabama tornadoes were fed by Gulf and southeast Atlantic air – period!

I am not a propaganda agent. Referring to my book's first batches of pages I have also produced authentic temperature records** going back 29 years, kept in the archives of NOAA, Asheville, NC. You have therefore the temperatures of major southeast sites and two from our southwest regions. Even though bathtub-warm waters of El Nino tucked up against California late 1982 and early 1983 please note it had no unusual effects of temperature averages in the southeast, including far west Las Vegas. However, Phoenix got a lot of rain.

*The same weather system that I drew for you in earlier pages describing what happened to Jarrell, Texas where 25 died, May 27, 1997. The same southeast flow from the Atlantic – warm, muggy becoming oppressively hot over Texas.

**El Nino buffs claim it's warm effects linger for months. A fact: Practically all southeast records herein show temperatures dropped in 1983 – the very year El Nino ravaged California. Me and Jack Webb have something in common except mine has a masculine and military flavor – in essence, "Just the facts, Ma'am."

Asser! C'est fini! My friend, Russ Limbaugh who I listen to for short spells (three hours would kill me) has thrown French expressions about but most amusingly with an accent no less – I suppose this implies sophistication somewhat. Might as well copy his newfound style. Judging from his followers, many from the boonies, they're probably having trouble coping with such newfangled expressions. Lord, the statements they make over the phone - God help us if they ever rule. Voila. The other day Mr. Limbaugh said, "Some folks are writing in complaining they do not understand the big words I use." Quite the followers.

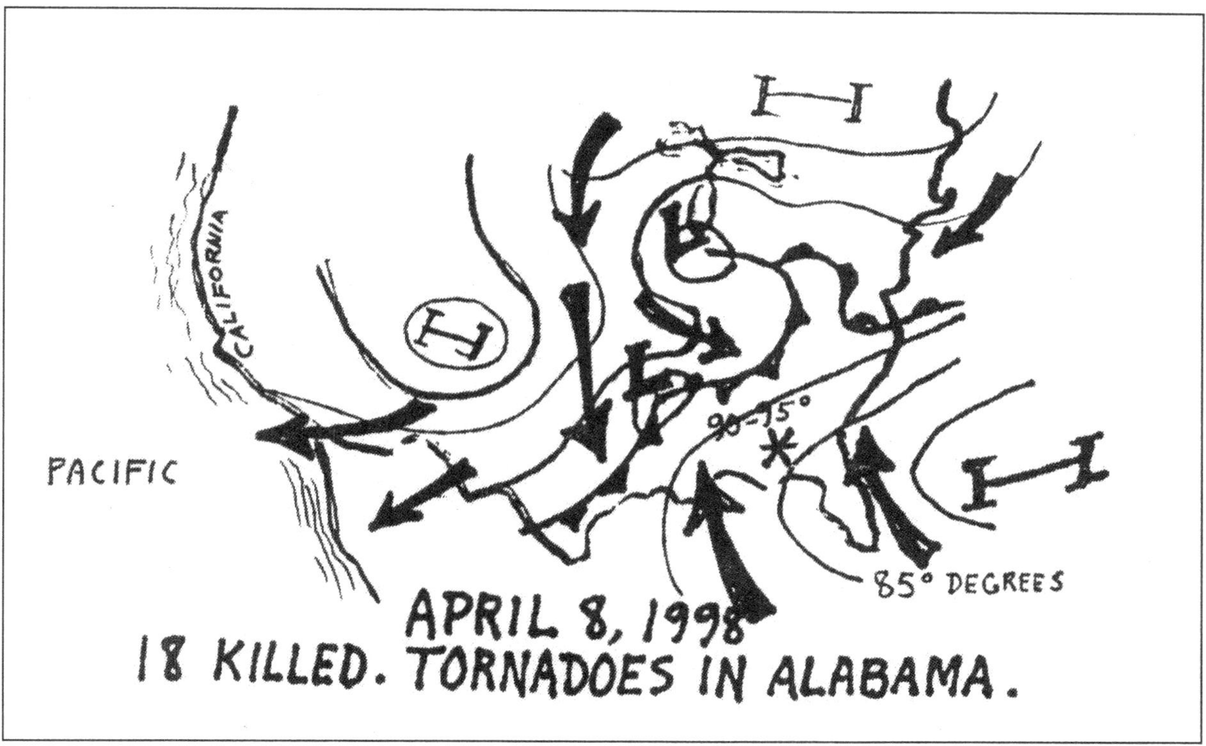

My radio – 11 PM June 3, 1999

Oh boy! Something New! Something Scary! And…A connection with weather – decisions and fatigue factor.

There is a new book out, "BrokenWings" by an airline attendant, Natamya Anderson, claiming that pilots are tired, or dollars before safety. Jim Bohanan interviewed her. Pilots called in and agreed. Jim Bo and others felt that the government should be firmer in view of June 1st Arkansas crash, etc… Some man called and said that the government shouldn't have anything to do with any business. Jim Bo called him, "Brain dead!" and

hung up. He then justified his curt reaction by saying, "Profit is fine but it is evil if at the exclusivity of safety." Airlines are squeezing pilots to work 14 hours and so on. Safety factors in planes have decreased.

More seats have been installed and not enough emergency doors, etc… My point of course is that there are a lot of brain deads! We are approaching an era of massive greed. There use to be 3 pilots, now airline companies have two fly their ships. Nobody is fooling me, that pilot was tired and wanted to land despite warnings of severe weather. I learned about flying fatigue 56 years ago.

Don't misunderstand me. You are not brain dead if distinguishing one cloud from another or what the temperature is in Milan, Italy (94 now) does not interest you, but I question your mentality if you believe that weather cannot kill you or debunk pollution maladies, or think that airline companies should not have to put up with tight governmental scrutiny. I'd hate to terrorize you by listing what I have seen go wrong with aircraft. How about a patched up tail riveted the wrong way and just before a landing approach a big chunk fell off. It happened in Japan. An American airliner (not a company name) crashed and over 100 people died. How did sloppy workmanship get by if maintenance standards had not been lax? The private sector is made up of human beings and I'm afraid there is a minority who prioritizes money over lives. Remember? Submarines off the New Jersey coast? Truck drivers called in also on pilot fatigue factors on Jim Bo's show and claimed that they too were being stressed out. Amazing, however, the American Truckers Association conceded government efforts had reduced road deaths (5000 in 1998). Preach I must. All we need is voter diligence and kick out idiots – but not based on knee jerk reaction because we want "Down with everything governmental." Shady characters don't like cops! One of my peers overheard two pilots talking while waiting at an airport. One said, "I'm not flying that thing!" From this he rearranged his choice plane schedules. Both he and his wife breathed a sigh of relief after an easily handled accommodation.

Something for Everyone – Disaster – A Warning

If you would like to fly your own small pleasure aircraft there is another facet of weather you must consider as threatening besides thunderstorms, fog and icing on wings diminishing lift and weighing down the whole plane. That is gusty winds, particularly on approaching or making a landing. Of course, I'm not dealing with children, everyone is aware of their vulnerability. All fliers know what they can do or cannot do. But, it can happen. My feeling about the Agadir tragedy described some pages back is based on that plane took off at 4 AM in the morning and it was possible the pilot was tired, not alert as he should have been and if an engine failed there are certain things he had to do. It doesn't appear he did – there I was stumbling around wreckage before 5 AM. Getting back to winds, it can happen, the unexpected gust can be costly. Still, in small aircraft, few die because of slow landing speeds but nationwide there are accidents and some hospitalizations besides crumpled planes. But, that as a subject is banal. Furthermore, everyone knows their limitations and are additionally forewarned by control tower personnel. Still, depending on where you field of travel is located I wonder if the reader, particularly the young, know the effect of hills and mountains on winds and just why heavy gusting occurs on their leeward side and goes <u>on</u> <u>and</u> <u>on</u> for <u>miles</u>.

Two Boeing 737s go Down, Both "Yawing"

A classic example of an airport having more than it's share of "tricky winds", quoting an article I read describing a baffling crash of a 737 there, resulting in all occupants dying is Colorado Springs, Colorado. But let me warn the reader the FAS (NTSB) knows more about aviation than I do and if, after an extensive investigation, still going on (I think) they discounted weather factors, although there is still controversy going on, then I have to accept that. But it is my prerogative nevertheless to hypothesize how such a place as Colorado Springs tucked up very close to the Rockies could experience violent gusts at levels, say, 400 feet altitude, but not detected because it is invisible and they are limited to measuring devices at ground level and none in the sky above. There were strong winds that day, a recording quotes the tower as reporting surface gusts to 45 mph and adding, "It's an adventurous situation."* A NSTB representative, shown by Mike Wallace had this to say, verifying also my feeling about people being infatuated over computers: "Many pilots today are highly skilled at computer interpretations in aircraft but are slowly losing skill in handling the "stick"." According to the article on the crash in Colorado a very frightened pilot yelled for help - that is not typical of any airline pilot. Reason: The aircraft on a landing approach began to "Yaw" and then finally, one right wing dipped down too much; of course, the left wing pointing upwards, all lift was lost – the inevitable happened – it plunged down out of control from 400 feet – 55 tons of aircraft. I just heard a pilot on Mike Wallace's TV show say, "It would take almost tornadic forces to move a 55 ton aircraft around like that." I disagree for one very good reason, he, like so many, doesn't seem to be aware things weather-wise are changing dramatically. "Yawing" means the tail section sways from left to right,

That means: Come in but at your own risk

while the wings dip down one side then the other, see-sawing. What caused this? Well, I respect FAA's NSTB expertise and the last I read they suspected a valve that automatically adjust or dampens the rudder's reaction to yawing (whatever) and of course, then the pilot can keep the ship on a steadier course. Boeing of course, also under heavy pressure insists they found nothing wrong with the valve. The argument goes on, lawyers are in on it. "Winds or was it the valve?" Therefore, let me limit myself to strictly suppositions with one purpose only, to impress on your minds the main theme of this book and that is, things are changing drastically in our weather!

MOUNTAINS, HILLS, VALLEYS AND WINDS.

I look at this beautiful USA map I have, three by four feet of it, and there, just to the west of Colorado Springs is Pike's Peak, some 14,000 feet high; just to the north of that mountain running west-east Trout Creek Pass, elevation 9,488 feet, further north in between more mountains, other lower passes. It is a perfect arrangement since that tragic day winds were blowing exactly from the northwest for powerful surges of an air mass piling up against Pike's Peak then suddenly released every so many seconds. It's inertial power boosted and more concentrated through a "Funnel effect" inside the mountain passes, bursts forward as a separate stream, much faster than surrounding air. I'll theorize, but certainly not question authorities, when that airliner's wings were tilted down as shown in the illustration below and supposedly a blast of air (invisible, remember and how could control tower personnel really know) slammed against that wing ---well, what do you think had to happen. The recorder also revealed a thumping sound, twice, occurred in that plane. An expert seems to think winds could have done that – they had to be powerful.

The problem: How can you know what the winds may have done at 400 or 500 feet altitude? Even if nature would repeat it would be very difficult to measure such an oscillating invisible force. You have anemometers at surface sites and even on top of mountains: an example our 5,000 foot Mt. Washington in NH registering quite often gusts over 80 mph*, but no such instrument in mid-air! You can't send balloons to be tracked on a continuous basis – not with planes flying around. Even if you could send one every half-hour, gusting is always intermittent. But, repeating again, I'll just assume NSTB experts are on the right track – they're quite marvelous detectives from my having followed previous investigations. All the above of course, has to do with effect of terrain on winds. Another Boeing 737 went down in Pennsylvania and also above hilly terrain but the investigation goes on. In the meantime, all you future pilots, "Pret guarde a toi!"

BACK TO TORNADOES

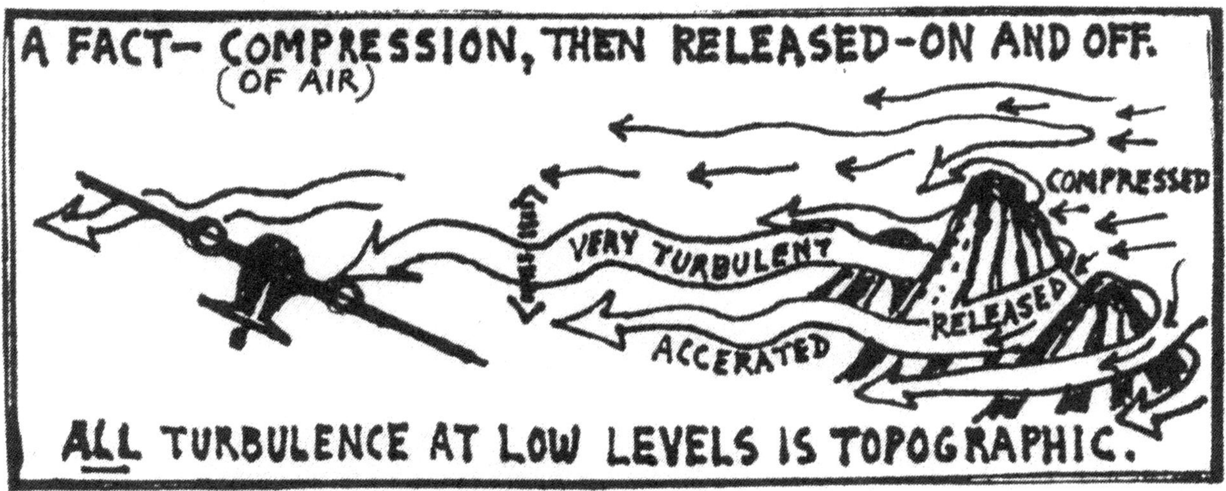

And now for the funnel. Let me try to explain why the description "Dipping Down" can be regarded as an illusion even though such a definition is practical in it's brevity. Along with everybody else I have always referred to such a dramatic onset of destruction about to take place as dipping down. Once I was asked how does the funnel remain intact. I didn't find it easy to give a satisfactory answer.

Looking at a funnel formation from a distance we may not see the initial stage. It is there nevertheless but invisible except for the hanging V-shaped dark segment of a storm cloud. There is at ground levels first of all a wide counter-clockwise motion of air but not strong enough to kick up dust and debris. The diameter may be anywhere form one mile to ten or fifteen – some experts have said as much as 20 miles. <u>There</u> <u>is</u> <u>a</u> <u>funnel</u> <u>nevertheless</u>. It takes perhaps but one second from this first stage to the final minimal diameter where maximum ferocity occurs along with the very visible whip-like action. Initially, a drop in atmospheric pressure inside the invisible vortex, as viewed from a distance, attracts surrounding air with it's normal but higher atmospheric pressure – a mass convergence toward it's vortex target takes place. But the vortex's outer wall is spinning and centrifugally deflects the rapidly encircling outside air charging in. It's deflected but while rotating about the wide funnel it exerts tremendous pressure on the spinning funnel wall. For the want of a better expression a squeeze action takes place and with that, the funnel's diameter is immediately reduced. Because of violently

Even when surface winds are half that much.

increasing centrifugal forces nothing from the outside can penetrate it but it is subjected to relentless encircled pressure. At this stage, in a flash, a powerful vacuum is created by the constancy of the immense cell above (I prefer to call it the Big Vac Machine – we laymen like dynamic expressions) at the top of the narrowed funnel. Remember, a narrower pipe and a constant suction by the powerhouse above, which may be several miles to six miles wide and 20,000 feet or more in elevation means spinning and rising air inside the vortex reaches unbelievable velocities along with maximized vacuum. We have to think in terms of maximum vacuum causing an immediate temperature drop. <u>It</u> <u>starts</u> <u>at</u> <u>the</u> <u>top</u> <u>of</u> <u>the</u> <u>vortex</u> since that part is nearest to the base of the powerhouse. Therefore rising warm moist air becomes a whirling wall, a thick <u>light-blocking</u> <u>wall</u> <u>of</u> <u>coalescing</u> <u>ice</u> <u>crystals</u> <u>and</u> <u>earthly</u> <u>particulates</u>. You now are seeing this first stage as a dark area <u>inside</u> the top of the funnel, the funnel having been already in existence all the way to the ground. The increasing vacuum from a tightening vortex, the process exacerbated by tons of ice crystals propelled into the cell then effectively completes coalescing condensates right to the surface level. You have not witnessed a dropping funnel – you have witnessed the formation of <u>condensates</u> <u>moving</u> <u>down</u> blocking out light as it does. Depending on how much loose soil, underbrush, straw, plus whatever remains of what is pulverized, what have you, pulled in the light-blocking visual effect varies from light gray to black. Tell me, dear reader, are you still convinced tornadoes dip down or are you not? But, it is just a technicality that's all.

I had good intentions not only to cease adding excepts but to exit properly from what I thought would be the final pages of this book with some nonchalance, savoire-faire or just maybe Que sera, sera but an article today, May 7, 1999, quoting some more scientists, by our dedicated Providence Journal, pleases me to no end! I can't resist commenting. It proves my point. Turbidity (I love that word) reigns unprecedented in this Ping-Pong game of environmental contributors but regrettably, it also plays right in to the hands of Carbon Club Members, namely the oil, coal and car manufacturers along with their well paid propaganda agents. Turbidity because today' group of alleged scientists in print having now decided, empirically they will have you know, La Nina is to blame for tornadoes of May 3 to the 5th in Oklahoma and other states. The beauty of this article, however, and I derive satisfaction from such statements made, is that it totally contradicts claims made by El Nino enthusiasts, now in a state of repose waiting for some more gaffs by politicians or whoever. All this morass being created smells of experts looking for more research funds. Excerpts below should make my convictions clearer. This article is another powerful example of attention-seekers unintentionally or intentionally generating turbidity – just had to use that word one more time. I promise not to repeat it again. As I have already written, logic is not guaranteed with degrees. However, I'll leave everything up to you – you interpret any way you see fit. Frankly I'm beginning to believe there might be some radiation from computers damaging a few brain cells judging from the slew of articles and suppositions flung about. Last May an alleged expert from a major Think Tank wrote to the effect that El Nino created tornadoes that wreaked great damage and deaths in Florida and adjoining southeast states. Others chimed in with the same view, supplementing their convictions with beautiful colored computer modeling. Today's article is the second proclamation that La Nina, the cold one, has caused Oklahoma tornadoes, etc… Part of today's article below, contradicts the opinions of last year's El Nino enthusiasts! "Wonnerful", as Lawrence Welk use to say!

La Nina effect blamed
for powerful tornadoes

Studies suggest tornadoes
were <u>reduced </u>by as much as 24 percent
in some regions during the most recent <u>El Nino</u>.

Reduced mind you! Yet too many times in the past year all you could find in print was El Nino being responsible for more tornadoes.

Overlooked of course is the fact that our supplier of ice-cold air, the North Pole and adjacent Arctic regions has been rejuvenated (already explained) and quite naturally, as this country experienced vigorous cold air mass invasions could only result in not only strong northwest winds in conflict with excessively heated warm air from the south but quite naturally an abundance of snow from storms in mountain regions, which we need anyway. Let me insert their opinion of this with the following clipping:

La Nina has generated a stronger, colder jet stream in the upper atmosphere*, while warm, wet air at lower levels has been pulled from the Gulf of Mexico into the Plains states.

It's part of the same widespread weather trend that dumped a record 91 feet of snow on

**Incredible! La Nina's cooler air does not rise! It moves west toward the orient!*

All of us should know lands east of the Great Divide need water that melting snow from the Rockies has for hundreds of years provided. If residents and weathermen fuss over snow accumulations, all I have to say is, water is more important than the nuisance of shoveling snow. I mention this 1999 snow factor because the article implies something bad and of course, La Nina is the responsible one. That's over dramatizing. It is also inaccurate. But powerful arctic surges of biting cold air, which were normal this winter, are not just shallow mobile domes marching southward but have some vertical dimensions also – therefore winds at 10,000 feet or 20,000 feet more or less reflect what is going on at the surface and of course, the jet stream wraps around the numerous major oceanic storms that occurred over the northwest Pacific and into our coastal regions. With no compunction whatsoever they are claiming La Nina, this new sensation even though it has periodically rose out of the cold Pacific depths for thousands of years, as causing a rise in tornado frequencies in our central Plains. Therefore, it takes the (intent) focus off warming trends, more carbon dioxide, and more pollution. But, such a claim may be just someone looking for publicity – that flaw in the scientific community called "hype" by not only other experts but also Cornelia Dean, Science Editor for the New York Times, excerpts of her statements in later pages. Anyway, here goes with the La Nina representatives – one sentence suffices. My graphs in early pages show this is a nice piece of fiction.

"The incidence of tornadoes on the central Plains is slightly higher during La Nina."

Remember some basic physics and also the prevailing easterlies out of the equatorial Pacific zone which would drive any air chilled by La Nina to move toward the far western Pacific and not over the USA. Furthermore, cool air at oceanic surface levels more or less stabilizes weather and in all probability provides a few degrees of relief from heat to various places such as the Philippines and so on into regions further west.

I am not interested in jet streams right now but I bow to others loftiness – my respect for their persevering with forces beyond our knowledge. What I am calmly protesting with you is the ailment affecting some experts – that is, a mild form of memory failure, or, is it "We need research money." <u>Completely</u> discarded in their well-written premises are facts already established. <u>We</u> <u>have</u> <u>measured</u> more carbon dioxide than ever, <u>we</u> <u>have</u> <u>measured</u> more and continue to do so, carcinogenics prompting increased testing of toxins. <u>We</u> <u>know</u> the ozone layer has been damaged and <u>we</u> <u>know</u> UV radiation is stronger. Unless thermometers are not being read properly by weather bureau personnel in the Caribbean, for example, there is no doubt in this simple practical mind of mine temperatures over a wide range of semi-tropical regions are much higher than say, 30 years ago. Either that, the scientists in questions are very shrewd and like riding the merry-go-round. Pray tell, what the hell are they looking for! Now, 60 years for industry to fully comply with necessary cleaner air mandates and their contributing toxins suspected of damaging the ozone layer is a sorry state of affairs. Certainly I am happy to read such a contradictory mess on previous pages but it is also sad. It leaves politicians and the public in a state of mind, well, "Que sera, sera" defines it. Worse, the Carbon Club has reason to cheer.

Continuing on with the article. They make a statement, a very questionable one, slightly more tornadoes during La Nina outbreak years. If the reader is the analytical type go back to pages 6 to perhaps 8, compare La Nina years which are shown with small circles on temperatures traces of cities, with tornado frequencies. Some years this supposition holds up – some years it does not – it's the same old morass again. Put it all together, dear reader. You have had a strong resurgence of cold air invasions this past winter. These powerful winds pushed more cold water from the Aleutians areas. Such currents sank as cold water will but in converging at the equator with those cold currents of the southern hemisphere also converging up arises La Nina – it's upwelling cold water, period. They call that thermocline inducement or whatever. The Arctic produces cold powerful invasions of air – meeting warm air hence big storms in winter and tornadoes too soon this year. That is the only reason, the Arctic and not La Nina. Let's put the horse in front of the cart, please. La Nina, it's rising fame (never heard of it years ago) in how it is played up is analogous, figuratively speaking, to the Loch Ness Monster (too much and steady consumption of Scotch Whiskey will make anyone hallucinate) popping up in the Pacific equator. While I'm at it, a tidbit of weather in Scotland and some history for what it's worth. You need whiskey in Scotland – it's windy and cold but hardy people live there. In fact so unpleasant for Mediterranean types, Hadrian the emperor decided nothing could be gained venturing that far north except pneumonia on a grand scale for his armies. He went halfway north in England, built the famous wall still there. Many years later, his legions not receiving any more pay, the empire slowly collapsing, history shows Roman soldiers settled down in England. When you see Englishmen with prominent noses it suggests some other influences besides Germanic. Just kidding about noses. As for Emerald Isles, it is true – I left Africa in 95-degree heat during the month of August. Arriving in Belfast wearing light Navy attire, all white, the 60-degree temperature was chilling but the extreme rich green of everything growing was a sight to see.

A story of pre-destined death – or was it? I didn't think of myself as being psychic that terrible day. Aviation was at time, between 1942 and 1951 or great fascination for me. As deliverers of death in war I quite naturally analyzed each plane for it's immediate impression of potential speed, what it had accomplished in the arena of conflict and most exciting of all, the power, the sound of blaring exhaust and high pitch whining propellers – combined vibrations filling the air. Take off – that was my kick. But there was one plane we called the Flying Turkey. Pilots during the Pacific engagements who flew this dual cockpit aircraft must have been super brave or were born of cavalier spirit and death-defying to boot!

It had clumsy lines, an awkward looking belly hanging down between the landing wheels within which a huge torpedo was housed. Worse, in attacking a Japanese warship they flew just above the water in a straight unwavering line and the necessary slow speed made this more than a very dangerous venture – the odds of ever coming back were against you. You can't drop a torpedo doing over 200 miles per hour. I'm sure you've seen kids watch stones they threw at lake waters – you can count how may bounces before it pops the last bit of spray. My guess the TBF lumbered in on it's target just a bit over 120 miles per hour and that made them easy to hit. How many were knocked down we will never be told. How many times I have seen documentaries and some crippled TBF smashes down on it's carrier deck? But here after the war at NAS Squantum we had a few Flying Turkeys. It was my imagination I know but going down that runway, noisy as you can imagine, TBFs were not inspiring. And landing, that was a display of a heavy bird plopping down – not graceful at all, hence the name assigned to it by our personnel. Behind our weather office counter I had a few copies of flight plans. One TBF was due to depart for Norfolk, Virginia and shortly thereafter a comfortable reliable two engined utility SNB – comfortable because it's insides were insulated. It was also destined for the same air station south. On a day off you could hitch a ride if there was room. Twice I went to Washington, DC. Even strolled through the endless halls of the Pentagon and took in other sights. This youngster that I knew, perhaps 19 years old, had asked me when the TBF was due to leave. How can I forget what I told him, "What do you want to fly in that plane for? Wait 20 minutes and ride in something good. There's a SNB going to Norfolk." He didn't respond, walked away and seemed tense pacing around an adjacent waiting room. We never did find out why that TBF crashed in Norfolk – we never were told what actually happened. I was troubled after that and wondered if the end of life is written in advance.

It can reach the point of monotony, my quoting or excerpting claims and comments by organizations, both federal and private environmental groups. But air quality improvement achieved (?) between 1995 to 1997, percentages given of certain toxins (unlisted) this time, May 17th for each New England state, encouraging figures produced by EPA, lauding cooperating factories and published in our Journal. It is all very cheering but is in disagreement with yet another report made available today the 18th. It is stated by EPA that our 1998 evaluations are not that encouraging. Quote, unquote. Today's focus was on Somerset, Massachusetts' coal burning power plant and several other electricity producers under fire by protesting environmentalists (backed by hordes) who say the air may be getting dirtier. I'll not annoy you any longer as nerve ends can be frayed with repeats of the same old ups and downs but in a 24 hours period we are again watching the ping pong ball bounce back and forth – quite a game.   Since my opinion pertaining to the <u>overlooked</u> factor of near normal winter weather mentioned quite a lot of pages back has common sense going for it I still have to leave it up to you. I am not afflicted with the teacher syndrome.

If you recall I did specify in the beginning, self conscious of repeating details, we are being treated to a respite by nature but that cannot last, alas, not with a growing economy and expanding population and a horrendous increase in vehicular engine sizes. Let's be for real! Unlike Russ Limbaugh (my buddy) who often leads off his radio show saying he is so superior claiming one half of his brain tied behind his back is all he needs plus the payoff, his "Talent on loan from God." I'm quite human and if at times I have prayed for divine help (especially when scared) I've yet to receive a personal message from the heavens above. For a man who speaks in an agitated hesitant manner just out of school and of late keeps throwing French expressions around, with a suitable accent no less so as to sophisticate the impossible to sophisticate crowd calling in, all singing the same song, their metallic tense voices reflecting an innate wish for the return of frontier days, not realizing the growing complexity of expanding populaces and world trade, he sure is a vain one. He allows no good to bad gradations. The expression Liberal is abusingly all encompassing by both he and his followers. Unlike their generously separating Republicans into different categories, "On the fence", moderate to slightly to the left, nice and perfect gentile Christian straight party liners, all liberals according to this Hooray for Mob rule crowd, wear the same stripes – they are not gifted with flexibility, and so on. But that is what dogma is all about. God-sent, my rear end. I've modified what I really think.

The following has to do with a derision thrown at we Frogs. Yesterday after reading the detailed report of two competing yachtsmen, Garside and Mouligne, I flew into a rage. Just kidding. Actually it seems to be rather a lighthearted description of Mouligne by Monsieur L'Anglais. If God put these people on a fog shrouded island and placed the French on what I know to be, weather wise at least, a beautiful land, He must have had a good reason. I'll return to Morocco after this. As I already wrote, each day stimulates a reaction that I respond to. I found their reports on weather in remote parts of seas fascinating. Evidently so did the Providence Journal.

Everybody wanted to be a hero! If all pilots I was in contact with for ten years were the paragons of gentlemen warriors there was one "Crazy Frenchman" who came out of a different mold – that was Lieutenant Tetreault from Lowell, Massachusetts. I heard that distinguishing remark made during an incident where our flight commander didn't expect him to come back from the usual 12-hour flight out to sea. This man never was one for the calm cool demeanor that I accurately cast-type naval pilots. He drank gallons of coffee, smoked and when something went wrong he exploded with choice expletives. He was out there this one day, in the Bay of Biscay, on the usual submarine hunt but created quite a stir with his radio messages. Suspiciously, he never answered certain firm requests to get the hell out of there. I remember officers right next to my little piece of business space in Agadir, talking in grim tones. "What the hell is

> …the Frenchman, had fooled Garside, the Briton. In a pre-race qualifier across the Atlantic, Mouligne had deceived Garside into making some bad moves. The Frenchman was up to his old tricks again, and Garside responded:
>
> "Three times…I've been suckered by a foxy frog. It has taken a while to dawn on me but now I know the truth. Behind the cool, charming, Gallic exterior of Monsieur Mouligne, there lurks a mind devious in the extreme…There's 5,000 to go before we get to Cape Town, more than enough to get even…"

the matter with him!" I heard that once also. But first, that body of water just north of Spain to the coast of France in winter was usually tempestuous and from our shaking view in a bouncing PBY, a near solid sheet of white foam blowing from wave to wave and clusters of fragmented scud-like clouds beneath a dark low overcast, mist spreading across plastic waist gun closures. My having been on one of those trips, I considered myself lucky for not being assigned again for another excursion there. I thought that this would be one hell of a place to fall down in. How could anyone find us, anyway a very slim possibility of surviving a forced landing? Not only was it dangerous since we were very near German occupied land, but shifting stomach intestines, up and down and sometimes sideways steadily for over one hour made me decide to feign illness the next time I was assigned to another trip like this one. This ocean area, notorious for rough weather drew our flights there however since submarines docked in France traveled to and fro across the Bay of Biscay. All of us, and I'm sure readers also have seen wing tips move up and down in turbulence, well, sturdy as PBY's were and still are in Australia, I wondered when rivets would start to pop. Well this darn fool Tetreault spotted a German four-engine bomber and if I can remember, a Junker or something like that. It bristled with shooting irons from our having to get familiar with enemy aircraft. Our PBY's had a peashooter 30-caliber machine gun in the bow and one 50-caliber gun on each side. Tetreault's sole message came through, real fancy like, with the assuredness of a Napoleon no less, "I am pursuing the enemy and intend to engage him" after describing the bomber. It seems she had the bright idea of strafing that fortress with his BB gun, if he could get close enough. I'm sure the German pilot would not have allowed the PBY to comfortably pull alongside and count how many 50-caliber bullets were headed his way. * Not prone to religious expressions, I believe that the other pilots assembled next to my caboose were thinking, "May his soul rest in peace". The bomber pilot stepped on the gas and disappeared, totally ignoring Tetreault. Had it been a clear day he would not have been so lucky since fighter planes would not have come out to say hello to him, I mean German fighter planes. Just a matter for the Junker radioing a nuisance to be rid of. Tetreault came back, everyone happy of course, but not him. Oh No. He had to blow off as usual. "Damn it, that son of a bitch ran away" and walked off in a huff. I think that the Navy should have been more considerate and made him a fighter pilot with one provision; have something that would prevent coffee from spilling when flipping a plane upside down. The French artist Paul Gauguin (1848 – 1903) lived in Tahiti for years and painted rather interesting portraits, which attracts droves of tourists after they

Every fourth bullet is a glowing tracer – easy to follow.

tire of sleaze shows. He died rather early. Must have been a typical Frenchman. Better to look at his portraits than generals on horseback in London museums.

Then we had the odd balls after the war. Signs of changing times after 1946 while still in the Navy manifested themselves in rather peculiar ways. For one, rules were more relaxed. One guy was unique and the butt of our jokes. This pilot flew back and forth to various points south from Squantum and always in the front cockpit his constant companion, a huge white dog of European breeding. I mean huge! It was about the size of a five-month-old polar bear. No, it didn't wear a helmet and goggles, but he sure behaved like an experienced airman. Up he would hop on the SNJ wing, climb into the cockpit, it's fluffy tail gracefully following and then sit down. This Lieutenant JG would come in, fill out a flight plan while his canine buddy rose up, plopped two huge paws on the counter, gaping mouth full of teeth, staring at me while reassuringly wagging his tail, showing great interest in what was going on. I wouldn't have been surprised had one of his paws slid the form over to his master after I signed it. Conversation began one day and the pilot just had to tell us what happened between him and his wife, as if this was very funny. Good thing they had no children. His wife didn't like the dog and we theorized he was too much the center of attention or she was sick and tired of throwing hamburger at this oversized creature. She gave her husband an ultimatum, "Get rid of the dog. It's either me or him." To which he replied, "I'll keep the dog".

Don't think this is a spin-off of chauvinism in disguise. I always thought, for example, that women were much more careful drivers than men were years ago. I also want more of them in politics. But today, the young ones are rather aggressive behind the wheel. An insurance agent told me that they contribute to half of accidents, unlike years ago.

Another day, a portly governor who was a known alcoholic, came in quite stewed but walking in a reasonable straight line, destined to be transported somewhere south. Quite a fuss from his entourage. Accompanied by several revering staff he went to our SNB but couldn't get up the high step. It was too high or he was too heavy.

A small crowd had gathered, including several ladies who obviously were from the upper crust of society. They pretended not hearing the governor blare out for everyone to hear, "Push my fat ass up this thing, will you!"

Becoming more and more conscious of dragging weather topics out I seek variety with a hodge-podge of childhood to maturity episodes of little consequence in a world of new values, but relief no less. What is wrong if I bounce from one time frame to another or avoid any continuous plot since this is not a story of a star? If I can extract a smile from you that would please me to no end. Or maybe pluck at your heartstrings?

Perhaps it was my having taught her in 1941 but my mother was never a good driver although in some 30 years she had no accidents worthy of mention. I made her quit at age 80. When they say "What red light?" after they run through one, it's time. Happened twice with me next to her. The last time an angry truck driver blew an air horn for at least three seconds. She was indignant and said, "What's the matter with him!"

Some of us are very interested in prewar cars. I believe the 1932 Plymouth Floating Power Four Cylinder was unique but reached it's demise as a smooth engine rather quickly. Uniquely different simply because in the hands of someone like my mother for example, the engine just asked to be destroyed. About six months before Pearl Harbor, at age 17, October would change that to 18, I talked mom into helping me buy a car. A patched up bike had become discouraging. In my mind this would be paradise of course and a sure way to achieve the ultimate triumph. I truly wasn't girl crazy but despite strict traditions in a primarily Italian and French community I thought one of the advantages, besides glorious prestige, mobility to take me everywhere, money or no money, would b the favorable impression I could make on some dream girl even though it was like panning for gold in the Sahara Desert. Such vigilance on the part of strict parents protecting their goddesses! Prom dances? Never heard of them. Before I come back to my 32 Plymouth I had, this one day, dreamily staring out the math classroom window of our all boy high school. There, but a quarter mile away, setting on a hill for me to gaze at, a beautiful high school filled with boys and girls. The math teacher didn't like it, trotted over to me and whammed me on the side of the head with a book. I do believe it may account for improving my thinking processes. Laugh if you will but those were the times, those were the standards. Drugs didn't exist. We smoked and on Saturday night the boys and I would help the beer industry but just for one night. The beer was ten cents a glass. Besides a dollar my mother gave me and a quarter that I earned in a gas station, that was it for entertainment. No crime, no policemen that I can remember except for one state trooper who we called Mr. Brown, so idle was he that we could count on him spending time with us standing on the corner but not watching the girls go by. They were all recluses.

My mother must have had an ulterior motive for helping me buy a car because shortly after this heavenly acquisition she asked me to teach her how to drive. My stepfather was opposed to all this but yielded to mom's urgings. I didn't know it but my treasure on four wooden spoked wheels painted yellow with black pin stripes was soon destined for the junkyard. It was beautiful, all black, four doors and nice upholstery. Purred like a kitten except you definitely felt the purring – a pleasant vibration in those days, if I may say so. I borrowed $75 to buy it but Mom made sneak payments as if Dad didn't know. First accident Mom had was a minor one but she thought it was very funny. It cost us a junkyard replacement wheel. She had put the brakes on driving over a patch of winter ice, spun around, banged against a curb and ended up facing the direction she had come from. She enjoyed telling everybody about this new experience. Thought it was hilarious and kept saying, "I was so confused." This four cylinder engine, mounted on three coil springs floated alright but only if you knew exactly how to let the clutch out and just so-so gas. If you stalled it, that engine reacted violently, slammed to the left then to the right with one hood-shaking banging sound, then snort once through the carburetor. When mother was trying to get the hang of it, I dare say that engine was pulling apart those spring mounts from so many attempts it was now a matter of time before the engine thrashed everything under the hood. Engineers in those days had shortcomings believe me. Metal was of poor quality. It was a common occurrence for axles to snap in frigid weather. The day after Pearl Harbor I joined the Navy. Two months later I came home for a few days leave and the car was gone. "It's a good thing we got rid of that car," my mother said. "It was a piece of junk."

When my oldest son was very young he once asked me if I ever had been shot at. I said, "Sure. Twice." And would begin to laugh. Two Moroccan soldiers, both sporting rifles and like the young German lady in Casablanca, harbored hatred for we invaders. Sometime in early 1943 I had left the beach from a long walk taking in surf and ocean air and began climbing a 150-foot embankment on the top of which a tower perched. Two shots rang out and while quite above my head the sound of whizzing bullets was less than thrill. They had no intent of killing me and knowing I was in no position to retaliate, they simply sauntered away. I was very lucky. In Port Lyautey there is a cemetery and rows upon rows of white crosses stemming from the invasion of November 8, 1942. Never got any press time either for this particular very important area which served to launch several years of effective submarine hunting. From 1940 to late 1942 Germans functioned in wolf packs, sank thousands of ships and many seamen perished but gradually we became the wolves and they the prey.

Between that girl in Casablanca who unceremoniously rejected me and two soldiers shooting bullets my way I must have been the scapegoat for our invasion of November 8, 1942 – a conclusion I reached later.

The brain's memory function when it comes to past acquaintances has to have two separate compartments, I swear. Think about this. I remember faces perfectly, every little detail but not the names all the way some six decades. Names must be stored in another section, so help me. Must be a malfunction unless I am not alone with this problem. If so, then I think I'll submit that one to some high placed medical society. (Joking again)

I'm not preoccupied with morbidity. I like to clown around. It' just that there is a need to break up the humdrum of climate topics and to give you flash backs of olden times.

Was it a premonition? I doubt it. My maternal grandfather didn't like the idea in the late 20's that cars could replace horses. Biddeford, Maine was a typical sweatshop town and no cars to be seen. Grandpa was killed at the age of somewhere around 50 by the first Model T to invade that town. Evidently the driver was anxious to show his new found speed of mobility but horse, carriage and grandpa went down – he fractured his skull. He had told grandma something bad was going to happen with these newfangled contraptions he despised. Probably knew he couldn't afford one. The horse ran off but had to be put down. Grandma, who lived to 91, had to fill me in on details.

Do you believe in imminent justice? I do. Oh, I don't mean death in war and all that – of course not. I mean individual louses. I was cruel at the age of twelve. I took a fancy to catching bull frogs thinking what I heard about it's legs being good to eat. That's why the English call French folks Frogs. They're really creative like that. I heard and read that expression used three times in one month not too long ago. There is this Mouligne Frenchman who competes in around the world yacht races and his closest competitor, an Englishman who he managed to beat in several legs of the race who, in their frequent E mail letters to newspapers (fascinating for me) said, "I'll not let that Frog beat me in the next one." Then some bit of a movie that I caught and switched channels had to come up with the same derogatory tag-some yo-yo actor thinking he was cool. When I lived in Louisiana, people of Canadian French origin were, behind their backs of course, called Coonasses. I'm supposed to tell you a story about justice and frogs but I have to let you know something first. If Affirmative Action is lamented in no uncertain terms it is usually by the very same people who are responsible for it being mandated. I got a good does of that when I was a kid trying to get a job carrying golf bags for old duffers pulling in with their fancy cars. The caddy master had a list, which you signed and would call you in the order shown. Came to me and he just skipped me – called the next kid. The next Sunday morning, after getting up at 4 AM, walking 3 miles to the golf course, I got the same treatment. He didn't like my name I guess. After all, Germanic genes are the best, n'est ce pas. They're such a thoughtful bunch. This pond I frequented had enormous frogs hiding under overhanging stones or sometimes their eyes could be seen behind a lily pad. I would flick a piece of red cloth, insect size, attached to a fishhook and sure enough, they would lunge at it. Then with much grunting, frantically using their back legs trying to kick the hook out, front paws clutching at it. Nah, it didn't bother me. I bashed their heads against a stone wall – took them home but didn't think much of what their legs tasted like. Aha, justice was due! I had placed my fishing rod against a wall, the hook dangling a few feet above the ground. Bending over to pick up something my ear got snagged on that hook but real good. Off to a doctor after Mom looked it over. It had to be pushed through so as to cut off the barbed end. The doctor did just that and after a noisy "Ouch" it was over with. He laughed. I never bothered with frogs after that. I felt silly that day, walking home, pole in hand, one hooked Frog. (That's an intended pun.)

This case was not one of predestined. We were in Hot Springs, Arkansas. They had a large pit with a dozen alligators. Two were enormous! According to a guide of sorts one was supposed to be 400 years old but he explained this questionable claim was determined by counting pointed ridges on their tails – whatever. In the middle of the pit there was a post with an attached sign, which read to the effect, "Watch your pets. Rudy the Terrier died here." The fence struck me as inadequate – it was about 3 ½ feet high.

Sometimes I think selected members of the animal and bird kingdom are protected by nature. When we lived temporarily in a small farm house in the middle of nowhere, the nearest village of about ten houses and one variety store called Barker, New York, there was every morning a male pheasant and his three wives (?) pecking away in the backyard. Without telling my wife why, I bought a used rifle I had designs on pheasant dinner some Sunday. Guess what? The very next day, no pheasants and they never came back for the two months we stayed there waiting for a home to be built a few miles near Niagara Falls. Dogs are very smart. I don't know if this is funny at all but a small museum next to the famous falls had all kinds of rigs crazy people had fashioned to float over and allow themselves to hurtle down. Quite a few died. This one barrel on exhibition had a hole drilled into it about the size of a nickel. The idea being to allow air of course. The trouble is the man took his dog with him – he just had to add a novelty to the insanity of it all. Attached to the barrel a framed description of what happened. The guy got knocked out from the impact but he died from suffocation. The dog had stuck his nose in the hole and kept it there while the barrel floated down the Niagara River. He survived, of course.

The things I have seen in my life – I could write a book but it's subject matter that is not that all exciting. For five years I rolled the odometer some 35,000 miles every 12 months. Lost my driver's license once in a speed trap in buffalo. Before you got to an intersection the speed limit was 35 miles per but after you crossed it dropped to 25. There, in line, other victims and huge cops of a predominant breed in Buffalo, writing up tickets. Same thing before in Rochester, NY. You roll along at 55 out in the country, there is city limit sign and then a 35 zone. No houses, nothing around and I'm following two other suckers. Bing. I'm in court and some lady judge is soothingly talking to a handsome 20-year-old and she says, "You come back here John and it's trouble – dismissed." With me, family and all, traveling to make a living, my plea meant nothing. She flipped over some pages in a thick book and said, "According to the law," tapping her pinky to a page, "I have to take your license away." I got upset and said, "If it's a matter of just a written law just why do we need judges, your honor." She glared at me and threatened to charge me with contempt. Well, I drove one year without a license in that state and I was so nervous I sold more equipment than ever. I actually had to go to a parish priest to get my license back. I saw so many things, death on the highway, a farmer laying next to a tractor that had rolled over on him and residents waiting for an ambulance. A trooper who doing a mad chase in the country ran into a hay wagon pulling out from some farm driveway. Loose animals on highways and some very beautiful scenery also. And snow, Oh Lord. Try Watertown, New York sometime. What a crock. To think that there are writers who claim it has always been warm – in other words, there is no warming trend. Gosh, it was cold in western New York in those days. I spent three days in Sioux City – those Indians are something else. I saw a sign in the back of a car window "General Custer got his." We had one in our barracks at Cape May when I was but 18 years old, just out of boot camp and this fierce looking Sioux Indian took pleasure in strutting around us while teaching how to throw a monkey's fist – he was a Cockswain, that's a deck boss of sorts. But, one night after coming back from drinking he woke us up by noisily plunging a huge knife through a sheet metal locker. Boy, I thought I best stay humble with this character, feeling very vulnerable sleeping at night. I forgot, a monkey's fist is a large balled up knot and some 30 or 40 feet of line – they sling it from a boat deck to a dock. The trouble is he flung it at lockers with a resounding whack. Passed through Phoenix a few times and even entertained the notion to live in Arizona. My brother spent two years in that state and told me to forget it unless you don't mind sand in your sandwich. I wish I could write like the author of "The Grapes of Wrath". I also saw the spectacular and the beauty of the northwest. The coast is beautiful. Write? I did say I'm not a poet – just a technician that's all. In August of 1971 my youngest son was killed in an auto accident. I fell apart and sought to bring my family somewhere else besides RI. I drove to California. One Saturday I went to Santa Anita to see a famous horse, Canonero 11 and the races. Such pageantry and ladies strolling about in silken gowns and wide brim hats, golden trumpets and an ancient chariot pulled by four horses – it was beautiful. Can you imagine 84,000 people applauding? It gave me goose pimples. Everything out there in front of that grandstand was regal. I decided this would make the perfect escape for me when I returned to RI. Five years later, strictly as a morning hobby I owned a racehorse and still hold a lifetime license to train. I enjoyed every minute of it. Every day there was something hilarious going on. There were two doctors and three lawyers who also escaped their jobs but most horsemen are dreamers. I don't talk about it – most everybody think of some character slinking up to a betting booth. You love those critters. They're bright and affectionate. I have seen men cry when their thoroughbred dies, such as I have seen in fires. Of course, some of the males who haven't been gelded like to bite – you have to be careful. One famous horse, Forego, had a reputation: "He eats hay and grooms." Until they neutered him.

I'm pushing variety in writing. The truth is that I want to connect with you. But, if you are an Internet Addict you turn me off a little. I just read a one-quarter page article in the USA publication by a teacher in New York City. But say it isn't so, dear reader, the opinion he makes is distressing. How can I interest youth in studying what this book is about, really? He maintains kids memorize a flock of facts from computers, throw it together on some test and then forget it. "They are losing the ability to think, do not want to bother with analyzing anything, hate to put together something written of a complicated matter – in short, they're getting stupid." Say it isn't so, please. USA is the leading seller followed by the Wall Street Journal. I'm glad the journal is paying a lot of attention to environmental issues – they know the money factor there and what awaits us. Yesterday, it was announced that HMOs are blowing up their premiums – they're all losing money supposedly. No wonder, our surgeon put some screws in my wife's broken hip that took less than two hours. Eight and one half grand! And the hospital? Yipes. $15,000. Greed now knows no bounds. The Republicans (here I go again) keep bragging about Reagan and how we prospered in the 80's. He eliminated government controls over industry – let the market reign but we set records for mergers – never has this mad rush to buy each other out has been equaled in our history. As a result, there were many more millionaires but several million lost good jobs also. If you criticized the government in Senator McCarthy's days you would be suspect of being a communist. Let me clear up something since I've taking jabs at our government. That's nothing but a temporary group of politicians and that is not what is meant by our country.

I had to find some gracious exit from troubling times awaiting us. Just think, despite some normalcy in Arctic invasions of cold air this past winter, the heat factor south is powerfully abnormal. Obviously then, why not concentrate on why. What damaged the ozone layer if not harsh chemicals from industry? I doubt if it is autos. Furthermore that is a health problem plus the irreversible* carbon dioxide problem. But, what if we concentrated on stopping the burning of coal and oil in power plants and stop the sulphur dioxide emissions, which also convert to sulphuric acid when contacting moisture. Redundant a topic as it is, such measures financed by our government would greatly reduce the frequency of tornadoes – at least it would be worth a try. But, the White House has now made a deal. I already told you of the 60-year concession.
Until we reduce burning of any combustibles.

SOMETIMES I FELT AS AN ALIEN BUT I ADJUSTED.

Here is a simple country boy story pertaining to the wisdom sometimes shown by animals over their offspring. I had to adapt to rural mannerisms in traveling Louisiana, Arkansas and Mississippi as a representative for an electronic firm – quite an adjustment to make. Besides strange creatures and different peoples there were quite a few sights to take in. Sometimes it was a crawfish festival or sitting in a famous old "beignet and coffee" restaurant right next to a 30 foot high levee, or something quite strange – it's an eerie sight to see the masts and stacks of a passing ship at a level much higher than where you are sitting – the Mississippi River flows at quite a speed just on the other side of that levee. And the coffee, New Orleans kind? Oh boy. Takes some getting use to but after four or five bouts you get to like it. Beignet is not a doughnut but a delicately fried something or another with powdered sugar – it could prove to wreck one's diet – you want to eat more than two. Before I get to my coup de grace, well, actually coup de nothing, about a bright wild pig that made my children laugh let me cover what you have to put up with in Arkansas and Louisiana. Needless to say, speaking the King's English was out of order unless bartering with some businessman. One early evening, in an Arkansas dealership the task of teaching five mechanics became difficult for a moment. They were anxious to share a story with me, somewhat embarrassing to one chap who protested while the others could not contain themselves from laughing. Their butt of a joke had been caught and I quote one joker, "You see him, he was trying to stuff a live calf in the back of his car – couldn't get the darn thing to fit in and the sheriff caught him (laughter) just got out of jail." More laughter. On the way to Shreveport one Saturday I was doing 70 mph in a straight two lane highway – scattered farm houses here and there and some cattle. Up ahead one trooper flagging me over. He went back to another motorist parked in front of his vehicle and finished writing a ticket. I had pulled up about 20 feet behind the trooper's car. While he was writing me up I saw his car out of the corner of my eye very slowly backing toward mine. A slight bump and the startled trooper looking foolish said to me, "Go on but watch your speed" but never gave me a ticket. Since I was in heavy Bible country all I have to say, this was a God send – I wasn't doing too good as a salesman down there – too many poor people. This was in the 60's and racial tensions were mounting. I was handed a notice by a Ku Klux Klan member one day and with that an innuendo, "You'd better come to our meeting" pointing to the sheet I was holding "Guys like you need to know what this is all about." Of course I became worried, these "aliens" wouldn't hesitate to hurt your family or burn your home down. What had I done wrong? I kept calling blacks Mister or Missus and befriended a few. While a Chevrolet service manager I had sent a black young helper to test a car we had turned up and told him to run it up to 60 on a nearby country road. A half-hour later here came my boy looking sheepish in the company of a short stocky trooper. He was a notorious racist. "This nigger" flicking his thumb practically in Joe's chest standing behind him looking miserable, "Tells me you told him to git on the gas pedal. Is that right?" "Yes sir." He left us and walked into the showroom after telling me he was going to talk to the dealership owner. Later, the boss told me, "Don't worry, I gave him a cold beer." Times have changed but how would you like to be in charge of 18 men and every day you have to hear the N word used. I got real disgusted after a while.

There was beauty to Louisiana nevertheless. All kinds of creatures such as a very large rodent called a Nutria who would be seen swimming across bayous. It came from South America. There were vast fishing areas anywhere from brackish waters to that of the Gulf and other outdoor activity. We had a 19-foot boat but we sold it. In nine years we attended three Mardi Gras. You think you're just going to stand there with thousands of others on sidewalks and stay calm. There is something crazy about mob reactions on the happy side. There we were jumping around like loonies trying to catch necklaces and bric-a-brac mad cappers on passing very original colorful floats threw at us. I knew quite a few doctors where I worked in New Orleans and they also got caught up in this insane behavior. The parade seems endless but I wouldn't think of going to see it at night. A touch of weather information now. Spanish moss grows all over, hanging from live oak branches but has no roots. It obtains it's nutrition from the air which proves there is something else breezing around. Some people eat anything! They told me Nutria were very good. A neighbor one day offered my wife two dead squirrels he had shot. She politely declined his offer but when I got home I didn't hear the end of it. "How could he do that?? Shoot those poor things!!" On and on, "How can somebody eat squirrels?"

TO ESCAPE WAS A MATTER OF LIFE AND DEATH AND NO ONE LEFT BEHIND!

Getting back to animals of a higher form at least devotion to offspring, for one attribute. We were on a dirt country road, just exploring, riding around this Sunday afternoon. On the side, foraging about, a wild female pig and judging from movements nearby under bushes, her family of piglets. Had my family stayed in the car she would have ignored us but despite our carefully easing ourselves out of the car she made a decision. We were a threat and it was time to round up her brood. She simply and I might add acting very calmly about it, laid down on her side. She sent out some grunting message and we had to interpret it as a signal that it was Dinner Time. A half a dozen piglets charged toward her and attached themselves to her food supply. That done, she jumped up, ran full speed into the woods with the whole gang right on her heels. That's what you call in military lingo, one well-organized retreat. She pulled off a head count procedure and an "All Hands Accounted For" routine so quickly that no sergeant-at-arms could equal.

> ■ A federal appeals court says the Environmental Protection Agency exceeded its authority in devising the rules.
>
> **Los Angeles Times**
> WASHINGTON – A <u>divided</u> federal appeals court yesterday overturned the nation's most controversial and far-reaching regulations that establish air-quality standards across the country.
>
> Declaring that the Environmental Protection Agency construed part of the 1990 Clean Air Act so loosely as to render it "an unconstitutional delegation of legislative power," the judges by a 2-to-1 vote agreed with the <u>American Trucking Association</u> and <u>other industry plaintiffs</u> that the EPA exceeded its authority in July 1997 in setting air-quality rules designed to govern levels of smog and soot.
>
> The decision, which stunned environmentalists, represented a major setback to Clinton administration efforts to clean up the environment, and officials said an appeal of the ruling was likely.

THE WINNER: MONEY!

This morning's paper May 15, 1999 has done it again with their daily bits of worrisome news! No peace for wearying me but no rancor for ignorance either. Is the EPA that dictatorial, that omnipotent, since hundreds of companies for the past 15 years have successfully defied this necessary government entity? Since the men in black are not in agreement I think this smells, like nitrogen-oxide, fresh from huge double exhaust stacks on trucks.

The following, as a hoped for system of long range forecast has possibilities but the concept I present covers a period of long ago when abnormal ultra violet radiation did not exist. I have to allot myself the privilege of presenting you with a theory. It has to do with the fact that we have had always cycles of floods except if we go back one hundred years or even a thousand the periods of innundations had to be more widely spaced although it wouldn't be something I would be interested in researching. Other than a little more knowledge it would not be exciting at all. For a moment let me come back to today's times and how a cycle of heavy continuous rains throws a monkey wrench in matters of pollution figures bandied about and unintentionally false claims spring up continuously. We are told by one group, "Aha, we have reduced X toxin therefore success in cleaning the air has made great strides." And then someone else parked on a hill somewhere or perhaps located downwind from a major industrial area comes up with the opposite – discouraging figures that contradict. "Oh my, such and such a toxin has increased, etc…" The former make the opinion with empirical sureness, claiming such and such a mandate pertaining to devices on autos, this or that enforcement since 1980 or whenever has reduced such and such a pollutant by 19.812 percent – exactly that figure of course.   While it's true some degree of improvement always counts and cheers, the fact of what our weather has done, let us say for the past six months, case in point, the winter of late 1988 and early 1999, is <u>completely</u> <u>ignored</u>. Even if it were not overlooked the whole business of throwing figures at us is too complicated anyway – not that we should cease analyzing continuously. But to insist figures accurately represent what is going on is not too realistic. Firsts of all our Arctic region came to life this past winter because the air it received had to be cleaner (already explained) therefore, healthy normal cold wave upon cold wave set off storm after storm. Snow piled up in the Rockies and major low-pressure systems marched across the northern half of our country – winds blew and so on. Naturally this activity rids us of a large amount of pollution – after all, all those vacuum cleaners contribute something else besides rain and snow and with it the reactive verbiage over accumulation figures. And so, the pollution measuring experts, not even considering such advantages simply are convinced, or want to convince, "We're on our way – look, nitrogen oxides have decreased 8.92 percent in our state" and so on. You see my point? Can we be exact over causes and effects? I doubt it. Therefore, there remains but logic and appreciation of the Big Picture. In short, Nature has an easier component that we can appreciate. * If you don't believe in some sort of cosmic arrangement it is futile for me to expect your interest in what follows. I did say that I'm theorizing however. One should believe there is a reason for everything. If the earth's axis tilt changes by some inches every year (I believe it's eight inches) it is obviously (my belief anyway – copying Einstein) a deliberate arrangement that has been going on for millions of years for the express purpose of changing the earth's surface. Lush forested and/or harvested lands become deserts – time to retire tired soil – considering beneath the Sahara there is hidden evidence of greener times, just twenty feet under the sand in some areas that have been explored. Then of course mountains rise out of the sea heavily coated with protein drifting down for milleniums while submerged. How many hundreds of years before our soil is exhausted? I have no idea. I just spotted a student in the back row – he's sound asleep. He loves Star Wars instead. The screen flashes, Poof, 200 people extinguished by a ray, their ship but specks orbiting about Planet Nincompoop. Careful of Zobra though – he is leaving Planet Barboza with 10,000 indestructible robots, all armed, unless the good guys can find a purple asteroid with magic properties that promises the building of time warp shields. Once Zobra penetrates this shield his body will turn to dust – the robots will go out of control and destroy each other… Oh hell, how does such junk appeal to so many? Seriously, bear with me. This does have something to do some day, you can invest in an umbrella factory because you and only you will know when torrential rains will return. I just had some trouble staying serious. Why does man killing man, all movies making money excelling in this appealing crap?

Pollution is not all man made. Some centuries ago it was fine dust picked up from deserts (still is) and arid regions. The atmosphere as a layer is very thin (pay attention next time to satellite photos) can absorb just so much dirt and it would be fantastic if we could determine when it is ready to switch on the laundry machine. Why do I say it can absorb just so much? If this were not a fact (theory or no theory) then excess dirt particles would absorb more and more solar energy – this would create more warmth at all levels – furthermore land heat would be delayed from working it's way back up to the stratosphere then to the sun, during the night. Hence, excessive heat buildup on earth. Worse, warmth at high levels stops precipitation producing instability. In other words no rising air from earth's surface, no clouds, etc… Total stagnant stability would prevail and the inevitable desolation. Life would gradually end. But, the plan does not allow that. Alas, dirt absorbs sea moisture and you

Unless you enjoy complicating it.

know the rest. This peasant has his own concepts. Remember that Jesus didn't go to some ancient MIT, he was but a carpenter. Ask your computer what it thinks about my theory. Periodically, nature unloads, it takes a shower, figuratively speaking but I wish I knew how often in ancient times, now in the present or in the future.

THE WOLF

Sometimes in childhood there is mental telepathy – just a glance – a hurt look. My friend Leo, when I was but 13…we silently communicated one day when his best friend, a great big German Shepherd vanished from his life. No burial either. I used to walk several miles through the woods to go to his house and on the way skirted a fenced off pig farm, which, when the wind was right turned off neighbors. Hundreds of pigs. But it was a short cut. I saw this police dog chasing a small pig. Reports had it that some dog was killing them. But I didn't know. From behind a tree an Italian farmer, blazed away with pistol and down went the dog. Walking over the man loaded up again and finished him. It shook me up, believe me! I played with Leo but not a word was mentioned over the whereabouts of his dog. I knew, and he knew that. Old Country Italians knew how to handle chicken killing dogs that they owned. They just tied a piece of chicken meat under it's jaw tight against the throat. After several weeks of smelling decaying flesh the dog was cured. But, it took a milkman to kill another police dog, who frequently turn wolf when they run loose, no leash laws and a habit people had in my younger days, to stop a female dog and her obedient son, their nightly habit, of course, dig under fences, kill two chickens and off they would go. They picked on my stepfather's chickens one night. The son, a male dog was busy digging under a chicken-wire fence while mom stood sentry and kept looking around. I had borrowed a BB gun from somebody and waited at an upstairs window. When I tried to slide the window open they ran off. This female dog had a bad habit of sneaking behind a milkman but one day he smashed her in the head with a full bottle. The neighbor who owned her said nothing. But as a night pirate she had some success with us a few times. Dad, as I called my stepfather, was very upset, looking at feathers all over. You see, I'm a farmer also, sort of. I raised all kinds of critters as a teenager. Everything intrigued me. My   Stepfather was quite the hobby farmer while he worked steady in a factory and owned a rather nice home on a hill. He would raise 500 chicks up to either food size or egg producers. One time disease destroyed all of them in 48 hours. But, he learned all kinds of ways to raise them successfully. I learned about mineral requirements too. Chickens turn vicious when in need of something and will peck another to draw blood. Dad would feed them charcoal and whatever and that would do the trick. Chickens are not "chicken" at all, I saw a mother hen with her brood attack a cat. This was all entertaining to me, of course. I'm sorry, it's all very mundane but I just love all animals, including certain insects, like bees for example. My wife goes into a ballet dance, flapping her hands around if a bee wants to look her over. I tell her to stop it, it won'' hurt her but...what a fuss.

MY SOUTHERN HEMISPHERE WEATHER REPORTS ARE FINISHED FOR NOW.

By the way Frog Mouligne got beat by Mr. Garside, the Englishman in the last leg of their around the world yacht race. Overall the Frenchman who lives in Newport and is in business won the total race by points but Garside beat him in the last leg from South America to South Carolina. He did pay tribute to Mouligne for a change. I followed this principally for the hard to obtain weather information in the southern hemisphere. Quite the messengers these two men. The hero of course is the Italian, whose name I don't have in front of me but gave up one leg of the race by rescuing this lady whose yacht had tipped over. She was lucky – some felt she drowned, trapped inside.

EXTRAPOLATING REALITY OF HUMAN BEHAVIOR

When some 40 to 50 people die, trampled and suffocated in a mad frenzy to escape, you know that they were badly frightened. According to a newspaper account, a Russian crowd, mostly the young at a rock concert, had been drinking but generally alcohol makes one braver. Even two policeman died.

Minsk is located on the 54.54 degrees latitude north and if you follow that line westward to a more familiar geographical location it runs across the northern extremes of New Foundland to the southern tip of Hudson Bay. All three sites from Russia to Canada are definitely on the chilly side, weather wise. You are not to experience big hailstones in such country whatsoever, but on May 31st it happened in Minsk. Now, in Texas the reaction might be "Dang!" or "My car!" but that Russian crowd totally panicked as globs of ice pummeled them along with the eerie sound it created. This was a new experience, something very strange, something that struck fear in their hearts. Ideologues, which I'm not, would be inclined to say, "That's more evidence of earth warming! We've got to change things around right now!" Cornelia Dean's statement in her article strongly suggesting more information from scientists and brings crucial matters up to debate but watch out for ideologues pushing too hard (words to that effect) I want to make it clear that I don't think we can accomplish anything overnight. But that is what I ask you to consider: Other than more effort to combat the health hazards of nitrogen-oxides, Benzene, mercury or what have you, you're already knowing what that is all about. You must understand also why I am annoyed over the lack of information either deliberately delayed or apoplexy has set in amongst our leaders concerning the most crucial problem of all which is a more powerful dosage of ultra-violet radiation. That translates into prolonged periods of heat, droughts and wildfires. Someone was quick to blame air conditioner Freon and aerosol sprays a decade or more ago as one reason why the ozone layer above our atmosphere is in a sorry condition. I'm not qualified to argue chemistry but it is my opinion that something of immense proportion, something vastly all permeating and of a steady voluminous quantity in the way of acidic effects has to be a fact since we are talking of a vast portion of our skies some 15 to 30 miles above us. Be that as it may, my opinion as it is, those scientists who felt Freon was one culprit must have had a good reason. I am not alone in my convictions adding to their <u>relatively small</u> <u>dosage</u> wafting into our atmosphere, the huge volume of industry produced sulphur-dioxide which converts to powerfully corroding sulphuric acid when subject to moisture, that smokestacks pump out in ever increasing amounts. Consider just how great this quantity of toxins is emitted when power plants ask us to reduce our use of air conditioners, which they did several summers ago. Just try to imagine in a growing population, more affluent each year and more air conditioners and refrigeration in use, how great, if I may resort to a figure of speech, the enormous quantity of coal being shoveled into the roaring bowels of power plant boilers. You ought to see the huge pile of coal in Somerset, Massachusetts where citizens have twice demonstrated over the filth produced there. I'm not alone over this issue as several times in newspaper articles it has been mentioned that the ozone layer is being damaged by the chemistry described above. That protective umbrella put there by God and/or Nature is the only means by which we are protected from being slowly fried. Skin doctors are now very busy treating cancer but the problem is that the average citizen thinks skin cancer is but lesions, some red spots and so on. I've seen what it really looks like in the advanced stages. It is deadly, in fact one of the deadliest forms of cancer and gruesome or not let me jar you a little. If many southerners are referred to as "Red Necks" derogatory as it is, it is also because their carelessness is obvious after you have been around them for nine years as I have. At one time I operated a service station in Louisiana – I like outdoor work – I like dealing with people and most of all I don't want to go back to the boring work of drawing weather maps having over 2,000 of them under my belt. My sons helped me and earned spending money. One day, one came back from a customer's car and told me he couldn't handle something. "Please Dad, go take care of that customer." Nothing bothers Blood and Guts old Navy Chief here, but that face behind the steering wheel really jolted me! A portion of his left side of his face was gone, just melted away, teeth showing including molars and he kept dabbing away at what was left of his mouth with a large piece of gauze. His wife (I thought she might be) sat in the back seat, looking miserable. And so, Mr. and Mrs. Internet see if you can find that in your screens. Of course I'm harsh. How else do you penetrate the average person oblivious to what the sun nowadays is all about! Why the silent treatment by Washington or scientists? That gets to me! Nothing, zilch about the ozone layer. Articles printed several years ago completely forgotten. Can you guess why the silent treatment? There's no other word for this negligent attitude. The answer is simple. I'll convolute over material earlier in the book: Don't disturb moneyed interest and foremost, electibility has priority over sanity! And those fat stacks of lobby money, wow!

Yesterday, June 1, 1999, an airline plane pelted by hail and rain and winds crashed in Arkansas after swerving off the runway – you know all those details of course. All over this world more and more weather problems. To slight the power of nature is suicidal. As for the sun, latently it is deadly.

All I have suggested is a piecemeal elimination of coal burning plants with the financial help of our government and leadership willing to push and prod other nations to cooperate and not put off everything for 20 years. Insofar as vehicular emissions I suggested ways* in letters to Washington hierarchs that would not imperil the economy. An example, stagger work hours so as to reduce fuel sucking traffic jams every morning and every late afternoon. Already we have an ozone alert warning here in Providence, June 1, 1999, "Free bus rides and leave your car home" but it doesn't work. If the ozone problem becomes too critical in the future then what would be wrong with a four-day workweek and stay the hell out of the city so it can recover? Oh no, moneybags factory owners don't want to hear it. What is wrong in our electronic day and age of having red lights able to monitor movement of cars through intersections? We sit there, dozens of autos idling away, waiting for a green light, watching one or two cars all by themselves roll by. Urban sprawl? Good idea but that is one terrible decade consuming project. And what do the legal fraternity in Washington say to me in their letters? In effect, "We're on top of this Andre,

*Many more suggestions than listed here.

don't worry, we're on the ball and always have been." Oh sure.* In the meantime hurricane-warning centers are bracing themselves but that isn't what troubles me most. I fear agricultural destruction. Secondly, red tide, growth of bacteria and proliferation of insects. Water reservoirs must be constantly treated more and more with chemicals that are even suspected of being harmful to us. Bottled water? Benzene is spreading in areas where it is mostly collected. Did you know that agriculture represents 10 percent of our Gross National Product (that is a huge piece of our business) and affects everyone from the farmer, markets, truck drivers, to waitresses and finally 250 million plus of us at the dinner table. But, Andre here is told, "We have 2400 scientists and 2500 economists, two Nobel Prize winners (of course) at our disposal." "No problem." "We are concerned however of rising ocean levels in the next 100 years." Additionally, the White House has just waved a Magic Wand. Nature must wait, that's all there is to it. It has to wait 60 years for the Carbon Club to get around to correcting their God-awful smokestack emissions. And to repeat, the Supreme Court decided that the EPA was violating the Constitution for imposing rules for industry to abide by. No debate as it should call for…Pow, one kick in the groin by lawyers cloaked in black. Real dandies, oblivious, comfortable in air-conditioned buildings.

The least they could do, if they really cared, is to diplomatically pressure China and others over their shipping tons of airborne garbage over our western states.

*Backtracking now. A June 14, 1999 Wall Street Journal article reveals an attitude problem, but doesn't explain it that way – that is my reaction. "The Clinton Administration is trying to postpone until 2001 an international convention to complete the treaty to curb global warming, etc…"

I was told that our government was right on top of the problem, remember? A letter from the senator, very confident of course – aren't all lawyers confident? A sample bit again of that letter: "We are concerned…" Now on June 14, 1999 another article by the Wall Street Journal, this being the 5th one in less than two months. Doesn't that tell you something? Who else but this popular newspaper scans the horizons with precision for whatever developments will affect our economy? This article reveals no one in Washington, DC is really that concerned. Whoever is meant by "Our government" has decided any meetings and discussions over worldwide pollution, in other words the Kyoto Treaty members, need to be put off until the year 2001. That is the latest.

My only wish was twofold, a program to enlighten the public and wake them up, certainly not by me and also, just hopefully now, Washington, DC will come up with someone who will try to persuade China, India and others to take more action over their filling vast areas of our skies (the world's) with pollution – much of it finding it's way here. A huge area of the sky over the Indian Ocean the size of

America has been found to be heavily contaminated with pollution, per another article by the Providence Journal. It's getting worse, year by year.

Right now researchers are relatively silent – maybe they can't afford to bite the hand that feeds them.

The Supreme Court has knocked EPA to it's knees. So much so, Carol Browner, administrator has adopted a hopeful attitude: "We will try to at least pursue clean-up mandates against <u>one half</u> of polluters, if we can turn the Supreme Court's ruling around." If? In the meantime Al Gore who is beginning (finally) to do some rousing campaign speaking, having been pegged as "wooden" and of course, aroused by Governor Bush's stirring successes here in northern New England, doing his best to find a new home away from overheated Texas, is being over analyzed by the likes of Russ Limbaugh. Today, June 16,1999 my ears buzzed with his usual excitably masticating to a pulp what Al Gore or anyone else's words truly mean when he claimed Al Gore's book clearly can be interpreted that w will all have to give up on the internal combustion engine in 25 years. Without saying so, Limbaugh and no doubt Republicans will jump on this; it's horses and bicycles for all of us if Gore becomes president. I interpret the vice-president's ideas quite differently.

I am not a cynic but cleaner air efforts have been seriously set back. The best laid plans of men, however, can go up in smoke. Nature has no political timetables. We are being conditioned to the problem on a petty scale whereas I'm very convinced that there will be more and more deaths and destruction, not only here but also worldwide.

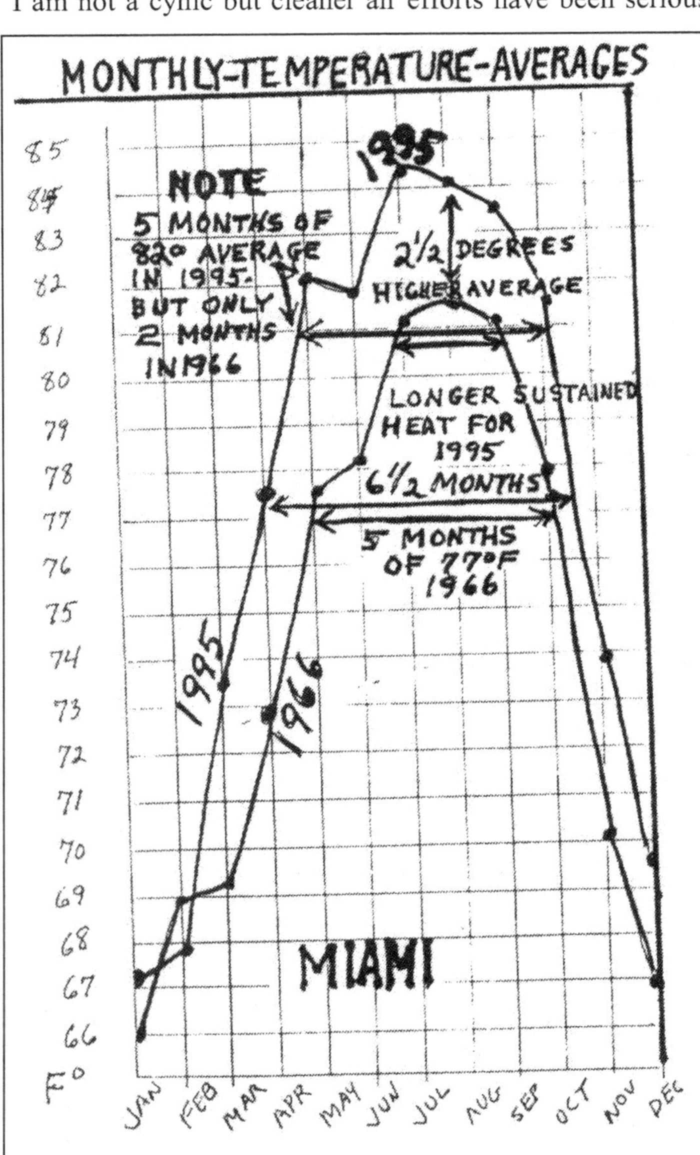

I gave prominence to recorded realities but you have not been truly exposed to that in any concentrated concerted way. Electability and money has outweighed reality. One possible reason is that bureaucrats in Washington, DC are obsessed about secrecy. I am reasonably quoting Jack Anderson and Jan Moller: "Instead of one thousand bureaucrats popping inked "Secret" on everything coming across their desks we probably have a million ego-oriented office managers wielding rubber stamps."

I'm glad that I live in Rhode Island – we are free of tornadoes and huge thunderstorms. We don't have to worry about grapefruit size hailstones either. Green Airport shouldn't have problems with such menaces to aircraft.

As for the profound wisdom emanating from the big intellectual centers, such as New York, Philadelphia, Washington, DC, etc…let it's residents consider losing three or four years from their potential lifespan since I read where many are not believers and pay the likes of Russ Limbaugh to say "Environmentalists are Whackos – there is no earth warming – that's a myth." Puny humans challenging Nature!

The whole world's climate is but a study in heat versus cold. Just temperatures – once you comprehend the interplay you've got it all. But, here is the problem. Nobody presents it on TV or in the newspapers in the manner that it should be, such as what you see on your left. A little explanation along with a blackboard session and perhaps 10 million viewers might say, "There is earth warming, isn't there." But you are treated to words and that's it. If then words do, what happens when some opposing force such as the Limbaughs and think-tank writers or the same on radio talk shows and they say, "There is no earth warming, it's El Nino, don't you see." What happens then? The once interested and the once ready to believe now begin to falter. That's what we have now – seesawing between yes and no. All Far Right spokesmen married to the Carbon
Club have it easy. All they have to do is toy with minds, most folks in a state of denial anyway, and they're very clever about it. Besides, how upsetting is it when experts say earth has warmed one and one half degrees in 100 years? Good grief! Really, who cares? I know a certain senator who wrote me is not losing any sleep over it and that goes for several hundred other House and Senate members. They'll never admit that but when they dilly-dally over objectives 20 years from now that isn't exactly indicative of paranoia, is it.

Earth warming is not fire and brimstone every day – it's slow, progressive and it is insidious. God doesn't want to fry us. In the graph above you are looking at a 29-year climb in monthly average temperatures. In your mind visualize warm air loaded with moisture having originated in equatorial waters, moving across Florida and the Gulf, then overland toward Alabama or Texas. From watery temps of mid 80's, in April probably more like 82 degrees but in traveling over our southland solar heat (5% stronger now) having baked the soil a bit, that same maritime air reaches temperatures up to upper nineties and near or higher than 100 degrees by the time it reaches mid Alabama or anywhere else from that point to west Texas. That is when tornado warnings resound over the air waves of course. But, the difference between the 1960s and the 1990s is the reason we have more tornadoes. For example, 580 twisters in 1966 and over 1200 in 1995. Has anyone ever told you that, I mean in concrete terms, statistics, charts or whatever? No. Referring to graph, April averaged 77.5 degrees in 1995, which means daytime temps were in the upper 80's and during the night in the 70's. Alabama experienced tornadoes that killed almost 20, April 1998. I include Jarrell, Texas previously explained as having the same weather patterns as Alabama. Now compare the 73 degrees average in 1966. That goes for almost all southern sites during that period. Quite a bit cooler both days and nights, therefore the reason why tornadoes in 1966 were much less frequent and tornado conditions were generally ready <u>later</u> in May. That's one point you should gather from the graph* - please note the longer period of warmth in Miami, being one of the better indicators of fueling air masses, in 1995. <u>Five months</u> of 81 averages as compared to just <u>two months</u> in 1966! Plus a 2 ½ degree warmer trend in summer, which by the way is not good news at all. More months of heat means more tornadoes – redundant of me but I want to be convincing. Additionally, although I can't jam this book with too many illustrations I can produce graphs for a lot of weather recording sites all the way to Juneau, that Alaskan city we have already covered graph wise in the book.

There are shrewd folks at work drumming up confusion. I do believe dollar bills would roll out of their heads if cut open. To my way of thinking they are all Louis the 14ths.

Just to make a point, Mobile now records 3 degrees warmer and Atlanta over 4 degrees higher than averages of the 60's. Late insertion: 78 degrees in Fairbanks, Alaska on June 13, 1999.

It would be unfair if I were regarded as someone constantly on the lookout for something negative whether political and/or weather wise with the sole aim of piling it on. Instead my experience tells me what to look for as

All data from NOAA records.

symptoms to share with you. For example, a major typhoon (it was called cyclone) May 20, 1999 spawned in overheated polluted air over the Arabian Sea not only demolished coast line villages of Pakistan while producing extraordinary winds up to 170 mph but some 700 hundred human beings missing, maybe more – it is now just the 22nd of May. Newspapers, at least here in RI gave it a few inches of space and as far as weather experts, not a peep out of them. What does it mean to me, this typhoon? First, it's out of season, much too early, therefore excess solar radiation and heat retaining dirt is to blame. Heat is heat and that is what hurricanes and tornadoes thrive on. Of further importance and significance to us is the fact Pakistan's coast is on the same latitude line that, as a reference line, encircles the world and crosses between Florida and Cuba. Remember, this equatorial zone wrapped around the world is subjected equally over every cubic mile of it to the same 3900 plus Angstrom frequency ultra violet radiation from the sun. We know this solar heat is stronger than it was, say, 30 years ago and we know the shield that should be filtering this force a little better is not in good shape. We suspect, and I can't see anything else that would do it, acidic type chemicals produced by, in their order of worst to less, coal burning power plants, diesel engines and autos. And so, weakening at the knees, worried about support from industry and elections coming, the White House, has simply allowed Carbon Club to have their way, to continue unintentionally or indifferently sickening and killing thousands with powerful toxins and just as worse, not giving the Ozone layer up there between 15 to 30 miles altitude a chance to recover! Contributing to my being perturbed is black robed lemming mentality in the Supreme Court who have declared the EPA is violating the Constitution in the way they are setting up rules for industry. Therefore America, standby for scorching heat*!

I've already worn thin my method of respecting thermometers all over our Caribbean for example and all over the entire world for that matter as an argument over phonies selling specious material but quite clever at literary arrangement and witty frills. But let me add a little more for those who do not believe there is earth warming. My family and I lived in southern Louisiana in the 60's as I already have emphasized. I asked an old timer fisherman down there "When in heaven's name is the hot muggy weather going to end?" No cool days whatsoever – it was always somewhere between 95 and 100 degrees in some spots, like the city of New Orleans. Most assuredly he told me the heat starts about June 15th and ends about the 15th of September. Sure enough, that's what I experienced. If there is nothing wrong, as the Limbaughs and hot shots from think tanks want you to

Predicted hot summer – nobody else did. It happened!

believe, determined to keep you misinformed and definitely reeking with ulterior motives, why then for the past 20 years or so the heat continues right on into November? Why tornadoes in January? Why the heat doesn't wait to work it's way up from the semi-equatorial regions for June 15th and already in mid May, 90's are being registered in the Keys and southern Texas? Lemmings don't think, they just go into a self-destruct phase, that's why and it applies to human beings.

We had a banana tree in our Louisiana backyard but the fruit was small. Summer didn't last long enough or probably a soil deficiency. It wouldn't surprise me if our agricultural crop sites will not be as productive as they use to be and eventually further northern areas, such as Canada, will be more suitable. My concern is food prices and other costs that warming climate will escalate to where social problems will increase. Am I a worry wart? No. I just know dubious leadership when I see it. Younger generations are bright, no doubt about it, but they are too young to have experienced the sun when it was more agreeable and less dangerous. I do remember. A fact: also a worrisome medical topic is Australian medical circles: Skin cancer rates are rapidly rising. I just heard some baritone-voiced TV ad-man say, "Stop living in the past" Over something commercial but that's what good weathermen are about. The whole business is one of referring back to memories of systems and their habits, the rhythm of seasons, remember some weather arrangement a few years back that proved to be dangerous and recognizing it is here today on the weather map. To think that systems behave in a constant manner and therefore fixed rules apply, or there are sure fire formulas to be had with a lot of research, is not to recognize that Nature is very fickle.

I have to insert a little more humor. When I worked in New Orleans I by chance listened to a very sad story by a broken-hearted Italian citizen who worked for their consulate. She, with the help of her hands, putting on the saddest look her big brown eyes could muster up, accent and all, asked us to help her find a car that was stolen from her. When she left telling us she would be back in a few days to see if we had any luck we could smile and joke about it. This car was exclusively the only one like it in all of New Orleans. It was a tiny Vespa and if it was gone she should have been persuaded how much safer a bigger car would be but we knew that wouldn't alleviate her misery. About the same time, speaking of small cars, this psychiatrist, one of our customers owned a tiny Sprite convertible but finally got rid of it after the back end got mashed two or three times. His reason: "This car is so low, everybody keeps running into it." Despite the sad story of the Vespa lady we thought it was very funny, especially her antics. Anyway, somebody stole it all right; drove it to the edge of Lake Pontrachain where the retaining wall is about ten feet above the water level with built in steps and with the help of others simply slung the Vespa into the lake. She got over it and soon was driving something a little safer. Maybe we were cruel but we thought this fitting end for a peanut car in a heavy traffic city was a blessing in disguise – we laughed when we heard of the Vespa's fate. We did like her – she was quite charming. Of course, we always put on our most serious look when she would saunter in and ask if anyone had seen her car.

If I were to be regarded as chauvinistic from what follows, humor thrown at a silly young woman, I would be pained. Let me therefore put up a buffer, just in case. I want very much to see more women in politics simply because too many men are wolves in sheep's clothing, power-oriented, greedy and basically the cause of wars and misery.

My wife, this one day, was caught in moderate traffic, waiting for a light to turn green – foot on the brake on an inclined two lane one way street. To her left a dental office building, it's large parking lot sharply sloped toward the street. A young woman, probably a receptionist judging from what ensued later, had parked her Beetle Volkswagen, stepped out but before she could enter the building realized her car was rolling backwards. The vehicle in front of my wife moved up a few feet but my wife hoped the VW would choose the increased space. Almost. It glanced off the front bumper, deflected slightly up the incline, bounced against a curb then demon-like, rolled backwards again and crunched my car's right rear fender. Hitler's People's Car had done a job on us. The girl trotted up to my wife, sitting there dumbfounded and said, "It's just like an old time movie – I don't believe it" and then giggled. A policeman arrived and after filling out some papers sarcastically asked the girl if she knew what she was supposed to do when parking. When my wife called the young lady's insurance agent as suggested the respondent man exclaimed, "Oh no, not her again."

It is now May 26, 1999 but I can't bring this book to an end, not yet anyway. Every day the Providence Journal (Pro Jo) has an article that helps my cause, namely to validate some aspects of my contention that there is a shaky relation of truth and sincerity versus titillation – oriented information the American public has been treated to, to the extent even members of Congress and the Senate that I have read about don't believe in Earth Warming while others do. Therefore the tragedy of delayed action or at least concentrated debates.

Don't misconstrue the following as another choice piece of criticism since I am not positive if a one year project that took place in one of our colleges (leaving names out) was successful or not. About a year ago ProJo had a photo of representative so and so shaking hands with a department dean while staff members stood by. They had been granted 150 thousand dollars, the agenda or curriculum to foster a better student journalist by working alongside attending scientists. The idea being those graduates would then be better disseminators of scientific information. I did elaborate on this need in past pages of my book judging from the absurd to masterful presentations I have collected. That is why the book is replete with excerpts, leaving names out, even though quite a few had vague origins, no names of experts given but others' nearly incomprehensible for public consumption, pedantic no less – the aggregate of which presents you the reader with an accurate history of a game I call truth versus showboating. Ten years of this Ping-Pong game raised my ire. But, in reading of this introduced program covered by ProJo I do not trivialize it's intent, please. It appears however, judging from two articles yesterday and today, that the project was not that successful. Yesterday our paper quoted both student and reporters who participated as saying that scientists talked over their heads, "We don't understand what they said." "They always say that they have a lot more work to do." (Oh boy, what did I tell you many pages back?) But the scientists were also quick to add; "Reporters are too impatient." But today's article, suggestively being more in depth but emphasizes the same thing; excerpts below. The NY Times Editor of Science, Cornelia Dean, makes me feel like boasting a little. Please note in her quoted opinions the word lethal underlined because that is a badly needed choice of words. However, I did emphasize I wasn't about to emulate Nostradamus. But, it is gratifying to pick up a needed boost for what I had already insisted upon before Cornelia Dean's views came into print and also we have some level-headed individuals not intoxicated by gyrating stock indexes.

I can't copy Limbaugh's "Talent on loan from God" or "With half of my brain tied behind my back, etc…" I'm not that potent a dispatcher of facts, but here goes anyway. Don't you think I've done one hell of a good job putting you in the position of not being bamboozled by spin-doctors? You now will have the satisfaction of interpreting TV's weather picture and predictions with astuteness, a penetrating acuity, (a reasonable facsimile-weather Guru no less) that you never had before. I have taken the mystery out of mystery peddlers, some blatantly promoting rubbish; some that I strongly suspect are manipulating minds so as to garner future research grants. At least you will not believe some trim-bearded showboater telling you they don't know what causes tornadoes – I never got over that one! Now you do and much more in depth than you think. No applause please, no accolades and I hate lecturing even though General Motors sent me to school just for that. I don't mind autographing though and all you young men, don't squeeze my hand too tight, I'm getting old. No TV makeup either – can't improve on my looks no how.

New York Times Science Editor Cornelia Dean called on scientists yesterday to overcome their reticence about dealing with the press, not for the sake of the media, but because their failure to do so could prove "<u>lethal</u>" to what <u>should</u> <u>be</u> <u>a</u> <u>vigorous</u> <u>public</u> <u>debate</u> on a number of vital issues.

All too often, she acknowledged, science journalism is "oversimplified, <u>overhyped</u> or just <u>plain wrong</u>."

I think that the article describing Cornelia Dean's opinions was the best! I asked the lady who printer types to frame the excerpts.

Going along with Cornelia Dean's aptly expressed "scientific hype" let me quickly review the roller coaster ride we have been taken on. Our newspaper, ProJo, whose principal chairman is an environmentalist, offers almost daily the amazing discord and morass caused by enthusiastic experts. One day, for example, air quality has improved over Maine and Vermont for this or that toxin and the next day's article bemoans over 100 wells contaminated by MTBE. Adding to the confusion: There is twenty times over acceptable norms of mercury in the

air in the same New England area, my having quoted a scientist frustrated over "lemming" skeptics and an inert public.

<blockquote>
They should not remain silent, she said, on issues such as high-tech reproduction, genetic testing, physician-assisted suicide or <u>climate change</u>.

Dean said scientists need to be more forthcoming, for the sake of a public that is all <u>too</u> <u>often</u> <u>uninformed</u> about a number of issues that emerge as hot topics for political debate.
</blockquote>

Now that I have leaned on more prestigious opinions than those of mine have, nevertheless substantiating what I have maintained here and there in the past pages, let me summarize one acutely accurate and very pertinent point. A point conveniently overlooked by a flock of alleged experts, whether they are shining on TV or masterfully* producing articles. Earth Warming is not the heavens falling down upon us nor is it some mysterious difficult force at play forcing several thousands of scientists probing a billion (whatever) cubic miles of our atmosphere (they're even up in the stratosphere now) while we are told over and over again that there is much to be done. That's exaggerating. I have shown in the earliest pages through authentic records, a sample I included of their method by NOAA at Asheville, NC where weather records are stored, key weather sites in the oceans that show summer heat now lasts 10 months when it only lasted 7 ½ months some 25 years ago. Do I need a master's degree from MIT to draw for you graphs that clearly make the unusual change obvious? It was just a matter of my knowing what to look for. In short, you have been shown what Earth Warming is about! Earth Warming means earlier** and later tornado seasons**, well out of season. It means long drawn out summer heat (nice for beach goers if they buy a lot of Sun Shield Goo). It means great harm to agriculture and dead cattle while dust devils whirl about over their rotting flesh. It means prolonged hurricane seasons and of course, borrowing from Dean's very strong statement, <u>lethal</u> results.*** Did you know grapefruit size hailstones fell in Texas May 26th? That requires stupendously powerful updrafts. How would you like that hitting you in the head or punching holes in your roof or maybe your wife's car looking like vandals had beat on it with a sledgehammer? Speaking of getting hit in the head (figuratively speaking) somebody in Washington D.C. needs it badly.

*If masters of literary finesse or experts have not made a peep or jumped into newspaper print over the above two indicators there is a lot of heat around when there shouldn't be, it is because <u>they</u> <u>no</u> <u>longer</u> <u>can</u> <u>blame</u> <u>El Nino.</u> Ultra conservative think tanks' media oriented members look for something in weather that they can use to punch environmentalists around. These two factors above have them probably retrenching – hence, no articles as of yet.

**Over 150 tornadoes, some in November and many in January of this year are very abnormal.

***Typhoons in mid May (Japan) and late May (Pakistan) are also out of season.

In a light hearted fashion and not as professional as I would like and yet substantive and interestingly as I could put together bits of the past as a semi-autobiography and what molded a philosophy of my own and perspectives in a day and age where ignorance could not recognize free-wheeling irresponsibility after which wholesale slaughter of human beings on a level never achieved in the world's history. I admit I have been weaving also a transparent veil, laced with humor through which dangerous developments I anticipate appear less than foreboding. It was, how would you say, my tiptoeing through the tulips with you in between technical descriptions of what is dismal but also warning you the dike ahead may collapse. That's another way of supporting Cornelia Dean's viewing quiet chaos in inner scientific circles and political merry-go-round tactics as potentially <u>lethal</u>. I said to myself after reading of the White House capitulating to the Carbon Club, giving them 60 years to meet standards of decency and worse, the Supreme Court temporarily taking the bite out of EPA, are these people in power nothing but Louis the

14ths? Quietly they intend to build up numbers of emergency shelters and even are caught smiling while examining devastated communities.

My Capitaine le Detective nose unearths contradictions in superbly written material by an author fortunate enough to be published by a most prominent organization, two pages devoted by a local gazette highlights the pitfalls of urban sprawl which as the author details will take us on the road to disaster. Leaving the sensible reasons he gives aside and without my copying a Russ Limbaugh "Aha" he points to the harm done for a decade following the oil embargo of 1973. I don't remember anything terrible following 1973, but apparently the author wants oil consumption to remain high and keep on going up. So; great, if that is so very important to him. But, he condemns the growing numbers of autos from coast to coast as producing an automotive slum. If the author would like to see a more sane arrangement of mobility, doesn't that mean we will need less oil? Didn't the publisher's editing department spot that contradiction? I did. However, the author merits our praise nevertheless as he envisions great changes for the better if somehow leadership frees itself from political paralysis.

Wednesday, June 9, 1999. An excerpt from a daily column by Jack Anderson and Jan Moller of United Feature Syndicate, Inc. is shown below. Of course Republicans and some Democrats too are guilty of promoting <u>polluter-friendly</u> policies! The title of this day's article is "GOP TURNS IT'S ATTENTION TO EASIER TARGET: AL GORE." I see Mr. Gore as a visionary rather than a skilled* politician but pathetically Republicans are going to work on convincing the American public that he is a menace to their lifestyle – a preposterous idea! Too bad we don't have a tribunal some 10,15 or 20 years from now and march those contributing most to destruction and deaths for an embarrassing and career wrecking judgement across cameras for all of America to witness. The Vice President was absolutely on the mark when, for once, he blew off and told the Congress "You have your heads in the sand!"

June 12, 1999

Gas emissions drop

WASHINGTON – Although the U.S. economy is humming and fuel is still cheap, the United States finally may have slammed the brakes on its insatiable use of energy and its output of the gases implicated in global warming, two new studies have found. For the first time in nearly a decade, U.S. carbon dioxide emissions did not increase drastically in 1998. In fact, they decreased slightly, concluded the American Council for an Energy-Efficient Economy, a Washington think tank. Greenhouse gas emissions dropped 2.2-million metric tons from 1997 to 1998, falling 0.15 percent. (Knight Ridder)

If this excerpt collected today the 12th of June but forcing me after contemplating finishing the book several days ago to insert it here out of sequence considering a diary mode of writing along with relief-providing items, represents good news, it typifies exactly why the public remains numbed-down, that expression more PC than dumbing. In this case it is not intentional however, but it must be clearly understood by you the reader, they're

claiming the USA has slammed the brakes on CO2 emissions, etc…is but a misconception – or a misnomer referring to the word "brakes". At present that is an impossibility. It is nature that has done so – not human kind, my having already amply explained in previous pages the purging effects of several years of widespread torrential downpours ending earlier this year, 1999. Everything that burns creates CO2 – you know that. Because growing populations, expanding economies and more and bigger vehicles than ever it is preposterous to think nothing else than more carbon dioxide can result! Absolutely ridiculous! Someone fails to understand major floods

A little more charisma is needed.

worldwide, immense storms over the north Pacific, many parading across our northern lands this past winter, all these rain and snow producing systems not only drag down CO2 to sea surfaces but also act as giant vacuum machines pulling up pollution and excess gases and from high elevations are transported to our polar caps. Core samples of Antarctica's ice reveal that. Furthermore an agitated sea during tempestuous weather, months of it, also absorbs CO2 since oceans are known sinks for this heavier-than-air gas. While I am not qualified to chemically describe the effects of churning ocean waters it has been advanced by less than positive scientists change occur in the way CO2 is absorbed during storms.

I am trying to express myself as a good lawyer might. It is important I spare no details and painstakingly present every facet available or otherwise my case, my argument would be tainted with comments suspect of contention. Of course, less CO2 is important but the upward trend in more of this gas has to resume sooner or later given the reasons already explained.

SPEAKING OF HYPE A FEW PAGES BACK, ONLY "BRASS" MATTERS. UNHERALDED, MOROCCO NEVER MADE DOCUMENTARIES.

That is very easy to explain. Historians need top brass to rant and rave over. Port Lyautey (now Kenitra) never had big-time warlords. Here was a country, other than Bing Crosby and Bob Hope (that was it) singing "On the road to Morocco" captured more or less after three days of fighting, some 1000 ships and the USS Massachusetts, a carrier, thousands of soldiers (I remember the 3rd Division with their blue and white striped shoulder emblem) involved, maybe 30 or 40 killed November 8 to the 10th, 1942. On December 23 another 1000 ships or more arrived in Casablanca. Belatedly, after they were emptied of human beings and vital cargo, Germany sent six bombers but didn't accomplish anything. Here was a base, this Port Lyautey that was the busiest spot in the world, hundreds of fighter planes arrived there and the least recognized, several squadrons of submarine-hunting PBYs. Six submarines sunk but nobody, but nobody amongst our citizenry has the slightest idea how this amount to the equivalent of three Battles* of the Bulge in it's effect on the enemy. Do you know why? Historians again. You are treated to films of Captain Eichenburger, (any old name!) sub skipper, bearded, hat tilted back of the head, swiveling his periscope and zoom, there goes a foam churning torpedo. Then, boom a ship explodes in half. Then you hear the following, "Captain Eichenburger is credited with sinking 87 thousand tons of shipping" just another statistic over the weight of metal ships. Nothing about thousands upon thousands of human beings drowning –zilch. But, that is not what I am leading up to. If Hitler had been less anxious to strut over conquered countries and had recognized the spectacular results that could have been attained by hundreds of submarines WWII would have dragged out and just perhaps, England in desperate straits for materials shipped there by thousands of ships, many never reaching her shores, would have taken a terrible beating. Logically, if war intrigues anyone, submarines in great numbers would have changed everything. Consider just what one submarine can accomplish in preventing a dozen ships and then some, from bringing vital material and troops to Europe. Therefore PBY pilots did their tremendously effective work but never have been praised – Morocco is nothing in the annals of military history. On two separate periods November and December 1942 some two thousand ships crossed the Atlantic and never had one torpedo to contend with. One reason, planes out of Iceland and other strategic spots along with whatever destroyers we could spare, devastated German submarine fleets. Therefore, just a tribute to a few dead soldiers in this photo taken shortly after the invasion of Morocco and what Port Lyautey's role was during WWII. Aside from hardly visible white crosses there were many wounded.

Port Lyautey, never mentioned in all the documentaries that I have watched, was the only spot on the other side of the Atlantic that provided the allied effort with the means to wage war in Tunisia, the first battleground against the German army – the number of soldiers temporarily encamped there and fighting aircraft was awesome. At least we had Ernie Pyle. He knew what it was all about.

One sub or fifteen could have sunk a dozen troop ships. 1600 GIs on the transport I was on – so thin hulled as most were, it wouldn't take much to sink it.

From the dessert to the sea
Thousands of voices sigh,
But a letter, you were alright.
You smiled from so far away.
Time and sand, wood painted white.
A.L.

First, a changed sky – can't leave you in peace. I cannot expect a great interest in sky scanning, a daily habit of mine although the few moments I do is not noticeable – can't have folks wondering. The intricate beauty of cloud designs reflecting purpose, motion and sometimes forewarning has little appeal for many and other than casual comments you are always treated to, i.e.… "Looks like it might rain" or "The sun is coming through" that's about it. Fine. To each his own. It's not interesting conversation material anyway.

Although the evidence of increased humidity (permanent, by the way) is stronger than usual at high levels (16 to 22 thousand feet altitude) I wish I could interpret that as badly needed rain on the way. I can't. In order to have

Little rain in June a major pain for many
June 22, 1999

BOSTON (AP) –Not a drop of rain has fallen on Danvers farmer Bob Connors' corn this month, forcing him to pour 5 million gallons of water a week on his fields in a 24-hour irrigation effort.

But the crop is still showing some withering as it enters a critical late June growing stage.

Conners' corn needs rain, fast. But the weather reports have refused to bring good news, so Connors has stopped watching them.

"It's kind of scary," he said Monday. "We're talking about losing hundreds of thousands of dollars."

The lack of rain promises problems for everyone from farmers such as Connors to homeowners hoping for green lawns.

Some communities have already enacted mandatory or voluntary waters bans.

needed ground soaking rain, other than passing showers or thunderstorms, spotty as they are, we need the action-generating quality of a colder Arctic region. The sun is too strong for that now. Without the contrasting surges of cool air rolling down from the north there can be no steady widespread precipitation producing systems. At the end of this chapter I'll explain why my heads in clouds above and what it means. As for the above, I view it as very painful.

Since what follows would classify me as a pain, someone who is trying to drive you crazy, let me dispense with a reiteration. You are now as I insist, going to feel the expense of earth warming. And that is very sad. I'm going to shift the focus away to someone else however. Froma Harrop, who is quite the writer for our ProJo, had an article May 11, 1999, that, if she were to run for a political position not too many would vote for her. She defames the sacredness of the American lawn. In words to the effect, lawns are a pain, tiring and very expensive. But, a certain lady added her interesting bit today, one Karen Asher, educational chairwoman of the RI Wild Plant Society. Inspired by Harrop's article titled, "My last battle for a lawn." Karen Asher also wrote an article (June 22, 1999) diplomatically stating it's time to shrink the size of our lawns. Good God! She goes on with what I call a guilt trip for all you people crawling on your hands and knees plucking weeds or whatever by emphasizing the tremendous drain on our water resources. On the east coast 30 percent of water pulled out of soon to be inadequate reservoirs is used to water lawns. Then, oh boy, 5 percent of emissions comes from lawn maintaining equipment. The title for her offering, "Time to shrink lawns" and goes on with some very interesting ideas, trees, bushes, gardens, build meadows and so on. Additionally her society has available books for all kinds of property beautifying projects.

I have to add here our reservoirs will not handle future demands, therefore it's just a matter of time where watering that green carpeting will be problematic. I also note in my daily travels that the sun's changed intense radiation (you young don't realize that, I don't believe) is turning lawns brown. To aggravate you some more Karen Asher, my concurring with her, insists we use ten times more pesticides for lawns than for farmlands. And while you're gritting your teeth over this let me again grind a point home – Most folks don't care about corporate smokestacks belching out poisons and damaging the ozone layer and obviously almost everyone is in love with gigantic SUVs, vans and trucks – therefore, start paying. After writing something like that maybe a bulletproof jacket will be needed. A lot of kooks out there!

Speaking of, well, not kooks, but a stupid idea, the former governor of Massachusetts, Mr. Weld has stated that we should open the doors for more immigrants. That's all we need! Jam up city populations which will drive out (per David Brudnoy) decent middle class residents but sticking with this chapter, more demands on a faltering water supply system. You, taxpayer, have to accept the probability of having to pay to expand reservoirs and aquifer potentials – but both also require normal climate with ample rain – more expense in other words. I believe that Mr. Weld, coming from a long line of successful* something or another (have to be polite) must have friends who own armpit sweat shop factories.

*Microfiber, a company owned by a family from Scotland just gave their employees $2000 for vacation money – something that scrooges I know have never done – or why they're rich.

This one may call for not only a jacket but also a bulletproof car. Sooner or later we have to get rid of 2 cycle engines, whether it be lawn equipment or boats. They do deliver more power than 4 cycle engines but I don't see why we can't have them mandated (oops) on everything – they have to be bigger that's all. The oil* mixed with gasoline consumed by 2 cycle engines is incredible and much of it is** unburnt. It ends up as a mist in the air or worse permeates surface levels of lakes and coastal waters. I don't think the oil family Bush would appreciate this.

On a calmer note Cirrus clouds at high altitudes predominantly had that swirling, lacy delicate look but now, more often than not, excess humidity from the far west, originating over the Pacific and sometimes our tropical south, manifest their heavy dense presence of moisture in their own special way. Cirrus clouds appear as thick splashes of white and also, watch high flying airliners and see longer trails of vapor they leave. But, it's useless moisture. It's headed back to the poles where some day, scientists taking core samples from Antarctica or Greenland will say, "Hmmm, must have been very damp - -."

I would like to suggest that you go back to pages 6 to 8 and look at Juneau's yearly temperature averages for 29 years and see the remarkable warming climb! We have had a break this past winter but it will not last.

IT'S WAR! LEGAL WAR THAT IS.

June 23, 1999. Something new. The NAFTA agreement allows a participating country to lodge protests and follow up with suits when a large sum of money to be lost is involved. In this case a Canadian firm, Methanex Corporation, manufacturer of methanol used in the gasoline additive MTBE is suing California for 970 million for lost profits. The president of Methanex claims banning of MTBE because it causes exhaust to spew out Benzene, a carcinogen, will result in loss of sales and therefore is demanding compensation. I'm sure you recall my calling your attention to Governor Davis of California banning MTBE, phasing it out by the year 2002. If I were to judge I would approve banning MTBE and accept the consequences of exhaust then spewing out nitrogen-oxide. We can gradually reduce NOx (I'm skeptical of that – however) but it is the lesser of two evils. Benzene

*As a _quantity_ of unburnt hydrocarbons.
**Why they steadily blow out white smoke.

contaminated water is a grave problem which cannot be cleared up for a very very long time. Rest assured that protracted arguing will go on and on. The argument will more than likely focus on money rather than lethal consequences. Traditionally, it is money first and then human welfare.

Of course I look for everything that is wrong. It is flaws that sink ships, it is deception that has ruined nations. Anyone can gloriously write on and on how wonderful we all are. This can be a savage earth – we shouldn't be playing a crap game with nature! I look for selfish culprits clouding up issues. I read scientific articles and point out what they are obsessed about and what they ignore. As a kid in the first few months in the Navy I was handed it's bible, "Rocks and Shoals". Or, how to avoid disaster and keep the ship afloat and stay alive. It didn't go into how wonderful it was to be part of the American Navy – it pointed out, figuratively speaking, how to avoid running aground.

If earth warming as a problem requires 2400 scientists and 2500 economists, denoting then, somebody somewhere in Washington, DC footing the bill with taxpayer money, must be very concerned. It doesn't require a brilliant analytical mind to come to that conclusion. But why the silence? Other than the usual, for example, today July 16, 1999, the EPA is accusing coal fired plants of exceeding certain limitations in the production of electricity without a permit (?) coping as they are with great demands, proof additionally we're cursed with the threat of shutoffs hanging over heads in so many overheated cities because this nation has a deteriorating infrastructure. On the subject of inadequacies consider the incredulous wonderment of Manhattan's nightmare, ten dead from heat and a blackout while we spend billions stacking up smart bombs and plaster this planet with over one hundred military encampments. Now they're talking of landing on Mars. What's another 20 billion. On the subject of heat before I go on we have had always heat spells but my book is about frequencies, not record high temperatures. I forgot something in my haste – can you imagine a blackout ruining computer records because the batteries died out? Please. 679 people died from heat in Michigan during the last part of July 1934, 300 of them in Detroit. No air conditioners of course. Earth warming means increasing frequencies, not temperature records. The silent treatment? Either they, the powers that be, think that we are stupid, want to keep everything "out of sight, out of mind" or really don't know what to do. Before I use someone else more astute than I in political matters to back up my protest passed on to you, that just maybe leadership is lost in the woods I must confess that I am confused by the EPA and they're threatening stiff fines against power plants.

Anyway, I would be tiring to elaborate on this latest but old news. Best leave the Wall Street Journal and others to explain it. Furthermore, my intent here revolves around an attitude problem, i.e., nothing is coming out of Washington circles to clarify neither anything nor what their intentions are, if any. You are not told of 1200 tornadoes a year. You are not told what condition the ozone layer is in although Nimbus 7 is whipping around and around the world and a slew of rockets and balloons launched by a dozen nations and thousands of research meteorologists playing with computers and so on. Think I'm clawing away like a loose cannon? From a major science magazine, the last paragraph frankly admitting scientists need different approaches: The first sentence saying it all: "It's one thing to play around in the lab with our climate models. But the bottom line is, those are just toys." Says Jonathan Overpeck, paleo-climatologist with the National Center for Atmospheric Research in Boulder, Colorado.

AFTER THE SMILING FACE, THE CORONATION; IS CONTINUOUS REVERENCE SMART?

I strongly suggest that you read Arthur Schlesinger's book, "The Cycles of American History." It does away with beliefs, myths and the overrated but praises the worthy. The man has a hands-on, right there with the mighty, and describes presidents as they were and really are. Quite often they do not know what to do. But, bits of profound wisdom having to do with what this chapter is about, are on record from olden times. One shining example is President Andrew Jackson. Someone in his private entourage expressed concern over the ease with which an ordinary employee working in the Oval Office could find out secrets of the presidency to which Jackson replied, "Good! The more secrets the public knows the more intelligently they can make decisions and the better off will be democracy."

IT'S IMPOSSIBLE – IT WILL BE AT THIS RATE.

In bold print today: World population has doubled in 40 years. A secondary line: "Population expected to reach 6 billion this week." Never mind next year or the year after. Then in finer print: "Three hundred million live in areas suffering from water shortages." I want to go back to Morocco for a moment and describe a compelling sight, quite common and yet it struck me as to how precious water is. We have our vacuum machine salespeople. Arabs there had water salesmen hawking this liquid commodity, so essential to life but imperiled by changing conditions. They walked about from village to village, sticking out from the usual robed natives, the sun's glint bouncing off brass containers and cups, a spigot held in one hand, the whole affair held up by shoulder straps. You would see an eager client, after paying with coins or a paper franc, watching his cup being filled right to the line marking his money's worth, all business and then up would go the cup and down the hatch, unceremoniously.

No ice cubes – take it or leave it – warm or otherwise. The Moroccan Arab that I remember was very thin and I do believe it is because their body content of water is much less than ours. I recall one Navy doctor bandaging an Arab worker who had a serious cut, saying, "These guys don't bleed like the rest of us." I got thirsty typing this and had to open the refrigerator and pour myself a glass of bottled water. Can't stand the slight swamp taste with the odor of chlorine coming from our faucets.

I can imagine the day will come when your water bill will rock you since this is the direction we are being taken on. It pours in southern regions nearest the equator and some rain north and over southern Canada while a huge belt of our country slowly dries out. As yet it is not a perilous situation and the reason already explained has to do with near normalcy for this year. But, I am not at all inclined to do any wishful thinking – I am convinced things will resume for the worst in due time, next year or the year after.

Of leviathan proportions some 800 gray whales have died in one year, washed ashore from Mexico, all along our west coast and further north. They appear to have starved. Experts are not sure why although several theories have been advanced, some revolving around earth warming and pollution eliminating food supplies, whatever. Even someone, as expected, had to include El Nino. Although these creatures forage for special sea bottom food, small crab-like creatures and tube worms, it seems some facts of changing ecological shore line aquatic life, a few species of crustaceans dying or disappearing, makes it difficult to connect deep sea bottom conditions, temperatures so different than shallow coast line waters where the sun's additional heat and pollution have more impact. The amounts of toxins found in their blubber where almost invisible particulates and toxins tend to collect (humans take note – soft tissues, fat, call it what you want) (can acids be spotted by electronic microscopes?) doesn't seem to of high enough level to kill (how about to make sick?) and therefore this part of investigation has been put aside temporarily. But the El Nino saga – give it a rest. These gray whales became near extinct at the beginning of the century. Harpoons were responsible. Protected they bred from then a 2000 total to present day estimates of 26,600. Despite recurring warm spells in the equatorial Pacific every 3 to 5 years (El Nino of course) their numbers greatly increased. It is theorized there is not enough food. As for El Nino, on and off for centuries, the overall ocean surfaces of the entire world show but a one to one half degree increase, supposedly for the past 100 years. Could it be a degree of inbreeding making some (adult females it is said) more vulnerable to food absorption malfunctions, loss of appetite from tons of pollution (have you ever seen the murky density of particles, whatever they are, drifting down through ocean waters?) raining down from the skies and permeating sea water from top to bottom. I hope this is taken as just food for thought, something to kick around rather than thought of as an adopted cavalier for detail. Is it not true there is mercury present in our oceans for example? But women die here in droves from breast cancer. Despite at the rate of 40,000 a year, despite chemotherapy, there isn't any interest in trying to connect air and water-borne pollution with this dreadful problem! You hear, you see, you read over and over again this or that toxin in the air or water (Benzene for example) is carcinogenic. But the subject stops there. Can't we put two and two together?

Go ahead and vote for Compassionate George while the Carbon Club shovels campaign money in his coffers already spilling over. You'll eat more mercury, you'll breathe more oxides and drink more Benzene when his crowd gets through with you. I don't care if what I write creates anger, maybe that's what you need. No subtleties in my makeup – right to the point – right to the keester! Rest assured the demand for chemotherapy will keep climbing.

Imagining they epitomize clear thinking successful types who do not need government I'm sure enthusiasm for tax cuts abound in the ranks of affluents, who represent a major voting block. Panting, anxious for more, I'm sure they see the golden fleece waiting to be had and do not understand the tendency in roaring times to chase up and up that rainbow with the pot of gold as the ultimate prize, beckoning, to suddenly vanish as it has done before. Six times to be exact. The more sober and more experienced fear inflation as only tax cutting in a booming economy can create. For one, a contributing factor, "les noveau riches" don't care about price tags frequently nudged upwards in any inflationary spiral. Businesses have always been quick to capitalize on this reckless frame of mind. How high can stocks go before a dangerous plateau is reached, the index watched by the more astute, the older and less excitable, the "Stop sign" for the smart, price earning ratios at unacceptable levels – when investors begin selling, weak sisters bail out and national confidence begins to wane? In reading an ancient book "The Seven Pillars of Market Success" patience marks the very wealthy and the very wise – 10, 15, 20 years means nothing to them in the ups and downs of markets. They do not fear depressions – it's crying time for the general public and buying time for them. That's the name of the game.

There are decent Republicans (I hate to admit it) and it is becoming obvious their ranks splitting as they are stems from the old school upsetting the more modern rational, moderates as they call themselves or are labeled as such. The Hoover mentality has never gone away and in his day he despised the more evolved members of his party as they despised him. Old stuff it is not – history repeats – the chase for dollars is an old well-beaten road. Two outstanding Republicans George Norris or Nebraska and Robert LaFollette (had to be a Frenchman in there somewhere) of Wisconsin declined to support Hoover for reelection in 1932. This callous unfeeling president approved government loans for seed and animal food but coldly downed his thumb on helping the unemployed. We're not talking about today's more generous arrangements to tide over employees – we're talking about desperate souls lost in the wreckage caused by a wild laisser-faire period. From this disgust, Norris, a Republican no less wrote, "Blessed be those who starved while the asses and mules are fed."

It is stupid and dangerous to propel the economy at an accelerating rate. The results of a rapid worsening of wage disparities combined with the additional burden of climate change calls for caution and not Gold Fever.

What a salesman George W. Bush is with his "It is compassionate to give back to Americans tax money." Three dollars a week more for 20 million or more working stiffs and millions for the wealthy. Peanuts for many and caviar for the rich. If you lived in an old house in need of repairs and you loved that home and unexpectedly you receive a big chunk of money, what do you do? Run around and give a few friends and relatives gifts and squander it for something you really don't need? This country has needs and I'm not talking about handouts.

THEY CAN'T SAY "GOOD BYE, A DEMAIN." IT WAS SO EXCITING.

The effect of El Nino pushed up against our west coast 1982 and 1983, the excitement it generated for certain people bored by routine or seizing on a chance to step out of confined stagnation (familiar as I am with the monotony of it all) and talk to TV cameras. And, of course, an opportunity for the Status Quo protectors, that crowd selling us conservatism is something new when it is but rehashes, to brush aside tried and true causes of earth warming – forgetting the usual consistent scientist insisting more studies are needed. Consider that the environmental problem reared it's unwanted head at least 30 to 35 years ago but finally there arose in this California El Nino episode an opportunity for conservative proponents of "Keep our air dirty – profit comes first." This then was a phenomenon they vigorously latched on to! They gave El Nino an exclusivity it did not deserve, stretching it's effect long before 1982-1983 and still pushing it.

PLAY IT AGAIN, SAM

It was the first time in centuries, that I know of, except the west coast of South America, such a phenomenon had expanded as it did (no reason why given for it's dimensional extending) and moved (no reason given for it's means of mobility – might screw up their contentions) all on it's own from it's usual homegrounds to lay siege on our west coast. Naturally long lasting massive storms beat on our coastal regions this one time. Since El Nino's temperature rises occur roughly every three to five years the storms created are well out over the equatorial Pacific. Despite recorded periodic warm spells out there in the Pacific I cannot find (other than 1982-1983 and 1993 and only over states along the west coast) any evidence of unusual rises in average temperatures and precipitation have occurred in American cities correlated with El Nino activity. You have to squeeze real hard to find a relation.

If it were not for the blessing, the enlightenment of a recent report El Nino changes the whole world's temperature average by a mere 1/8 of a degree Fahrenheit, then only temporarily, plus the common sense of understanding there is always a balance going on between stable ocean values, the cold poles and the boss of all bosses the equator, but "puzzle sellers" sidetracking comprehension or at least familiarization, I would be hard pressed to devaluate a mental fixation or a persevering promotion, that El Nino is everything – nothing else matters in regards to our unusual (quite often) weather here in America, referred to as whacky on occasion. The <u>massive</u> <u>vast</u> <u>thermodynamic</u> differences between poles and the ever warm equator, creates churning powerful kinetic forces, vertically and horizontally, very violent here and there, that guarantees a comparatively small amount of extra heat such as El Ninos produce is effectively dispersed, diffused and well mixed in a trillion cubic miles of churning air. Heated air has to rise, fuel systems and then shipped to the poles. If it wasn't so life on earth would be unbearable. Only cold air, heavier as it is travels for any great length of time overland. Another rising center of attraction being La Nina, the chilling one has had little if any effect on American weather. Someone already has leaped head first, stage-bound, claiming, although it moves away from us and toward the Orient, it would bury Buffalo (an example put in newspapers) with record snowfall and frigid temperatures. I lived there five years, as you know. Always kept sand and a shovel in the trunk of our cars – got stuck in Watertown for two days one time, snow bound. Since nothing of the sort has happened caused by La Nina this man's shot at recognition or maybe supporting anti-environmental factions was singular and not particularly endorsed as viable and died as an idea as it should.

THEY'LL RESORT TO CHOICE PHRASING TO HAMSTRING EFFORT, I.E. "THIS IS JUST ANOTHER MOVE FOR THE GOVERNMENT TO TAKE OVER"

The Bug-a-Boo standing in the way of environmental cleansing innovations, or whatever, is oil and coal interests unwilling to accept reductions in consumption. Entrepreneurism is a wonderful thing. Sometimes the government should lend a helping hand and sometimes let the private sector develop something and arrange financing through the market system. If we, temporarily, unless someone brave rises up, have no practical visionary outlooks I nevertheless am very encouraged by two developments briefly touched upon below. However, there is always, lurking, waiting to agitate against air cleaning efforts resorting to instilling in the minds of many this is but infringements on our liberties and an impetus for more government, therefore, my hopes for bright innovations, new businesses, new vehicles (on and on) has with it reservations. The play on the human tendency to be nervous about change is a well-practiced one. But, we need both federal participation (the money, the power and the skilled are there) and the private sector.

John Todd's development, proven to work, the "Restorer", a floating platform with various ecological devourers of what is unwanted and filtering cells (too complicated for me to explain-not that I know a lot about it) rides ponds, lakes of toxins and bacteria-proliferating animal waste or what have you. But funding is needed. What would be wrong with federal help or perhaps towns and communities with polluted drinking water problems or condemned lakes and ponds (just happened here in Lincoln, RI) finding ways and means to pay for such a fine useful development? I don't know if the sales potential of the Restorer would appeal to the private sector however. Maybe it would.

BUT THIS ONE – OH BOY!

We have electric companies polluting while antiquatedly struggling to meet demands. Along comes a company, CFIC, that has really something exciting in the works! This company, located in Troy, New York, picking up on a development of Los Alamos, has a home heating furnace, circulating heated water-based, using natural gas that also produces 500 to 1000 watts of electricity, depending on its size. Of course, homeowners would still need the services of electricity producing companies. The cost? About 50 percent more than standard units. Just imagine if this was mandated (bad word) as required in new homes, our building them at the rate of 85,000 a year. No more oily smoke, chimney scraping and less coal burning. The unit pays for itself in three years. In the meantime, this Troy company is going to northern Europe with it's furnace because electricity there costs two to three times more than ours. Who knows what will happen from there. I do hope it will be a very successful venture and eventually adopted here on a massive scale.

Before the somewhat depressing appraisal of where climate change is taking us I would like to explain that there is a great difference between a two to three day high temperature record set, say way back in 1926, and five or six weeks of destructive hot dry weather. The latter figure, 96 degrees in Providence, in all probability occurred when a specific condition existed conducive to temperatures soaring up, such as: A major high-pressure center parked over our area, for example. Clear skies, no wind whatsoever and a bright June sun. But, the system moves off. That's the difference. However, one huge stale and enormous belt of high pressure over the semi-tropics, bloated to tremendous dimensions north, south, east and west, and going nowhere – that's a different story. We had that condition all over south and southwest in 1998 and it lasted seven weeks.

Excess and prolonged heat is the worst possible condition that can afflict the human race yet, little information if any is being disseminated to the public. Are we being told why? No, at least not thoroughly. I was told 2400 scientists are at work but do I read or hear anything worthwhile, anything instructional? From the few bits in newspapers, going back a year or more might as well consider everything is normal. Furthermore I question their allegiance, varying from pro-environmental to throwing monkey wrenches to delay comprehension. As for all those scientists, we're talking of well over 96 million dollars a year in salaries although not all work for the government. Plus a fortune for equipment. Couldn't Cornelia Dean pursue answers by badgering our elusive and too secretive government or are they just indifferent?

Aside from violent phenomenal weather, my having already overdrawn on the subject of more tornadoes, hurricanes, huge hailstones, even cool areas such as Minsk being bombed by ice, some 40 or so people trampled to death. One lady survivor saying she could feel the softness of bodies beneath her feet as she struggled to escape, terrorized and primitive reacting to the unexpected head-pelting accompanied by a low decibel din, visibility blocked out, but one thought; imagined heavens falling, there is also the proliferation of bacteria and disease-spreading insects. Right now, June 6, 1999, there are government workers in Florida battling African bees. As a matter of fact, Puerto Rico is infested with these prolific migrants from Africa, also finding our climate quite to their liking. Ticks abound here in greater numbers. Lyme disease is a problem. How can you enjoy the woods? We couldn't when my wife and I lived in North Kingstown for three years. Lovely wooded area and beautiful homes with no population congestion. We looked at the foliage – that was about it. Two acres my daughter and her husband owned but, our grandchildren were told not to venture beyond the yard – woods were out of bounds – wonderful! Drinking water reservoirs, (if they don't partially dry out this summer or the next, or don't develop patches of green scum, or hyacinth growth doesn't flourish) now have to be heavily dosed with chemicals. Bacteria thrive on heat – we all know that of course. But, are chemicals good for us? I doubt it. Bottled water? We spend over three dollars a week on the stuff but now we are told Benzene has ruined over 100 wells and of all places, Maine. The state is sparsely populated but it is now heir to big time producers of toxins west of there and heavily populated regions anywhere you look direction-wise, south, southwest or west. Internationally, need I add on and therefore overdo all this negativity by pointing to two million annual deaths from malaria and typhus – both transmitted by insects thriving on heat and moisture? A fact: It's all because of the sun. A sun that is stronger by just 5% is in the long run quite harmful.

I reported the facts. I did not speculate or twist the truth. You draw your own conclusions. I feel no obsession to witness great changes - experience conditions the rational thinker to be patient with hopeful ideas while trusting good will prevail over bad eventually. In economics, complexities require sages that are rare and if change is called for, the human impulse to grasp at quick fortunes deters and discourages foresight – the wise man, or woman of course, may drop by the wayside until bitter lessons teach and something new arises out of demoralizing periods – it has always been that way. Arthur M. Schlesinger, Jr., wrote a fine book, "The Cycles of American History" defining the to and fro of idealism, wealth, depressions, corruption and of course progress. Impeccable writing is not my forte but this nation and the whole world for that matter is faced with a very different kind of cycle – the socio-economic disruption looming ahead due to climate change. It is not perceived that way. There is no sense of urgency whatsoever. Inflation, which has to be exacerbated by climate change, is merely viewed as a problem of booming times. We are, borrowing the essence of economist Albert O. Hirschman's views, (his book "Pendulum Swings in Moods") prone to periodic alternation. We experience eras of private interest prevailing and then public action through representation. First of all certain scientists and experts belong to exclusive clubs and when they do publish a report it's aim is not necessarily to enlighten the public. If it is, then they ought to be made aware why Jesse Ventura became governor of Minnesota and try plain English. The American in general needs impactive imploding expression and not airy Shakespearean exercises or excess in lengthy English compositions, the delight of eggheads. How pleasurable! What should take 20 words to express is blown up to three paragraphs. Critics rant and rave but how many books are sold? Some non-frightening sense of urgency should be cultivated so that the public would be more open to consensus. Instead, everything of corrective steps is put on hold while the reality of expanding populations and economies, therefore increasing contamination of our atmosphere, as perilous and as increasingly more difficult to reverse, is not cause for worry. One month of nice weather and all is forgotten.

During the week ending June 25, 1999, again the Wall Street Journal has added two more articles. What did they boil down to? And, less than two million Americans read them. In short, the powers that be, namely EPA, Supreme Court, the government, the auto industry, truck organizations, oil refineries and coal consumers are playing chess except they hassle over who should make the first move. The auto industry is asking* the government to impose a requirement on oil refineries to produce gasoline having little or no sulphur claiming they cannot improve emission controls nor fuel economy (can't lean mixtures any more) because sulphur damages and/or renders catalytic converters less efficient. The oil man responds with "We can't afford it." And Detroit, how really concerned are they? The biggest tanker, the incredibly thirstiest fuel-guzzling leviathan vehicle that man has ever been stuck with or stuck behind their massiveness now roll off assembly lines in record quantities.

China, India and others, the Pacific sky loaded with their industrial particulates drifting over our nation – do you hear of any diplomatic effort, maybe an appeal if that is not beneath our dignity? Mr. Schlesinger makes a point of a gradual decline in the art of diplomacy – I have to refer to someone else making such statements. Frustrated but not giving up despite the ignorance of non-scientific types, (our Supreme Court) justifying as they did, curtailing EPA efforts on the grounds of "Insufficient scientific proof..." (therefore unconstitutional, that's the latest!) the lady in charge of EPA, Carol Browner, announced that she will try less stringent rules aimed at coal burning plants, plants that must deliver more and more electricity, year by year. Consider increasing heat (No proof? My God are they that removed from reality!) and 40 or 50 million (I have no idea how many) air conditioners consuming 100 to 300 dollars a month in electricity every month for six to seven months, never mind the four months of Yesterday. And you, dear reader, who ever presented proof in graph form as I have? No one. Let me refer now to where we are going. Let me resort to typing word for word what a very realistic Arthur Schlesinger Jr., has to say, page 46 of his book, praised to high heaven but probably not read by the average citizen whose support is badly needed:

"There are enormous potentialities for disintegration in contemporary America –
the widening disparities in income and opportunity; the multiplication of the
poor and the underclass; the slowdown on radical justice; the structural propensity
in inflation; the decline of heavy industry before competition abroad and the
microchip at home; the deterioration of education; the pollution of the environment
and the decay of infrastructure; the rotting away of the great cities; the farm crisis;
the mounting burden of public and private debt; the spread of crime and violence."

They don't want government it seems but are quick to appeal for federal mandates.

In fact Mr. Schlesinger goes on with the opinions of others: "One can be certain that neither public purpose nor the free market, will do away with these anguishing problems. This leads two of our most acute diagnosticians, Walter Dean Burnham on the left and Kevin Phillips on the right, to pessimistic conclusions about the future of democracy itself." In the same book I want you to know the <u>record</u> of 11 percent unemployment was set during the Reagan administration.

I'm sure, after reading my book some will react with "He sure is the life of the party, that one!" But, Mr. Schlesinger, praised on top of that and not bias toward any party, certainly tops me – that is tough material he has punched out! And with a superior style that I envy. I do feel better though.

"The accumulation of discontent will subvert the traditional political order and rush American politics into new and dangerous times." Kevin Phillips. Two problems: The few million readers of profound literature such as "The Cycles of American History" are not asserting themselves. Critics lavished praise on this book but...I do not hear or do not read of any approval – certainly not from Congress, what with so many cool lawyer youngsters holding down seats. Secondly, the average sophisticated reader lives quite remote, quite detached (I don't even think they're aware a problem exists) from stewing immigrant crowded cities. There will be no 1934 violence that I witnessed as a child – it will be creeping very expensive aggravation.

INCREASING INSANITY AND MORE AGGRAVATION!

A young black comedian, whose name I glanced at and overlooked, but seems well on his way to the top said, "Whatever happened to insanity?" He was responding to the many psychologists giving reasons for people killing each other and of course, the rash of murderous behavior on the part of the young. Sadly, mental illness quite often does not manifest itself until one is in the late teens – usually that is the case. Something to do with body chemistry changes I suppose.

The advertising of late to sell massive vehicles having dragster power has with it an irresponsible inducement to encourage running amok on streets and highways, exciting as the thought is for the immature itching for more expression. Some don't quite grow out of a juvenile stage, they remain the hoss who wants to boss and abound in our society. I hasten to add here that I'm not the typical grandpa, frowning on youthful activity – in fact, I like autos very much – always have and find today's cars quite beautiful and I don't mind cruising along at 70 mph. But besides immense power plants pushing 6000 pound vehicles, spewing more of what we the more rational are repulsed by and don't want more of there is the near demonic attitude beyond thoughtlessness becoming more apparent of auto execs promoting hot rodding with the bulkiest vehicles ever conceived. Where are they from anyway? In one auto promotional page of our newspaper the latest incredibly souped up pickup about to grace showrooms. Two later pages that followed, an article on test reports over the dangerous tendency of such vehicles, including SUVs and vans, to tip over much easier than expected. These vehicles accounting for 9700 deaths in 1997, the majority in such high topped and centers of gravity autos; 47 percent compared to standard autos' 22 percent. While it's true drivers can be blamed such massive tankers give a feeling of security and less speed than what the speedometer is really showing to the unwise who have yet to see smashed faces, bones protruding through skin and pools of blood. Three teenage girls died recently after losing control in a big pickup truck. Weight and design determines the critical margin of negotiating curves at reckless speeds. Granted this is all elementary but I am motivated to preach through my typewriter at what I recognize as the gradual return of the loony fringe that only the older can remember in the auto industry. No matter how high placed and educated I've seen a few polished whackos. I vividly recall an impressively appearing company executive, tall, well dressed, a T-bone fed countenance, masterful orator who, to my amazement, would stomp his feet, tears springing from his eyes when he couldn't get his way about something. Another character thought it horrible if we didn't wear black socks with garters at factory sales meetings. My boss told me, "Whatever you do, don't cross your legs and expose any skin – so and so can't stand that." So now the thoughtless and irresponsible are on the rise, the madness of the "dragging" 60s is returning and again are encouraging highway exhibitionists to buy super-powered dangerous behemoths 4 by 4s.

I'M DISTURBED. WANT TO KNOW WHY?

Ford in the promotion mentioned has got a dandy of a new pickup weighing in at 4600 pounds but extravagantly powered. (Hope you can stop it at 75 miles per hour.) It has a supercharger (not the late kicking-in turbocharger) that crams air and fuel into a hungry 5.4 liter V8. But that metric dimension is too innocuous appearing for me – that's a lot of engine, 360 cubic inches of it! There is explicitly only one purpose for such a supercharger. It is to give tire-scorching acceleration, 0 to 60 in 6 seconds and that's it, plus about two times more pollution as Gung-Ho roars by you. Now, be with me please. Why this dragster? Then Pontiac has a computer imaging of one of their specials moving at extreme high speed, trees flashing by on both sides (of course, it's a country highway, not a house or anything alive in sight). Words flow to the effect, "Feel the power, this is for you to buy – give your life some new excitement." It happened, this insane campaigning in the sixties. The death rate soared, especially in California. If not surf boarding they were hotrodding. I was out there in 1972 and I saw what advertising can do. This was the effect of money-grubbing assassins of common sense and irresponsible advertising.

I guess that I'm not with it, having driven over 800,000 miles and actually seeing something like twenty crushed bodies in my life on the road encompassing quite a few states. I am disturbed more or less, nothing to rob me of sleep but I remember too well my son killed as a passenger in a factory souped up car that failed on a curve. Youth quite often go through a stage of wanting attention by acts of bravado and Detroit knows this very well. Well, they're gradually bringing back the 60s, believe me. We have also the Philistine types who never grow up. Just what the guy needs – this new pickup can do 135 mph!! Yes, that is what is advertised, 135 – bait for machos. A lure for the lunatic.

I don't know how long a federal function (I think it's federal) now sporting a title I never heard of before, the Ozone Transport Commission, has been in business. Back in 1994 my conviction graciously (yes, indeed) passed on to DEM's representative over the phone and then by letter to Al Gore and two representatives, that it was <u>impossible</u> for less than one million residents in RI to cause so much low level ozone formation. Other than medical oriented organizations protesting and legislators arguing, the public, as usual, kept on yawning. I did point out this typical summer hazardous condition was exacerbated to a very unhealthy level (cruel to asthmatics) some days by prevailing winds bringing in the pollution ingredients and/or ozone formation from heavily populated areas (about 40 times more than RI) such as smoke-pall-blessed New York City, northeast NJ, and western Pennsylvania. Anyway, this fact, requiring little mental effort on my part nor profound wisdom, was politely rebuffed by DEM. Now, we have the Ozone Transport Division. Can it be? Did I unintentionally influence (well, a casually presented opinion) the structuring of another costly bureaucracy that will accomplish little, except to show their worth with inter-departmental exchange of much paperwork and an occasional announcement to the public? "Yesterday, the 95 degree temperature and a west southwest breeze, etc…etc…elevated such and such a city's (Let's use New Haven or exec territory, southwestern Connecticut) surface ozone density to a critical level, etc…" All it takes actually is one meter sitting in some spot that some college student can read and report to newspapers as we have here; i.e., such and such an ozone value occurred, etc… By the way, did you know a New Haven suburb, Hamden, July 1989, was struck by a powerful force four tornado that injured 40? As for being denied twice, I'm back with the humor bit again.

Emotionally damaged by this blow to my pride, no recognition of my genius in blaming major metropolises for most of Rhode Island's ozone and now this sad story coming up. The second time mind you. When the ozone level gets too high we have free bus service. Obviously they're trying to have folks leave their cars home. It doesn't work. Do you want to know why? Well, of course, most folks don't care anyway but some do – I have to try to be nice. But one reason is, it use to take me almost two hours round trip and jumping aboard six buses just to commute from home to work and back a total distance of four miles (one way). Disturbed one day I wrote a letter to a charming lady in charge of our bus service and let her know I felt I was living in Ethiopia – I wasn't trying to be funny either. That is an example of why I feel strongly about improving our infrastructure, obsolete

as it is, although I have to be reasonable since we're always involved with huge military expenditures and of course several other urgent problems. By the way, federal acquired tax money is paying for free bussing. Getting back to my esteem shattering experience, sitting there watching TV's Weather Channel – I got hit with the second Whammy. You surely must recall the amazing originality of my predicting the gradual Africanization (exceptional choice, a new expression) of our climate here in America. Well, twice now, those TV weathermen with not a smile on their faces, referred to Arizona's heavy downpours lately (July 12 – 15, 1999) flooding and all that plus southern California experiencing mudslides from…guess what they are calling those conditions? The Monsoons no less! It's over for me. I know how Dick Tracey must have felt. It's new this word for the USA, Monsoon, and my Africanization goes out the window. I can't use Indianizing, it doesn't sound right. Besides, I'm not a plagiarist.

American politics seem to be increasingly exhibitions of guerilla warfare and policies becoming fragmented – there is no programming – party structuring is confusing and no talk of the future. Instead, two years before each election we are treated to steady dosages of who is going to win. Too many politicians now are neutral despite publicized forms of "savage wolfishness" frequently asserting itself by corporate downsizing with intent to show an attractive fiscal year end profit, thus illuminating the inevitable glowing "For Sale" assets while the dispossessed collect unemployment and are forced to grab at some lower paying job. Reaganism did just that, remember, a record 11 percent out of work.

If a power shift as I believe is going on now, favors control of America by a relative few then democracy is in danger. Although our economy is prospering and other nations buy and sell with us and are doing quite well it is also a wonderful time for tycoons to sweep up more power, money being more potent than any weapon. Before and after the Great Depression, the insider's dirty works covered up by phony patriotic historians unintentionally helping Hitler to create hate against Jews and others, even though they were but a minority in the power enclave of Wall Street or what was once called "Insiders" by more honest writers. Because it would never do to be identified with their scooping up assets of desperate businesses the wealthy rats* of those days sneaked around

*Not a nice comparison but their methods, justified by believing it was their amoral right, fits well with "search and seize".

but today it is brazenly done, as if the big guy wants the whole world to know he's the boss hoss. Did you know huge rats get that way from cannibalizing? They don't bother with venturing outside old structures to forage for food but simply eat the young.

Because of moral myopia, those not concerned as to what happened before, the "modus operandi" being quite different today, because it promises increasing the value of mutual funds held, the successful young feel no need for governmental monitoring – a sure recipe for future chaos. In the meantime politicians sit back, chastened by the attitude of this powerful voting block. What the affluent public fail to recognize is that weak governments invite abusive influences and in fact <u>history</u> shows <u>us</u> that impotent leadership have capitulated to ambitious organized penetration by power obsessed individuals. What is the difference between a Hitler takeover and a half a dozen moguls owning everything and in charge of our destinies? Furthermore the Far Right and I'm sure they're not liberals, will see to it that we periodically wage war – it is very profitable. It is the reason why they want more defenses than is really necessary. If anything that is making Russia nervous is talk of our building up the Pentagon's budget. Of course, other countries will accelerate their arms buildup when diplomatic efforts to stop this insanity should have priority over everything. But we have the media and surely they will help calm these growing passions for wealth and power by a few. And unlike one radio pundit and a minor one here in RI believing voters are stupid I think the American public will recognize sooner or later an alarming trend in "consolidating monetary power". Regrettably we average less than a 55 percent turnout of voters at elections while Europe, having a thousand years or more of despotism enjoy somewhere between 75 to 85 percent active voters.

But politics is for more sage than I. I should stick with a subject I know best. I reacted to the Supreme Court cutting down EPA's power without sufficient debating. For sure the public had no say so in the matter. It is the equivalent, this ruling, of preventing policemen from issuing traffic violation tickets and limit them to just firm chastisements. You also now have a good idea where the power is – the White House succumbed to the might of Big Business. It yielded without much of a fight and gave a 60 year concession to the Daddy Warbucks now flexing their muscles who don't have to get going on reducing their destroying our environment and our water supply for a very long time. "Take your time fellers, we won't bother you." That's how it has to be translated.

When the cost starts to hit home and you react as you should, I can only suggest don't phone representatives – you naturally will be too gentle, too polite and your complaint will soon be forgotten. Write and make it firm. A letter is much more effective.

Remember, even if you feel immortal those little nodules called lymph spread around your body are quite busy collecting garbage you are and will be breathing more of. Thanks to a class of people who don't give a damn about you. One of my twin daughters just came back from a week in Louisiana visiting her "clone" (a little smaller, that's all) and with much grimacing said drinking water is horrible. Lots of oil refineries down there you know. We plan to persuade the other daughter to move back up here! More on physiological long term effects next pages.

We sue tobacco companies, rightfully so in a way, although the smoker has been warned but there is the possibility some day there will have to be some reckoning called for over cancer-causing agents freely flowing along with the very air we breathe. I want to detail into your mind the growing invisible menace to your health-it is here now and not a manana topic making sure something impactive is passed on to you. I spent many an hour studying physiology and while I don't stand in front of our bathroom mirror searching for evidence of expected decay, as would some hypochondriac it gave me an advantage. My wife broke her hip and for now she has a plate with screws but according to medical authorities, and believe me, it was difficult to extract four reports, exactly 25 feet of internet printouts from England, Australia and America; it seems that there is a likelihood of my wife undergoing the placement of an artificial hip some day. The death rate from this surgery is only 3 percent. I went one step further and wrote a congressman who obtained the latest report from orthopedists and it was very disturbing. You may wonder why all this detailing that generally afflicts the elderly. Well, it has to do with air pollution just the same. I ask your patience. It may only apply to the elderly but two outstanding athletes, Cam Neely, hockey player and Bo Jackson, football and baseball player both have had their careers wiped out from fractured hips and last I heard Neely

> **DR. PAUL DONOHUE**
>
> Most people are unaware of the lymphatic system, a second kind of circulatory system. Lymph fluid bathes every body cell, bringing nourishment and washing away germs and foreign matter. Lymph vessels, similar to small veins, vacuum lymph fluid and return it to the circulation.

needed an artificial hip while Jackson already has. But, the report from the government deeply troubled me – therefore no artificial hip for my wife! It seems surgeons are concerned over the long-term (maybe 10 to 15 years) effects of plastic parts within which a metal hip head (femoral head) rotates while one walks. Minute particles of plastic begin to circulate into the body's blood circuitry and eventually lodge in the lymph nodes. The life of the party here has something special to tell you though. If you don't get eventually upset over our government, who, despite the ridiculous tirades put out by ardent fans of mob rule (Lord help us if they succeed!) is there to protects citizens, killing year after year procrastinating, in a state of denial expressed by some Congress of Senate members, or fearful to speak out, the likes of the far right ready to pounce anyway on environmentalists, calling Al Gore the Tree Hugger and all that; if you don't do some serious reflecting as to the insidious damage to your system going on, then there is something very wrong with your thinking processes. Ask yourself this: "Do I want to live to a ripe old age? Why am I working hard, building up investments if not to retire both financially secure and possibly at a younger age - notwithstanding, remaining healthy as long as possible." Caring about your children's future goes beyond money and education. Forget how long grandpa and grandma lived – theirs were pristine days. It's 1999 and populations expanding while economies boom along fills our air with greater quantities of foul emissions. Like the math teacher who whacked me on the side of the head in high school I'm bent on leaving you thinking what this hidden olarchy, this assembled minority of hell-bent for power moguls is doing to all of us. Wait, lymph node topic coming up. No matter where you live if this government whose two parties are in a state of dissolution, therefore the personality TV cult requiring enormous funding reducing our voting opinions to who is the nicest guy, doesn't begin to talk up rebuilding our infrastructure while promoting cleaning the air, protecting water, etc…then don't think it is just lungs that are affected by carcinogenic substances in the air. I think I've got pretty good lungs and so do millions of others but you still are pulling in junk with every breath, deadly junk into your system irregardless and whether you have cast iron lungs or not, an excellent breathing apparatus or not, where do you suppose chemicals finally end up? In the lymph nodes of course. Why? They're there all over your body to collect bacteria and oh yes, particulates and other alien substances. Special blood cells in the nodes devour bacteria to prevent them from circulating in your blood. But, those nodes can overload; they can't handle mercury, benzene, soot, and oxides if the accumulation rate is too great. * Maybe not at all. Or, why so much cancer and leukemia in America? The result of just one type of lymphoma (cancer) called Hodgkin's has increased by 73 percent since 1973. By the way, it is the most rapid increasing cancer in the western world. Shouldn't you ask why? Pity the health buff (I'm not, haven't seen a doctor in 15 years) jogging in cities, traffic and all – My God. Sure his or her lungs will

Experiments on rats show benzene causes tumorous cancers.

have more capacity and be more efficient than ours but they're inhaling in a much greater amount of pollution and keeping their lymphatic system very busy. My only objective with all this is not to spoil your day but start thinking about bothering your representatives.

FOR THE IMMORTAL. FOR THE INDIFFERENT. FOR THE BODY BEAUTIFUL TYPES WHO MAY MYSTERIOUSLY DETERIORATE AFTER SIXTY. I AM A WITNESS.

It would be gratifying if I could make people angry! If this that follows is irritating and also disturbs your peace of mind (I hope) (a perfunctory Sorry to boot) don't take it out on me – try writing instead to your representatives who are quite busy sweating out next election. Firmly remind them of that. Turn the heat up! Stamps only cost 33 cents. They need a little shoving around. Sixty years for industry to finalize emission corrective measures?? Kyoto treaty meetings delayed and delayed??

HOW MANY BODY ORGANS ARE THERE?

There are two types of lymphomas. One already named and the other, afflicting millions is called Non-Hodgkin's Lymphoma. All organs are susceptible and vulnerable to cancer if the lymph system deteriorates and can no longer do what Dr. Donahue tells you what it is supposed to do. Pick any site in your body and be reminded, while you quite naturally concentrate on building up your retirement kitty, what one of the richest men in the world, Paul Getty, once said in his final years (he didn't live all that long). "I'd give all my money if I could have my health back."

I'D LOVE TO THROW SOME QUESTIONS AT GEORGE W. BUSH!

Oh hell, not about his father who I admire simply because he was a naval pilot having helped his son avoid going overseas during the Vietnam days, but what are his intentions in matters of environmental pursuits, if any. "Are you willing, governor, to scale down defense from the block-busting high Republicans want and spend a relatively small piece of it toward helping industries modify their plants?" I have about ten other pertinent questions that would make his backers, would I be a powerful figure, go into mild convulsions or maybe a touch of angina.

HOW CAN I GO ON AS DO SKILLED WRITERS WITH "ATTENTION-GRABBING" EARLY CHAPTERS, "CONTINUITY OF AN INTERESTING THEME" AND "CLIMATIC ENDINGS" WHEN SO MANY TRAUMATIC EVENTS ARE NOW OCCURRING?

Rest assured I'll connect most with weather! But jump from one topic to another I must. This first sentence is an approximate quote I am borrowing from the now deceased John F. Kennedy, Jr., which for brevity sake I will identify as JFK, Jr. "Politics is inextricably woven into the fabric of our existence." As for my view, unimportant as it is, what a shame so many Americans, other than a glancing look at newspaper items, more or less believing it is all futile to analyze and furthermore one vote less will make no difference, stay home on election day. Accessibility to what goes on in Congress and the Senate on a day to day basis, who votes for what and so on, is also not easy to come by. Besides, newspapers by virtue of what sells are not inclined to devote much space to a multitude of details. Important bills passed, what they represent and the 3 to 1 (whatever) Congressional approval is all very well and good but do your know how each politician voted? If you are a voter do you know what your Senator or Congressman or women are doing?

SOMETHING GOOD FOR A CHANGE

I am pleased at a recent addition to the Internet, a new web site USADEMOCRACY. COM which now, I have been introduced to through the expediency of Jim Bohanan's (he was very enthused) radio show and on-line guests, has a service promising House member visibility, badly needed. Frankly, in areas of what will prove painful from my point of view, the blind leading the blind, this new web site can change that. At least you can rapidly discover where your political choice is taking you. They also have a Chat Room arrangement for you to join in and exchange political information and opinions. If I have hinted at disdain for Internet extraneous information, much of it not that reliable, my not being too interested in fanatic "on-line" soldiers, this "something new and very intelligent" has all my appreciation. The promise made to you before elections garnering your vote may now, under this new benign electronic spying, keep your choice honest.

CONSERVATISM GETS CARRIED AWAY

A new slant for public consumption – earth warming is a blessing. The August 1999 issue of Reader's Digest has a unique article suggesting scare mongering environmentalists should be ignored while convincingly (if you're stupid) promoting the strange and incomprehensible idea that a warming trend in climate is beneficial, even suggesting more carbon-dioxide is desirable. Several quoted experts have a supplementary attraction to enhance the image of warmth promoting splendors awaiting us in the form of a photo – one very beautiful tropical type flower. Readers, if of the insecure "noveau riche", scared, liberals want to snatch their mutual funds and ground their SUVs or are just plain naïve, will now add to the ranks of anti-decency who unreasonably spit out nonsense without consulting with history's recorded monetary improvements under Democratic presidents. For example, John Kennedy's upping our gross national product 5.6% after Ike's sluggish economy and three recessions, thanks to several tax affected changes, etc... He also kept inflation from blowing up, limiting it to 1.3% and bringing down unemployment to a record low 5% - quite a record for a lousy liberal! But this Digest article is harmful. I must elaborate on it further. First, let me highlight how President Kennedy reacted to US Steel defying his plea not to raise prices – I think this is great and there it is in the annals of history: "My father always told me that all businessmen were sons-of-bitches but I never believed it till now." He did prevail. But just between you and me, not all business (it depends) people are bad – we all know that. Of course, a lot of them are running off to China, the American flag they vigorously waved for all to see, left behind. CEOs and managers will operate factories there and employ cheap Chinese labor.

In the meantime, the government haters, encouraged by the article I found preposterous will join hands with the Carbon Club. And why not? A lush semi-tropical paradise awaits us – trees will burst skyward in accelerated growth from heavier doses of C02 (steroid theory) (rain not important it seems) agricultural expansion will fill our
markets in unprecedented abundance, you will not need to buy winter clothing, balmy breezes caressing your tanned body. You will use less heating oil and gasoline since winter is anticipated to disappear, therefore outdoor barbecuing, swimming pools in use year round, glorious sunshine, maybe palm trees thriving everywhere and so on. Dear reader, do you believe this? Do 1000 kids at this year's Woodstock concert, prostrated from heat requiring the assistance of medical emergency crews, now believe the sun, more of it, and stronger, is so great? In this article, malignant as it is to our welfare, they resort to the old standby, pointing to climate changes thousands of years ago that ended the empty bleakness of cold snowbound regions stimulating a change to productive fertile lands.

Still wondering what possessed Reader's Digest to publish opinions of alleged authorities on weather, so far out I had to jump in and denounce this as preposterous. One of the least mentioned side effects of heat, presented as a blessing already mentioned, is the accelerated churning cellular electrical discharges, the interaction going on constantly except when we sleep, in skulls whose contents have been damaged by alcohol and drugs. Temperatures in the upper 90s and touching 100s flush out the innate insanity, erupting after hours of simmering heat and relentless sunshine penetrating the hairiest of heads. This year's Woodstock was marred by vandalism, arson and nasty behavior by a crackpot minority. It may account, this blessed and desired heat, for African massacres that defy comprehension in view of mobs swinging machetes, butchering anyone not able to get out of the way, poverty of course and anger contributing to madness. If Arabs are calmer it is because they wear turbans and loose clothing.

HOW CAN SUCH TRIPE BE DISTRIBUTED ACROSS OUR NATION?

Such an endorsement to leave things be and no doubt delightful reading for the Bush assembly, pumping millions into his campaign fund also sways public opinion to continue on in a state of denial. No mention is made of the nasty part of prolonged hot weather or how temperate climate zones have healthier populaces, no mention of where progress leaped forward these past three centuries, springing up out of the middle latitudes and _not_ in hot damp brain deadening, body functions just poking along typical of semi-jungle climates, hammocks and indolent natives laying there, listening to jungle sounds, beating off hungry flies during the day, hiding from ravenous hordes of mosquitoes at night and occasionally responding to a needed task with their most uttered word, Manana. Bush talks Spanish, why not me, everybody knows that means tomorrow this Manana thing. Unless two inch long cockroaches are acceptable, some used as pets in heated Oriental locations, responding to meals (cheap too) that being a possibility with the heaven promoted by several very bias experts. Then, the fascination of new pests moving further northward bearing in mind heat accelerates growth and numbers, enough having been said about bacteria. I had several Praying Mantises in a business I operated in Louisiana and although short-lived it was fun to watch them snatch insects – just think of all the distractions you can look forward to if this dream world envisioned by people who indirectly accuse environmentalists of being whackos. Where but hideous diseases such as leprosy prevailed if not in heat, glaring sunshine and endless rotting dampness? Where but in African climes do you see bloodshot eyes in every face from relentless ultra-violet rays? Cataracts too, any doctor can tell you that. Yet none of this is featured in that article. Instead the Garden of Eden is in our future. Apples won't grow and any future bikini-adorned Eves will have to lure Adams with bananas. (I had to put some clothes on her.) Adam will need a potent deodorant also.

I pause – calmness returns. Something bothering me, are camels good eating? It is also my prerogative to assail this nonsense with caustic comments and then top it off with some frivolity. Would they, these dreamers call the FBI and have me watched if I suggest someone toss a can full of fire ants in their backyards? You know, give em a dose of what semi-tropics are about.

But I'm not mean. Did I not tell you tales of lost loves long ago that didn't quite work out? There were a few others but I didn't want to bore you. And Murph and I, joyously horsing around bouncing from one Atlantic City bar to another after he waltzed his way into sudden riches – five minutes to make what took over three weeks to earn? And my lovely graceful stork? A delicious meal of stolen prunes? Oh, how can you not envy my dancing in New York with the cream of show business?

With this lymph node thing, appealing indirectly to women (or was I too obvious, too blunt?) I wanted to continue on to a firm finish but there is this unforgettable interruption, a plane crashing and three lives snuffed out.

WHAT ALTITUDE?

Gone forever, John Kennedy, Jr., wife and sister in law, a stunning heart wrenching ending as it was, my wife coming to tears without sobbing because a TV camera, closing in and staying too long focusing on a crying child's face, up close to see welling tears, I have to put something to rest. Theories will fly now, tabloids will reek with insane interpretations and I wonder, if it will approach the nonsense of Bermuda Triangles? Unless the National Transportation Safety Board finds something wrong with the plane, there is no telling how wild suppositions will fly. Pay no mind. The most inexperienced pilot knows what an altimeter is for. Each day that you fly, before you leave the ground, they must be set, as simply as adjusting a watch. The surface atmospheric pressure reading, available and always on call at every airport or the nearest facility is used. Whatever the pressure, assuming there are no rapid changes coming the altimeter is set to zero altitude. Whether JFK, Jr. could see or not see what lay ahead and decided to descend quickly where perhaps he felt visibility might be better or island lights or landmarks would be clearer or even became disoriented I cannot understand the tentative belief he didn't know how low he was flying or if he was unintentionally going at a higher speed nose down, when all he had to do was look at the altimeter and the airspeed indicator. Then, supposedly a plane was seen flying too low by a fisherman but that is hard to validate. Since the plane's wheels were down (allegedly) it doesn't seem he was having engine troubles nor did he radio he was in trouble. One fundamental rule taught to pilots in training calls for leaving wheels up and out of the way when engines fail and try to skip along and/or flop down in the water.

I heard a TV meteorologist most authoritatively discount bad weather as a factor. Careful, bubba. But, I don't say they should know precisely how much visibility could be expected at 9:30 PM or whatever. In general, if I recall, he said weather was clear and visibility reasonably acceptable. That was from his location here in RI or Massachusetts. I really don't know where his office is and do not car. Now I am reading where pilots of small planes all theorizing what could have happened along with a flying witness or two saying they would not have ventured out on a hazy night to fly to Martha's Vineyard. All very well and good but the most important questions of all have not been mentioned. Now, I don't know how many times naval pilots phoned or asked me what was the exact atmospheric pressure for the express purpose to calibrate (a simple adjustment) the trusted sensitive altimeter. The questions that should be asked, and I also recall the severity of naval investigation procedures, are: "When was the altimeter of JFK, Jr.'s plane last calibrated and by whom?"

By no means am I irresponsibly searching to blame somebody. Surely, JFK, Jr. was indoctrinated in something so very basic as reading an altimeter! According to reports his sole communication with a New Jersey control tower was to advise them what his destination was. Did he, that day, phone anyone for weather information? What I suspect rather than described reasons for possible diving, plunging, spinning in, volunteered by well intended experienced small plane pilots it appears to me JFK, Jr. did not know how low he had descended and if he did glance at the altimeter, that not requiring an instrument flying license or expertise, it is very possible the needle could have shown 100 feet altitude when he was just a few feet from surface waters. I use this figure 100 feet off the top of my head because I want to describe precisely how altimeters react to pressure changes and that figure simplifies the minor arithmetic involved. Remember someone saw a plane flying at that low level but…is it hearsay?

ALTIMETERS – OR COUNTERING TABLOID NONSENSE COMING UP.

Atmospheric pressure is read in millibars – let's leave tenths out. At the surface it averages 1014 millibars and at 5000 feet altitude it registers 850 millibars. Per one foot that comes to .0328 pressure drop from surface to 5000 feet. If the surface atmospheric pressure was 1021 millibars 3 days before July 16th (just an example, I know of nothing definite) and the altimeter was set at zero altitude but inadvertently overlooked three days later, that fatal day and the atmospheric pressure had naturally changed and was, let us say 3.28 millibars lower, or 1017 mb, then the altimeter, if not looked at before takeoff (I can't imagine that at all!) would show 100 foot altitude and of course, later, over a hazy and not discernible water surface at night, the altimeter would still be showing 100 foot altitude when in reality the plane would be but a few feet above the sea surface. This is but a study in the minor arithmetic involved with how altimeters work and not in no way advancing theories as to what happened Friday night. Again I cannot imagine JFK, Jr. not looking at the altimeter if he had decided to drop the plane down (possibly in a hurry) anxious to get a better view, bearing in mind approaching his destination at some 125 to 150 mph, confronted as I'm sure he was with thickening haze. Because colder bottom sea waters (I think in the 50s in this case) upwells up sloping ascending sea bottoms as the tide rises, and that is, coming up the only fact I know, peak high tide was due at about 11 PM then it is very probable a west to southwest wind against Gay Head, water along that southwest coast region could have been cooled a few more degrees since at 9 to 9:30 PM the tide had to be rising and there had to be some upwelling. We have to explain why that island area has haze bordering on fog rapidly developing at times. If the plane was in a steep dive as it is imagined but JFK, Jr. did not know it, how can that be? The speed indicator would show that, would it not – it would indicate 200, 250 or whatever miles per hour.

I have just heard over the radio an announcement by what should be a reliable network, federal authorities are investigating the possibility meteorologists gave out poor information to JFK, Jr., suspecting (not clearly defined) he had made a 1-800 phone call before departing New Jersey. This is unfair if one understands the complex limitations of coast-based weather personnel tied down to anticipating their own local weather. You would be surprised how involved they are with various industries requiring advance information. Furthermore, if some NY or NJ expert merely reviewing hour by hour weather reporting from Martha's Vineyard told JFK, Jr. that visibility was 5 miles (whatever) due to haze at 6 PM (whenever) he or she, or perhaps a recorded opinion, may not have been that astute, that fastidious, that worried or that experienced with the peculiarities of that island, such as rising high tides (it was at night) colder bottom waters upwelling and exacerbating the degree of haze bordering on fog formation over the island's coastal waters. Were they asked what would conditions be at 9:30 PM or what? Weathermen are not supermen or superwomen, (the latter scaring me off when I was younger) endowed with computer-like compiled evaluations of every square mile of oceans and distant lands, whether it be but 150 or 500 miles away. I have visited New England's best weather facility and was impressed with their highly technical school knowledge but I didn't see one person older than 35. It takes experience and brain-busting, hair-losing effort and diligence, how would I say, to be a great forecaster, albeit it is not possible to be infallible. It is not a profession for those who seek a sure thing in job accomplishments.

WAIT AND SEE

There will be examples of soaring imaginations, pending structuring the most impactive story tabloids can concoct over the Kennedy tragedy. One chap (really a screwball) called a radio talk show and insisted, panting over what he was anxiously about to reveal but becoming slightly agitated after the host disagreed, that ocean waters do not allow bodies to stay intact after 5 or 6 days, therefore no bodies were really found – dissolved, gone, as he believed. He vocally emphasized, in positive tones the navy retrieving bodies, the autopsy and ashes deposited at sea were all faked! Just a big cover up, that's all. It gives you an idea of what mentalities tabloids aim their unique renditions at and why they sell.

1998 AND NOW 1999. IT'S HAPPENING – WILL THEY NEVER LEARN.

Again, China, flexing it's muscles while Nature turns up it's cleaning machine, beaten down by heavy rains, a major river flooding and only in it's beginning, already 150,000 are being evacuated. Go ahead China and keep those smokestacks pumping out filth. In due time you'll come around to realizing this can't go on and on. And in America, some major wildfire out west while the earth begs for water, slowly roasting as it is. I have to see here in Lincoln, RI two huge tanker trucks specializing in hauling water parked on the side of a main thoroughfare, police directing traffic around them as they load up with perhaps 60,000 gallons of water that a five inch hose is taking in attached to a fire hydrant. Wonderful, what community is in trouble and needs water?

HIS FIRST LOVE – IT WAS NOT TO BE

King Hassan II of Morocco has just died. A very useful man negotiating peace in Israel while reactionary forces tried to kill him six times. How brave can you be!!?? But I must add the usual supplement, my two bits worth of "lost loves that didn't turn out right" aimed at the ladies, since many of you love novels while I dearly would like you to be more attentive to both politics and our technical world. There is something very romantic about a brief period in this Arab monarch's life, newspapers carrying on with his not protesting that I must tell you about. But first, don't be too in agreement with criticism of a tendency of France being recalcitrant when asked to fight with us against Arabs – there are close ties, business relations and survival in mind. Once Hassan II invited a French soccer team to visit Morocco. They were treated so lavishly one friend I know there thought this to be fantastic an occasion. King Hassan obtained a law degree in France, even joined the French Navy but this story I'm about to tell you is not a case of "L'Amour est un oiseau rebelle" that set him back emotionally exposed to French ladies as he was in his younger fiery days, studying in Bordeaux. Instead it was, "Au revoir ma belle, je t'aime a la folie, mais --." "My heart is broken." Beat that Russ Limbaugh, go ahead. The then prince fell in love with a pretty French actress, blonde and diminutive. Not very tall himself if he was to prove the strong one later. Prince Hassan was enraptured. Trouble is, by monarchist Arab decree, heirs to thrones cannot marry out of their nationalities.

OCCASIONALLY BUT VERY BRIEFLY I AM BITTER

I believe after six decades, maybe a few years longer an active mind has the ability to observe with things as they are. Based on history, boom and bust, several conservative presidents believing the only way to slow down inflation was higher unemployment figures which John Kennedy totally disagreed with and Reagan restoring that philosophy. It is not some unimportant personal suffering, unimportant to millions of others whose interests must be respected, albeit I've thrown some denunciations around. But, you cannot convey erudition without command of expressive language and yet, written intellectualism on higher levels dear to a minority does not reach the masses who in the end tend to vote for charismatic personalities. Wit and charm sells but can hide harmful aims or irresponsible attitudes. For example, President Hoover once said, "The poor must rise out of their poverty by spurring themselves." The whole world had collapsed and a woman like my mother and millions of others were

supposed to kick themselves in the behind with spurs so to speak and rise up above it all. My father did spur himself on and took a job nobody else wanted in a company owned by a Mr. White and died five years later after inhaling zinc fumes.

"SOME DAY I'M GOING TO TAKE YOU TO A BALL GAME"

The only thing I very vaguely remember is a few words he told me one day, printed above. I'm sure I was less than eight years old. We were sitting on a hill nearby. Hunt Street had a long series of tall tenements, side by side but to the north there was distractions. One dump infested with rats, a large gravel pit, forbidden territory for children but we sneaked a few adventurous excursions, climbing up to the top and sliding down, gravel filling our pants. I remember sitting next to my father looking down at a grass-covered gully, our Central Park, about 100 yards of it. There were even a few anemic blueberry bushes, if you could get to them first. "Some day I'm going to take you to a ball game." That's all I remember of him, the product, and the final outcome of wild and wooly untethered economics. Well Dad, it sure has been one hell of a ball game but I don't know if the umpires of that period were worthy of being called fair and just, a lot of them of Machiavellian described mentalities whose ancestors unscrupulous, wore horn-bedecked helmets, having contributed to the collapse of a system on an enormous scale. Even some Republicans were nauseated and let it be known – in fact, one, LaFollette helped democrat FDR become president in 1932. No one has ever, as they do now, compiled suicide rates and death from malnutrition in those days when vested interests filled their pockets. When I hear "We should privatize" nasty movies parade across my mind. I'm no socialist but give me a balanced system of government and free enterprise. In the meantime you'll not find me scowling and talking to myself – I enjoy laughing and joking.

"AMORE" – BUSH'S SONG

Perhaps he will legalize bullfighting. I watched his reaction when he was told JFK, Jr. was no more. His eyes flickered from left to right, his demise unchanged, no pain showing whatsoever, simply acting out, sorry was he over this news. I would like to explain why I dislike too much pandering to just one part of American society but I can't. I have nothing against any nationalities but a certain segment of his loved ones dislike Americans and do not want to speak English. Why does he encourage them?

BEWARE KISSES THROWN AT YOU AND FLOWERS AT YOUR FEET.

Once in the 50s I did remarkably well selling electronic equipment. Mr. Ralph Countryman, former vice president of Firestone, then manager of our company I prospered with for five years sent me a letter praising me. I knew what would follow a month later – an increase in quotas desired. I told my wife, "Take a look at the flowers sent to me." Scrutinize aspiring candidates for major positions and beware those who advocate too much private sectoring. Beware of aspirants who avoid issues and tell us over and over and over again, how lucky we are to be American. I don't need that, do you? That's sophomoric – like the teenage boy who whispers "I Love You" and has no intention of treating his girlfriend, as he should once he accomplishes the objective. Same thing. Right now, George W. Bush as radio topic material is stirring up radio media drumbeating over who, between Clinton and bush, were ducking military duties. This is the hot topic, not issues. Who cares! I don't. I don't know to what extent the olarchy will control Bush but I've heard it said over the radio waves OPEC and the Carbon Club now own him. Or, why did a campaign staff member boast "We have a lot more than 36 million dollars." And Al Gore? He's now wary but couldn't he elaborate on what he meant about mobility systems of the future? Or my pet project, help finance converting coal burning plants to natural gas? Where is the fire and spirit, the wisdom of Franklin D. Roosevelt, although far right Republicans have lambasted the New Deal but make no mention of great projects that advanced this country? Cut taxes? How about using that money for improving the country's deteriorating infrastructure? That would put a lot of people to work, don't you think? We will desperately need ways and means to increase our water supply. Do you hear that from any politician? All I heard from Bush is a hint of more defense spending although all he said was, "We are surrounded by insane dictators." What does that tell you? How inspiring!

"WHAT WE DO TO WATER IS WHAT WE DO TO OURSELVES." JOHN TODD

Again a quote from the same man: "Water is the vessel through which all life originates and flows. What we do to water affects the other creatures that share the planet with us." And, what do you hear from our government? Where is it with this imperative issue? "Water stewardship* should be a central theme of our culture." "It is raining pesticides." "Water raining down on us is slowly becoming unfit to drink." After reading two pages of
Do you actually believe the private sector is willing or capable? Jim Bohanan scoffs: "Don't tell me the "Market" will solve the water problem!"
Mr. John Todd who is a scientist working on ways and means to ecologically clean up ponds and lakes I feel I'm not all that bad of a gloom spreader.

FROM THE ROCKY MOUNTAINS – A REAL HIGH

A scientist in Colorado has just advanced an eyebrow-raising contention (my reaction, "Oh no") the validity of which from my point of view having the annoying habit of collecting easily acquired simple facts, disturbing* to bureaucrats who revel in special jargon and masterful juggling of figures, needs just a little less empirical presentation and a little more dissecting. Furthermore before I am accused of being an agitating spoiler, really directing my opinion and expending my capital of accumulated facts in the process, toward hope-for masses for their assimilation and judgement, some of this gentleman's scientist peers have politely sent notice they're not completely convinced of what truly is a shocker if carelessly accepted at face value and intend to study the matter.

A Double Shot, Mind You!

Young folks may have never heard of the fabulous detective, Dick Tracy, but if there was one character he was after who frustrated him most, it was "Double Whammy" as he was called. Just when Tracy was pulling his gun out with his right hand and handcuffs in his left, he got zapped backwards into a dazed helplessness from a beam shooting out of highly specialized evil eyes** - just a single shot mind you, but with a warning, "You're asking for my Double Whammy – I'm warning you Tracy!!" Of course the scientist in question is not a criminal and perhaps he's very sincere but his two-bagger hit may have a negative influence on public opinion and worse, the suggestion environmentalists may be responsible for increasing the earth's temperature. How about that! This most current emanation typically keeps a premise within a narrow boundary and by not including all factors that affect atmospheric heating, this being his central theme advanced, the effect is climactic – I mean a jolter. Such reports (certainly not all) reveal a technique I've overly described already. This latest claim adheres to the routine of never mentioning what we know to be true while highlighting some exclusive idea and making it exciting material – material that if accepted wholeheartedly suggests alleged efforts at reducing sulphur-dioxides (SO2) in our atmosphere may do much more harm than good. That should delight the Carbon Club!

**Remember? My friendly tiff with DEM*
***Don't laugh since "Star Wars" is just as crazy. And the goofy make up!*

He makes two impactive statements. One, due to our "dramatic rush" (quite an exaggeration – colorful no less) to clean the air we have succeeded in reducing SO2 by 12 percent. Two, SO2, supposedly, reflects the sun's rays and therefore we may have a more rapid rise in temperatures in the future than what was originally estimated. In other words, we may be on the wrong track trying to cut down one of the worst emissions there is, SO2. Imagine that. My first argument has to do with expressing our cleaning effort as a dramatic rush contrary to what has been going on at a limping gait, per, for one, the Wall Street Journal (and others) with it's probing of facts motivated by economic concerns, and of course, the "Yes there is more toxins" "No, there isn't" reports from all over the United States, keeping everybody concerned running around in circles. Secondly, while no mention in this scientist's report is made over the fact of our ozone layer temporarily (I hope) deficit in it's function the gentleman in Colorado may not be aware nature did most of the air cleaning and not human kind considering the catastrophic quantity of rain (only about 40,000 died) that was dumped on both land and sea (unmeasured out there but tons, no less) if we consider numerous hurricanes and typhoons this past decade. The argument which should be clear to you is acid rain and what it represents as a way and a means by which nature tries, if I may express it that way, to keep us from destroying her planet. Additionally to the point of repetitious boredom, efforts just in the USA, specifically curtailing (a moot question) SO2, coal burning being the leading producer, then diesel engines, possibly including autos, heroically described as dramatically rushed, does not take into account much of our air is continuously coming in from the Pacific, the Aleutian area and of course busy China and other major coal burning countries producing a zillion cubic miles of sulphur-dioxides. "Nope, we did it our way!" apparently, when in reality only nature did most of the cleaning.

I am a zero when it comes to chemistry; therefore it's possible SO2 does reflect the sun's rays but back to where? What happens to this detoured heat or is it bounced back heavenward? If this heat is kept away from us and supposedly congregates in our upper levels that should stifle the formation of tornadoes and hurricanes also. But, we'll have to wait and see.

As for our present heat, past and future, excessive as it has been and will be I'm an "ozone layer", "particulate pollution absorber of heat and collector of moisture" and "Greenhouse Effect" man myself. Those are the Big Three.

Maybe the government wants to remain silent, a bad habit and perhaps derelict also or not as sharp as they should be, while paying one or two thousand scientists (who knows how many) (who say little) because the ozone layer is a big secret and is a factor in missile warfare or whatever. In the meantime nobody seems very worried, stores running out of air conditioners and fans and too many imagine there is nothing we can do about it anyway. If freon and other aerosols condemned as a cause despite the puny quantity these gases represent and damaging the ozone layer plus punching a big hole in it there must be something extremely more quantitative in man made acidic products having such a major effect since the ozone layer up there, 15 to 30 miles above us, must represent a trillion or more cubic miles of ozone. We are in a sorry state of mind for not having done much of anything, relatively speaking, to push for immediate action rather than pay lip service to noble treaties in the future that promise (When – God knows) an attack on the problem of increasing production of destructive acids worldwide. I'm sorry I don't think much of world leaders in this matter right now. Regrettably the human race has shown a strong propensity to do nothing much until it is hammered to it's knees by a crisis – some deadly – some causing much suffering.

In this matter of impending dissolution of stable socio-economic world arenas I feel there is a deplorable lack of leadership but nevertheless very sophisticated and dedicated at killing while huffing and puffing, all clamoring for markets. We are digging a big hole for ourselves and it may take a very long time to extricate ourselves from it.

It's possible you may see bumper decals on giant SUV and truck bumpers such as these: "If it's too hot, blame EPA." "Buying economy cars is anti-social." "Long live diesel engines." "Burn more coal – it cools our planet." I'm waiting for the big "Aha!" though, aside from this sarcastic bit I just had to include. And where does compassionate Bush stand?

I'M ALL FOR FREE ENTERPRISE BUT STRONG GOVERNMENT ALSO

Do you know why there is more and more persuasive talk of private sectoring? That's easy. In 1981 the average CEO was paid 30 times what the average worker in his company earned. In 1999 **it's 320 times**! Power corrupts. There is now talk in Washington of removing tax exemptions and certain welfare for corporations wallowing in this Bonanza waist deep in greenbacks. A major firm here in RI and having branches in other parts of our country, flourished from defense contracts that you the taxpayer financed. Last year they dumped several thousand workers and just a month ago the president retired with a 43 million-dollar pension plan. A smaller firm just gave it's 600 employees pink slips and unashamed announced they had a contract with China to build a factory there and operate business as usual. George W. Bush pretends it is compassionate to give a tax break. He should be talking using expected surpluses to rebuild America, a country that can't even handle electricity requirements. But I know where he is coming from!

It must be the heat – I'm overreacting. The murder rate here has jumped up lately especially when it is hot. More cops are needed. One SUV Bronco owner flew into a rage because a woman yelled at him as he sped by. He turned around, rammed four cars then tried to run the woman over. In one month, three banks and two stores have been held up and robbed by youths.

Some good news for a change. Desalination is catching on – corporations are showing great interest spurred on by increasing water shortages and fear of underground water contamination. Naturally, now that they have contributed most to the gradual destruction of our water supply they must be smelling the sweet odor of big bucks – selling desperately needed water in the future is a sure thing. As you can see there is no disjunction in my mind. My feelings about the momentum heading for more corporate power and the Democratic party having lost it's sense of direction I'm not apt to write kindly of their increasing passions. But first, a gentleman in charge of a Massachusetts desalination project explained the high cost of filtering salt water but that it will be the preferred choice over the sickening idea of using waste water (sewage – don't retch) even though recycling the latter would be a lot cheaper. It wasn't necessary to tell us the public wouldn't approve! I may be going off on the deep end with this one but I think if Franklin D. Roosevelt were around he probably would get together with the Army

Corp of Engineers and study the more practical (my theory) idea of channeling off some of the greater rivers in America by carving out canals and piping water to various deprived areas. But, ole, ole, down with government and march on with the private sector. There is a surprising number of companies building desalination facilities. As for contaminated water supplies it is sad testimony to what we have done in the name of the almighty buck. Economically we charged forward while ridding us of some government supervision, stampeding for the golden fleece but our leaders beginning with a period before Reagan and worse when he became president had no comprehension where this would take us; water messed up, climate change, red tide (a prettier name, algae) disappearing species and so on.

Unwisely, those who already have too much greedily eye the surplus pie of Federal funds. The fever for tax cutting is rising and to hell with the future. For 30 million folks, it's peanuts.

On a more sedate note, fish wandering from their own swimming pools some 1800 to 2200 miles strikes me as an excuse to dump several hundred scientists who meander around with questions as to whether or not the ocean is warming and where it is warming, making sure it is all so very much a mystery to be researched for the next, well, at least 10 years. A new species has arrived. It has been spotted by Scuba divers in Connecticut coastal waters. The Sheepshead Porgy, (quite a name) quite comfortable here after leaving it's home in the Gulf of Mexico, a rather freakish behavior. So, you want to know if it's good eating, I know. Seriously, it is a relative of our Tautog but it has black stripes which means there is no chance of they're being attracted to each other, if I know fish. Very bias that way; (reminds me of certain folks) I hope it does not wander into the western boxed-in end of Long Island Sound where experts fear all aquatic life will die because of oxygen deprivation. The water is too warm. I hope that they're wrong.

No ecclesiastic drama meant but sometimes I wonder what God has in mind if anything. Maybe I should ask Russ Limbaugh how he managed a loan from God as to his purported gift of talent. Then the real dandy, he calls his program the Institute of Advanced Conservative Studies. Please, all that means is bring back the old ideas stale and callous as it is.

For the young.

Devote an occasional moment to seriousness – memorize Wordsworth's 1815 prefaces, in order:

OBSERVATION,
"THE ABILITY TO OBSERVE WITH ACCURACY THINGS AS THEY ARE,"
"TO KNOW WHETHER THE THINGS DEPICTED BE ACTUALLY PRESENT."

REFLECTION,

WHICH TEACHES "THE VALUE OF ACTIONS, IMAGES, THOUGHTS AND FEELINGS;
AND ASSISTS THE SENSIBILITY IN PERCEIVING THEIR CONNECTION WITH EACH OTHER.
IMAGINATION,
"TO MODIFY, TO CREATE, AND TO ASSOCIATE."

In reading choice pieces of wisdom by the likes of ancients, Alexis Tocqueville, R.W. Emerson and our current Arthur Schlesinger Jr., who it seems repeatedly quoted the frog* just named (I guess they're not all inferior to us here) it is clear one characteristic of the human race is to stand by and take comfort, (in fact almost an intractable attitude) in custom and resist change. Therefore, idealist as I am I have no illusions. Politically, if one wishes to stay intact and prosper at election time it is best not to promote disturbing ideas – the public responds best to promises of good times. Regrettably it takes a crisis for Homo sapiens to turn around and reinstate what they voted against. We cynically bounce from shades of conservatism to liberalism, sometimes overextended. We discover also presidents are not supermen and go for another.

But that is a subject for political analysis and the study of history – my view although having a connection with decision-making politicians has to emphasize weather. Putting humility and political correctness aside I sought in this book to make it clear why we are in trouble. Determined to debunk flawed interpretations of El Nino's ways, serving to shroud truths for some, I turned to weather maps and recorded facts since it's influence dominated thinking much too much. I have heard comments from the "El Nino farce" to "El Nino is responsible for heating climate change." Neither one is true. Neglected from convenience or ignorance (yes, that is very possible) is the forces behind El Nino's move toward our west coast, twice traumatically (only winter) in the past 17 years that propelled such a body of warm water beyond it's usual boundaries. In 1993, the previous beginning years of the 90s marking the onset of accelerating tornado frequencies never experienced in our history despite El Ninos occurring every 3 to 6 years for centuries it is obvious, referring to temperature records, unexciting as that must be to pseudo-sophisticates, the hue and cry went up on the west coast; "Get Ready! El Nino is on it's way!" Unlike 1983 this body of water did not literally pile up against our west coast; I cannot find temperature rises of any significance referring to three California cities, San Diego, Los Angeles and Eureka, all three having the Pacific waves impressively beating at their doorsteps year round, so to speak. But a powerful expansive high-pressure area had driven El Nino closer, weather-wise, nearer than it's usual centuries-old territory quite distant from California. If El Nino, claimed as it was to have parked against our west coast as it did 10 years before providing

Sorry, I heard a mother-son team radio hosts commenting on Armstrong winning the Tour De France. Wondering how the French feel even though they cheer him: "Oh, they're just frogs." Their ancestors had to be dock workers in Liverpool.

heated waters making life less chilling for surfers (it didn't in 1993) the thermometer records show no variation from previous non-El Nino years, before and after 1993. Nothing! However, rain quantity doubled and even tripled in contrast with general averages. But, don't anyone say, Aha! Oh yes, in that winter (1993) floods, mudslides and structures collapsing kept movie cameras busy and film footage presented over and over again, even now, July 1999, six years later, adding in words "El Nino the horrible, look what it can do!" Failing adroitly or miserably (whatever) from a meteorological standpoint, and if I also hammer away again, it is because I am aware the media and goggle-eyed showmen and baritone-voiced narrators, sounding as if they were witnessing people dying of the plague, definitely are more influential than I. The real facts are never correctly explained. In the winter, surges of colder air driven southward by the now dominating northern Pacific system meets air warmed by El Nino. It is also driven further northeastward by the southern blown up high, and consequently there is a squeeze play, an increasing gradient, as it is called, and a natural arrangement for huge storm formations. Such storms as in 1993 perhaps 500 to 1000 miles west of Los Angeles, therefore prevailing westerly winds at high altitudes drove these storms, one after another into California. Put the Spanish one to rest, it is not an entity propelling itself. Other than 1982 – 1983 and a lighter treatment in 1993 it takes vacations lasting 3 to 5 years.

WE HAVE EARTH WARMING, THAT'S IT!

I have shown you graphs of what warming trends are about. It will not go away. Don't dream, there is more coming. Even the editorial column of our Providence Journal reveals resignation, they had an article last week to the effect, "Give up on lawns, they've got to go." Yipes. Now, let me add one more graph on the next page. Let's have a look-see at a city located in northern California, Eureka. This trace as the others represents monthly temperature averages, faithfully recorded every hour, 24 hours a day, 365 days a year. Or, "Just the facts, ma'am." If a distinguished literary figure, confidant of presidents, lunching with the mighty can quote different personalities and use bad words I don't feel out of order by reminding you, pay no mind to the bullshit thrown at you by folks with ulterior motives. If you don't mind rats over a foot long, mice looking your cat in the eye, big mosquito wings drumming the air all night and bouncing off your screens, a zillion robust aggressive wasps trying to make you abandon your homes, thriving as they do on 100 or more degrees of heat (they love it), rattlesnakes moving further northward and everything else imaginable that proliferates in protracted hot weather just go along with the anti-government crowd. Want an example how far right Republicans think? Russ Limbaugh, today the 26th of July, arguing the merits of a big tax cut and denouncing protecting Social Security and Medicare as if his generation would never get old, callously blurted out: "Those battleaxes out there waiting to collect their social security checks and Medicare!" He'll go the way of all unfeeling demagogues eventually.

The world is not about to end but it is changing for the worse. I am not an alarmist – I am a realist. Do not think the graph's 1.99 temperature change is insignificant. It is 24-hour averages, 7 days a week, 365 days a year and not one afternoon's max temperature compared with another. I write with abandon –m there is too much happening! Orderly writing serves fiction well – mine is the truth and I have responded to day by day news compelling me to comment as best as I can but orderly disorder it is! If you wish to call my efforts intellectual arrogance I would be pleased – but my intent was sanitation – to purge deception and flawed reasoning. That requires bluntness and get to

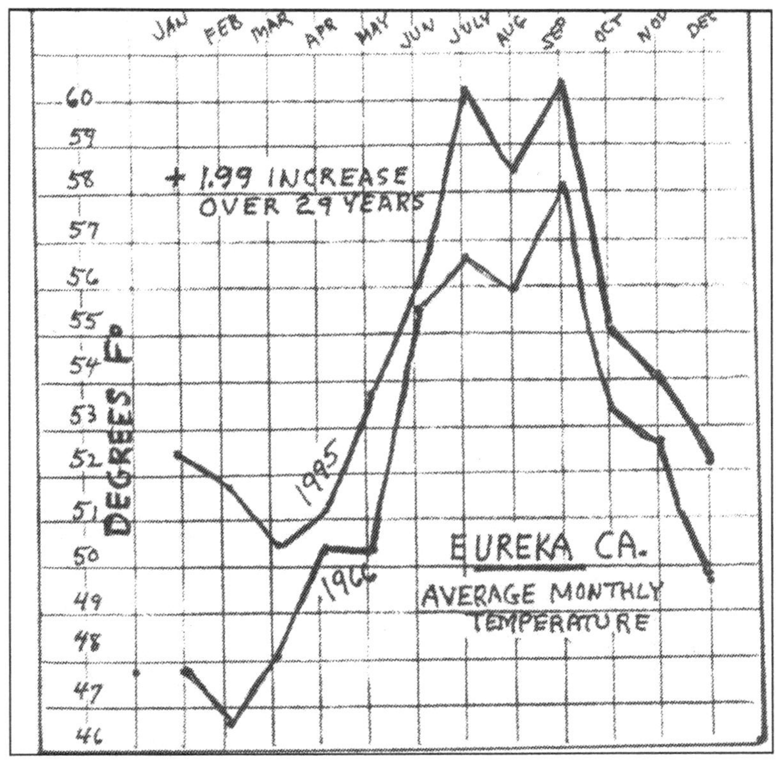

the point and reaches more rather than the less. If lulling and gratifying pedantics pleases a minority regaling over commanding phrasing and maximum use of every page of the thickest dictionary let the results of Jesse Ventura sink in.

I cannot maintain artistic continuity in day by day writing, even ending with an effort at smoothness not easy to attain, not as long as every 24 to 48 hours some news, some radio announcements has to be dissected by yours truly. Furthermore, I'm incapable of pretty, organized fiction, preferring one-two punching. July 30, 1999. We have had heat and droughts before, the mid 30s, 1968, 1998 and while I point out frequency and duration (seasonally) of violent weather and painful hot dry spells as an imperative matter for our attention, what is very significant is that heat records are being broken day by day with an alarming rate of recurrence. If cattle and poultry are dying, secondary to the tragedy of human deaths, they're being staple sources of nutrition as they are, farmers unable to grind corn to feed livestock because nothing is growing then I know I am justified, although concluding this some time ago, to point out there are two huge discrepancies in our thinking. Social tranquility is at stake with the expected rise in the cost of living but it has yet to be mentioned – it worries no one. Secondly, we should be more aware of our aging infrastructure, one aspect of it, such as our power plants cutting off electricity last week in Cincinnati and forcing hundreds of businesses to close down. Additionally, you are not hearing any talk of irrigation, water works and what have you, "compassion" being the slogan at the forefront of promises. Instead the money that is needed to propel us in the right direction is eagerly sought by advocates of cutting taxes, "across the board" no less to the delight of the "Insatiables" and the cult of "gimme more" who care less about 20 years from now. Many indifferent, flush with imagined "savoire faire" in money management, in regards to saving social security, not believing they will also be the old goats and battleaxes as the present generation of the elderly has been called by certain far right spokesmen. When I saw notoriously Republican hard-nosed Senator Phil Gramm of Texas arguing with Allan Greenspan, the federal man fearing spiraling inflation, I know where we are heading and it sure "ain't" paradise. That politician is another Hoover. Ventura used "ain't" successfully – I thought that I would copy.

I listened to an author, Jordan Goodman, his book "It's your money" catching the attention of Jim Bohanan. Because we are vacuuming up gasoline out of thousands of gasoline storage tanks, I mean the 200,000 gallon kind sitting all over our country, painted white, at an amazing rate and because Asia's economies have turned around and prosperity means more gasoline burning, it's cost per gallon to you is rising, such was this innocuous

sounding event coming on stronger over a more intelligent media but meaning little for now to the strutting "nouveau riche" but a worry no less to an astute radio host and a few of us out here above the part time thinker level, (well, not the majority anyway). We have to understand a businessman's concern calling in to talk to Mr. Goodman, reminding thousands and perhaps millions of listeners this will cost us a lot more money to buy food. Thanks to you exhibitionists and prestige hunting in your neighborhood backing out of your driveways with huge vehicles I will have to buy canned kidney beans from Argentina! There ain't gonna be any one dollar for a dozen ears of corn, it's up to two green ones now and maybe more. And cows don't produce milk nor do suffocating chickens lay eggs.

However, I am not that vain that I think destiny calls for my issuing wake-up calls. But the house that America is, in a poetic sense, has problems. If you see saw dust accumulating in your cellar at the bottom of wooden beams on top of cement walls (did you ever see that?) do you wait and wait to see and verify whether it's termites or carpenter ants and delay taking action? Do you procrastinate? Time marches on – we play games as if Nature will wait. Must we fit into R. W. Emerson's unflattering view of man – imbeciles?

I WANTED TO LET IT GO, BUT SOME BRAIN DEADS
MADE CRUEL REMARKS ABOUT THE FATE OF KENNEDY, JR.

Let us, you and I, try to recreate what happened. Let's treat this on a protective bent and think of John F. Kennedy Jr. as being intelligent and not just another wild one, that being mild as to what I heard over my nighttime radio – I didn't notice but the moon was full.

You are flying at 2500 feet and there to your left is the first sign of Martha Vineyard's land mass, however misty it's outline, some dim spots of lights, scattered in the haze. It's night, there is enough visibility from that altitude to make out blurred pin points of light but at your present rate of airspeed, probably over 125 mph, you had better drop down in preparation for a landing – the airport is only 12 miles or so away – maybe 6 to 7 minutes left at the most before making that runway approach. You let the wheels down. From this point on I would like to give JFK, Jr. a little more credit rather than supposing as it has been publicized too much that he lost control, dove down, spun out and all that hypothetical reconstructing going on over the last two minutes that he, his wife and sister-in-law were alive. He dropped altitude in a hurry – no big deal. I've experienced that. It's quite common a procedure if there is no other planes around. Why not. Young pilots don't mind that. I did mind, especially on clear days, once out at sea with fishing boats looming up at me or homes in Maine getting bigger in a hurry – besides my stomach and not the pilot's was not use to it. Like the driver in a speeding car he doesn't feel the same as someone sitting next to him. When JFK, Jr. was leveling off at a lower altitude he had to be looking to his left and not at the sea beneath him because it could not have been visible what with haze bordering on fog being thickest at surface water levels – he needed a reference point. What other direction would he be looking at then? Furthermore, unmentioned in news clippings and TV narrators of course, the tide at 9:40 PM, (time of crash) had about one hour and 20 minutes left before peaking. There had to be (from that island's characteristics) a thickening haze caused by colder waters rising along coastal expanses. Now, seat yourself in a plane, join me as would-be pilots, oh say, 500 feet to 700 feet altitude, the haze you are now in is thicker. Will you see those scattered island lights or any geographical reference points on your left clearly? Not very well. But you certainly would be more intense about seeing something to guide you, would you not? Was JFK, Jr. so preoccupied, staring at what little he could see of land and specks of blurred lights and didn't realize he was flying lower and lower? A fisherman supposedly saw a plane flying at 100 feet. Who is he, where is he? You can't see the water below you – a film of water vapor covered it along with the obscurity of night. Did he glance at the altimeter? Was it accurate if he did? We'll never know. If he could see the island and it's lights, did he make the mistake of judging his altitude from that? Did you know naval pilots forced to bail out at sea sometimes plummeted to their death from great heights after releasing themselves from their parachutes too soon – guessing as to how near waves were below by looking at the horizon? Did JFK, Jr. try to judge whatever direction or height his plane was at by the confusing results sure to occur looking at some vague dark outline to his left? I'm not quite sold on his being disoriented as present theories suggest – I think something was lacking in what was taught to him in flight school. * Yes indeed. Remember, the wheels were down and he did not radio he was in trouble, and that fisherman story, that raises questions. The percentage of private plane accidents in overall aviation is high – I think 38% - ask yourself why.

If I have gone to great lengths over this tragedy it is because it has some weather factors to think about, besides it rekindles the feeling, not overwhelming, but delayed, then enduring for awhile, that quite naturally bring back detailed memories of beautiful nice people suddenly no more. I see planes in motion – a propeller slinging sod, the look of a pilot who made it – special kind of people, my kind of people – noble too. When something traumatic happens I dive right in emotionally aroused but now I express it as best as I can in writing – there is, you must understand I'm sure, foremost and always the <u>human connection</u>.

I heard over the radio and of course one muscle-head journalist had to add his well written two bits worth as to why the fuss over JFK, Jr. insisting (a bit hard I might add) those three departed never worried about us, etc…, three columns no less. I should remind him it is quite human to be interested in figureheads, even actors and actresses – hard to explain, it goes beyond news. Maybe a song "People

Who Need People" (if that is the title) touches on it, this kinship thing that separates humans from animals, this feeling of something we could call group

How deep do they go into the danger of weather?

attachment, something subconsciously going back to primeval days, a member or members of our society suddenly no more. Certain animals mill about, touching a dead member but it is over with quickly. We, emotionally speaking, also mill about and it lasts much longer.

Is it a delayed afterthought or whatever, I have this feeling about JFK, Jr. periodically taking up my thinking time, whatever that is. How true is the report a plane was seen just 100 feet above the water? The visibility was not zero as best as I can find out. He probably could see the water beneath his plane. But, that is extremely dangerous to use as a basis for estimating (quickly, of course) how high you are flying. If you have but one to two foot waves as I believe there were that night, maybe some white cresting, that could be interpreted the wrong way – it could mean to anyone that the surface was, deceptive as it is, at a lower level. The theory I have read and heard is something to the effect that the plane dove in. If so, why one wheel quite distant from the spot where the engine and fuselage was found? If my hypothetical opinion and that's all it is, visualizes a plane suddenly skimming just over the surface, then that would account for a wheel and an attached component ripping away. Be that as it may, it won't bring back three people, our hearts having gone out to them.

According to Arthur Schlesinger, Jr., who I respect and probably hundreds of more astute than I, as a thorough analyzer of good and bad traits of our political structure and it's captains at the wheel, dismantling as he does, myths or crediting virtues, none of which raised the ire of critics, his views marching on unchallenged, the environmental problem was manageable a long time ago, the era of this unproductive procrastination sometime in the 50s. Be that as it may, economic reactions being a rather flighty component of the "by-sell" decision making process, peaking stock fever suddenly chilled by the slightest change of one index or another, there is nevertheless, stock reaction or not (might do good if it shook out weaklings) more than ever the need to come to grips with a grave problem that will not go away, worsening at an accelerating pace. If there is a flaw in our fluctuating psyche make-up (very human) it is to close our minds to the need for change, carefully administered of course. The American tendency to see-saw from churned-up hostility toward government to leaning toward accepting the need for leadership suggesting more controls is now in a state where politicians are partially paralyzed, choosing to stay neutral wanting to please everyone imagining I suppose increasing prosperity will overcome all and re-election is not at risk. From my point of view, it is an uninformed public, prone to be indifferent, that should shoulder much of the blame, vulnerably open therefore to the best snake oil salesman, suspicious of sincerity if not accompanied by charismatic buoyancy. Can you imagine, a 20-minute debate on TV, just a few words and millions make decisions from this. I just heard a political analyzer, John DePretio say G. W. Bush in his campaign oratory is copying to the word the same page Ronald Reagan used – obviously Bush knows what worked before should work again. You know, tax cuts, less government and so on. But, the public is coming up with some surprising reactions to efforts at big tax cuts and according to a radio poll taken by Myrna Lamb, a lot of people are against it. And they were not the elderly calling in either. No, DePretio is not a liberal, definitely not!

Issues are being avoided. Nobody knows what Gore, Bush or Bradley really represent. Exactly what is needed? If a car is deteriorating in performance and reliable durability is desired, do you spend money on a new paint job? For one, redundant as I can be, not being Kissinger or some other elitist, I want to emphasize again, the whole of society must stop being selfish in a way, and devote a little more time to take a good look at what is happening to this nation. Three months ago, as simple as it was in determining, I wrote, "Stand by for scorching heat." It follows droughts come with it. Nobody else wrote or advised that over TV or radio, not a soul. My point is, let's not dream things will turn out rosy sooner or later, in terms of years to come. Therefore, take a good look at everything. That is one problem but it is overshadowed by obsessive concentrations on building up future security (while nature has bombs waiting) and furthermore, irritated by what they imagine as a threat to their goals manifested with the usual "Go away government – we're smarter than you are." (Sometimes that's true of course, but) slogans of this nature whipped along by vested interests who will, if this rush for the Golden Fleece propels us into the inevitable (with the eventual loss of some Greenspans) "What goes up must come down" roller coaster, definitely walk off with all the marbles ass they always have.

Visionaries, there being none in sight would find it politically dangerous in these dollar chasing times, TV bombing my senses with endless commercials on getting rich, drug store book shelves loaded with Secrets of the Rich, etc…trying to promote worthwhile future major projects what with skittish voters fearing change, the opposition singing the same song "Liberals want to waste your tax money" and "Government is excess baggage" knowing full well only a major federal entity has the means and wherewithal as in the past to accomplish what the private sector couldn't possibly do, nor do they have an interest to even think about it. Business lives for quarterly profits, as it should be. Except for farmers whose numbers diminish, who, in America would be interested in a network of irrigation canals or pipe lines, tapping into rivers that are both a blessing and a menace, prone to flooding at times. Desalination of salt water, a fine idea will help water needs of growing populations of nearby cities but certainly not that of our agricultural expanses. Consider the wasteful Mississippi dumping rich mud into the Gulf of Mexico to the extent New Orleans gradually sinks from the sheer mass and weight of it while to the north the land tilts up.

Idealism is for dreamers and dreamers are a nuisance to a society not really interested in anything of national importance that goes beyond a few years from now.

In conclusive acceptance of what trend we are in although I am not the total cynic, always hoping just a wee bit, my increasing pessimism over the lack of foresight, clearly evident in a runaway climate change, the whole world at fault and no promise of action forthcoming, taxpayers here growling "Leave us alone" and filth producing Asia preferring to let it's people drown or smother, scenes of handkerchiefs held over faces in Hong Kong for example. I am in hope of merely offering a Home-Education course for what it's worth.

**GOO UP YOUR SKIN WITH SUNSCREEN ALL YOU WANT
BUT DON'T STAY IN THE SUN TOO LONG, BECAUSE:**

President Clinton whose state of the nation speech fired up applause but now as time goes by his one proposal I am most keen on may be set aside simply because there is a big pie of surplus money and all the kiddies are clamoring with their forks to dig into it. To financially help power plants convert to less climate deranging fuels struck me as urgently needed. The sun powers on, so strong that skin cancer (reports from Europe also) increases at the rate of 4 percent every year. More evidence noticed by yours truly, the sun has elevated coastal water temperatures in the Gulf of Mexico (not everywhere) to a 90-degree temperature! I follow temperature trends in the Caribbean. Why repeat what I see. It is without resorting to dramatic expressions, somewhat predictably unfavorable. It's influence is far reaching and extends heated air much further northward and for protracted periods. My having beat to death details of NOx and SO2, the latter impossible to curtail for several decades I still must insist something powerful and continuously being produced is not allowing us the benefits of normal sunshine! One chap I knew had left a battery in his auto trunk too long. It tipped over, the sulfuric acid leaked out of former vented removable caps and within one month a huge chunk of the steel floor disappeared – completely ate up. That is one of the most potent chemicals produced when sulphur-dioxide created by coal burning plants (and China – horrors!) and oil refineries comes in contact with moisture – the conversion taking place to a point where you can smell it under different weather conditions. What else would b wreaking havoc with the ozone layer? It has to be something very powerful and produced in enormous quantities. What do we need to do, wait 20 years while scientists fiddle around with experiments?

I worry about farmers. Without them this country would not be complete and self-reliant. Right now, August the 3rd, 1999 much talk of loaning money to hard pressed farmers being ruined by drought conditions, from government figures – politicians arguing how much, etc… It's fine to say "Here's $100,000" but a loan means you are $100,000 in the hole. It is a loss first of all – a hole you must climb out of. How many youngsters will take over this needed occupation if the prospects of normalcy get dimmer by the years?

By the way, the blessing of earth warming we should look forward to according to experts evidently approved for public notice by Reader's Digest has a catch, if I may facetiously find humor from our Journal's article today. There you are, sitting on a beach chair, say here at Barrington Beach as a few swimmers walk back from the waters you see them scratching themselves like crazy and then your eyes notice also, torsos not lying still no more, scratch, scratching away, like seals infested with sand fleas. Well, it happened here and doctors got their share of worried bathers suffering from "Swimmers' Itch" as it's called, the cause being a parasitic agent. Everything unwanted multiplies and thrives in heat of course. The beach has been closed.

But Aha! There is a ten-year-old boy, an entrepreneur who is making money thanks to abnormal heat. In some town in the Midwest, according to a radio report, grasshoppers have multiplied one hundred fold and devour flowers, plants and what have you, cultivated by homeowners. The young lad advertises he will clear your yard of these pests for X amount of dollars. He then goes home with cans full and feeds them to his chickens that enthusiastically function as disposers of a problem. Well, in one way, the Reader's Digest cannot be all wrong, can they? The boy is quite the capitalist, fills his pockets with money while reducing the cost of chicken feed.

I am not the keeper of the environmentalist's flame. I thought Al Gore was but he has been squelched, at least temporarily. He needs a sock-em, rock-em approach and define in optimistic rousing terms how the public should perceive change.

Isn't it amazing! For years millions have been sent to Israel and so, aside from defense needs, our friends there have converted semi-desert lands into productive arable soil while we dry up and grasshoppers take over.

Then we have those wishing for a big destructive hurricane to bring us water. What the hell, a piece of Florida destroyed, maybe Louisiana or the Carolinas, or Texas and even some Central American country, why not. That's smart isn't it?

WANDERING SCIENTISTS – LOST IN SPACE

August 4, 1999. Not malicious bent at all I want to indicate to you the detrimental effect on clear thinking caused by conservative politicians, haughty think-tankers scratching out impressive anti-environmentalist themes in a convincing manner along with snickering radio personalities specializing in ridicule. We now are told a 7 billion-dollar aid package for farmers is about to be passed by the Senate, classified as due to the <u>unexpected</u> destructive weather. 2400 scientists with their usual "More research is needed" (too many on the dole) looking for secrets of the universe, that part of it hugging our planet while all that they had to do is read thermometers and join little old me (remotely disassociated as we are) who certainly cannot accept "<u>unexpected</u>!" Swayed by conservative efforts to derail accurate interpretations of major factors, sticking out like a sore thumb, i.e., a heated equatorial region

for years now building up it's expansive air mass while the Arctic region clearly has lost it's needed cold-generating power during the recent past (two decades or so) so called experts are lost in their own high-falooting world. It is becoming costlier with time just because ignorance fostered by deceivers prevailed over logical reasoning thanks to campaigning far righters selling disrespect for environmentalists labeling them as "whackos". This peasant is concerned – potatoes up to near 3 dollars for a small bag from $1.99 and also milk per gallon – up and up we go!

Irritation is getting the best of me – I'm losing my composure. What illogical man promoted to a position allowing him to make public statements, very questionable ones, by coming up with this announcement a few days ago, August 2 or thereabouts, transmitted via radio over the airwaves: "This drought should not affect food prices because there is an abundance out west"? I don't know but with four states in the east, big producers of vegetables, it's farmers throwing in the white towel knowing a scorching sun has done them in and President Clinton pushing for a 7 billion dollar aid package, furthermore, my having observed* shipping containers, lettuce, cabbage, tomatoes, potatoes and what have you coming in from Quebec and lands south of the equator I think he has never set foot in supermarkets and seen prices inching up week by week. Remember, dear reader, a producer-shipper in California** who keeps 40 thousand gallons of gasoline in storage, phoned in to Jim Bohanan and said a 10 cent rise in his cost per gallon means shipping will be more expensive and therefore you can expect food prices to go up. There is being maintained a sort of dream state in this nation to the effect "Oh, things will turn out alright eventually, why worry." That attitude suits the middle class alright but there are 20 to 30 million workers in the bottom rung of salaries who will think quite differently if not already. Again, let me repeat, at least present aspiring candidates for the presidency should have the intestinal fortitude to talk up issues instead of giving us the old circus "Snake oil" pitch. I'm for some tax relief but not something preposterously designed to boost incomes for those who least need it. Something has to be set aside for future needs and that is the job of the government with public approval of course. It appears there are a lot of Americans, who think like I do, very gratifying to say the least but I don't know if they can prevail.
*Always scouring my little universe.
**That is the "productive" west is it not? I think that optimistic messenger should spend more time listening to intelligent radio talk shows and broaden his perspective. Aside from climate effects doesn't this man see fuel prices rising? Diesel trucks deliver food cross-country and the increasing cost has to keep going up.

A Mesmerized Following – Reminiscent of Pre-World War II Techniques That Succeeded – "Repeat, Repeat and be Vocal!"

Had he not attacked environmentalists I would not write of him. To what low level of vilification the favored son of far right Republicans, Russ Limbaugh, the ultimate berator, measured restraint but frequently erupting diatribe, blurted out to make a point designed to create a schism between the haves and have nots who he does not tolerate, will stoop to, bent on agitating the middle classes and the wealthy into feeling Democrats are abusing them. Can you imagine him, today, July 28, 1999, insisting the poor should be taxed so that they too, finally, can experience what the better off are putting up with. Yesterday, old people were indirectly labeled parasites, the majority, women, who he with some anger to his tone, called "battleaxes" because he resents entitlements tentatively standing in the way of tax cuts. As for the poor, not bothering to separate welfarist from hard working nobodies he had to add, "Let them stop buying junk foods and beer" as if 10, 20 or whatever million low level laborers were careless bums. The poor according to this modern day Louis XIV (I dare not call his wife Marie Antoinette – I don't know her), speaking vehemently and intensely as did Goebbels in WWII, are non-producers in our society. The rich provide jobs for them while they leach from taxpayers. Their existence is a burden for the well paid. Unlike Hitler's propagandist promoting hate and extermination of the unwanted, this man wants to reduce poverty-level workers to the equivalent of serfs by virtue of reduced take home pay after taxes paying more to the government so that he and his crowd can keep more. It's that simple – why make it sound just, as he does, having identified himself (evidently successful at it) as the intelligent authority protecting the besieged wealthy and well off affluent from persecution by liberals. The poor (I speak for workers) do nothing for America, if my ears are not failing me, engrossed I confess, feeling I am listening to the ranting of someone selling not just a political choice but cruelty. Should his concepts ever be realized (I doubt it) there would be a society plagued with covert rebellious acts. It is already in motion this reaction by malcontents but not on a crisis level. Completely unmentioned, workers produce the <u>very</u> <u>goods</u> <u>that</u> <u>enrich</u> <u>his</u> <u>suffering</u> <u>followers</u>. I have a large family and I know several of them work very hard, make survival wages and one, doesn't even get paid for certain holidays. If you want to break away from seven dollars and hour as one daughter-in-law did, choosing 11 dollars instead, you have to work in a sweat shop plagued by unventilated lint polluted air while producing at a frantic pace. She bothers me but…that's her choice wishing to buy lovely this and that. By the way, is Limbaugh a German name?

Lord, what a harsh man! He does not understand the need for a balanced society. If his attitude is not a pretext aimed at garnering approval of Republicans and he is, as harsh as I imagine, I view him as an effective anti-environmentalist persona capable of influencing politicians. I don't think I'm wrong if I have him pegged as 100 percent pro special interest and not caring much about the general populace. He knows where the money is. His sudden passion for golf typifies some who, not athletic at all, wish to belong to a special society and imagine they will soon be members of highly esteemed groups when all they really are doing is poking a little white ball along and talking about if for hours.

My wife and I spent 5 days in the Bahamas, about 10 years ago, Freeport it was. We were at some breakfast room buffet style in the hotel. Not many people, too English for me, spoiled by the liveliness of Europe, blacks in charge everywhere in this tourist heaven, all very polite but I wouldn't try for a smile – they seemed rigid but in all fairness exceptionally courteous and neat – well trained by their previous occupiers. Four middle aged duffers sat next to us and all I heard was the intricacies of their game, infatuated as they are over golf.

peaking of a balanced society we've come a long way. Did you know what philosophy those folks back in the 1800s, when social discord created prominent spokesmen who influenced thinking, unsuccessful revolts and all that in Europe, promoted much to the delight of lords and landowners? One, David Ricardo, lived in London and his famous contribution "The labor theory" lasted until pre-Great Depression days. It was: "Just pay workers wages that will allow them to barely survive. If you pay them more they will proliferate and become nuisances." It 's economics true, but it explains why the far right doesn't care about working stiffs who will feel the brunt of climate effects.

SOMEBODY – GET OFF THE POT!

Couldn't we the public, finally recognize the game played on and on by a determined few, indifferent to potential disaster and suffering selling infallibility through doctorated spokesmen (haven't read of a woman yet – more honest and sensitive as a rule) belonging to so called genius loaded institutes, over the issue of environmental problems? I've resorted to including excerpts after excerpt to show you a range of opinions so as to reveal the age old battle of truth versus lies for one, confusing information for another, i.e., "Yes there is" "No there isn't" "There's more" "There's less" in matter of toxins, by whom and where they prevail. What does this nation need to come to grips with a very serious problem? Another dust bowl? Unprecedented death tolls from heat? Huge tornadoes and more of them? Softball sized hailstones? Come on folks! And wildfires? Even New Englanders now very

> ■ The Great Smoky Mountain National Park had dirtier air on some days than any city in the South, except Atlanta.
>
> The groups said that in the 32 states examined between April 1 and July 25, the levels of ozone, or smog, exceeded federal standards on 85 days in some parts of the states — a total of 2,743 such instances.
>
> "That's slightly ahead of where we were last year. Last year was pretty darn bad and this year is maybe a little bit worse," said Jayne Mardock, director of the Clean Air Network.

worried over bone dry forests, leaves shriveling, some falling, green weeds turned dead yellow waiting for a cigarette butt to start it all off.

Another excerpt now – my final. Jayne Mardoch is not a financed liar! She doesn't belong to the Carbon Club! We must be dreamers if we think this climate problem is temporary. Or, the nonsense: "We have made great strides in arresting foul emissions." I use her statement in the hope (am I also a wishful thinker?) it may stir a few ostriches to pull their heads out of the sand. We're getting nowhere contrary to claims. Our approach needs immediate evaluation and not what a selfish minority has set out to do, delay action for years and years. Insofar as scientists you can expect "More research is needed" to carry through another generation, at least.

Let's pretend that I'm just talking to you as a friend. No railing, no rancor. I did register disgust, true. Consider where the lack of resolve on the part of leadership has taken us and the reasons therefore. Perhaps it is confusion. Too much controversy so far – it has had its effect. Or are they afraid or not willing to take chances very much aware there has developed a down side to the repeated alternations in public moods, a cyclical thing I wrote about some pages back borrowing from analysts? The political animal, meaning intuition and instinct does not want to stick his or her neck out – safer to test the winds. One faction of voters, an influential one, has made it clear it is not keen about supervisory figures and more laws. Even believing government is predacious. Perhaps they would scoff at anyone suggesting (as I did) we're taking a long walk on a short pier or even if they are beginning to realize there might be something to earth warming in view of repeated episodes of unusual weather. Consider a tornado* today, August 8, 1999 in Long Island. Not a big one but damaging no less and I'm sure a lot of people were frightened. Twisters are spreading. I made that clear. But the attitude that may prevail is to build up revenue and sail right through hell and high water – "It will end, I'll not be fazed, I don't need anybody." Do you know what the author of "The Seven Pillars of Market Success" a retired broker doling out advice but here and there sharing a touch of guilt feelings over the Roaring Twenties, while explaining mass psychology, said pertaining to the wise and the foolish? What segment of investors are prone to brashness, overconfidence and failure to understand a nation's welfare and its moods being a rather sweeping recurring tidal swing affecting everyone? Don't be offended – it is in general the 25 to 40 year olds. Let that be, although nothing has changed in our history of up and down economics. We're not mentally cloned. Besides the information age is upon us. But it is a generation thing and that is both justified while also risky, if you disrespect wisdom of the more aged, risky if you feel it's your turn to guide the ship all by yourselves, figuratively speaking. 250 million people – one group, if that was the system, without the older members of the Federal Reserve (loved and hated) could endanger, could inflict much damage if their youthful passions were not occasionally reined in.

But our fireside chat (me doing all the talking) is not just about the young versus the old. I was just building up to a fact or more politically correct, the psychology of confusion and lack of resolve, the fault not altogether that of leaders who must after all, in a democracy, make decisions based on what the majority wishes.

Reports are not that definitive – but damage nevertheless-even Martha's Vineyard was in line for what is very unusual – also some damage.

Let's go back to 1992. 150 nations participated in the Rio Earth Summit. The idea was to work out a plan and take concrete steps to curtail the buildup of Greenhouse Effects, the focus on carbon-dioxide emissions. Fine. Based on the recorded density of CO2 in the atmosphere during 1990, two years later a good reason for the Rio meeting, we, the United States (biggest polluter of all but understandably so) made what we could call a promise to achieve that lesser figure (1990) of CO2 by the year 2000. It was clear to all concerned the upward trend of this emission would continue and not level off. It was just a matter of using less fossil fuel. But, what did we do beginning sometime around the very year we made that commitment? Without any compulsion whatsoever we began producing more carbon-dioxide and the amount has been climbing ever since. And you know very well why.

Do you know what country and its industrialists is doing the most toward this goal of cleaner air? It is Germany. I don't barge forward with opinions empty-handed – I have a lot of data from the Society of Automotive Engineers but I shan't bore you with engine developments and the efforts at improvements going on. I just wanted you to perhaps look into a mirror if you feel someone else is totally responsible when occasionally, borrowing from TV shows, Nature's Fury beats down on this or that area or some city official tells you your lawn has to be given up. Are we still friends? Irregardless, je vous aime quand meme.

My not having written anything particularly pleasant calls for reporting anything cheering, whenever it shows what side is gaining. For example, I read where the Growth of Goodness (actual title of an article) is evident – donations to various causes on the rise. If it keeps up I'll have a hard time finding someone whose neck I would like to spray beer on.

AUGUST 11, 1999

Imbedded with fixed beliefs taught to them by professors, who, it appears (I know I am convinced) remain faithful to old theories, either through discomfort over questioning them or an unwillingness to venture into new approaches fearing (I had never anything to lose) criticism, perhaps a contained snicker from crowned leaders, the top climatologists still fumble along. At least they are honest about what I know to be very difficult (see excerpt). For one, you cannot predict what part of our nation will be subjected to a drought, nor predict how many hurricanes. My suggestion is to cast aside the ideas that have developed into fixed beliefs – there is no secret equation to be had, which has kept scores of scientists searching the skies and the oceans. Then, come back to earth, keep it simple and start reviewing fundamentals. First things first, find out what is the sun doing, intensity wise. From there consider with more seriousness Nature has balance in mind between the poles and equatorial heat and the means by which She conducts her affairs. But Nature has a problem. And of course, pollution as to its accumulation and the solution (partially effective) resorted to by this power however temporary. I am glad, in the same article, an admittance, El Nino has been an excessive media hype affair, pleasing to harmful groups and detouring logical reasoning. In our present state of procrastination we are the fat man who has put off dieting until January 1st.

Today's paper, August 13, 1999, has an article praising a super battery being developed. The advantages; a list of uses very noteworthy has with it the usual belief electric cars will bless our environment. But, no mention we would be swapping one toxic emission for one much worse as long as we have coal fired power plants.

Meteorologists: 'No good tool yet for forecasting drought'

By SETH BORENSTEIN
Knight Ridder Newspapers

WASHINGTON — Just five months ago, the nation's top meteorologists warned of a worsening drought . . . in the Southwest.

The rest of the country didn't have to worry about a shortage of rain in the spring and summer to come, National Weather Service officials said.

Now, the forecasters concede that they failed to predict a drought of historic proportions because they were led astray by a combination of factors, including a relatively wet winter.

The current East Coast dry spell — the worst ever in New Jersey, Delaware, Maryland and Rhode Island — is so bad that President Clinton yesterday declared all of New Jersey and 34 counties in New York disaster areas, allowing farmers to get

*Turn to **DROUGHT**, Page A-9*

I need to find humor in life. If adults are not familiar with infants I soon discovered they saw nothing funny in this. My 15 month old grandson became very angry and loudly protested when my oldest daughter, babysitting, interfered with his plan to eat a huge black beetle, quite alive and kicking. What limited teeth he has were very busy chomping on such a crunchy gourmet bit. This, of course, occurred in the backyard – you can't let these walking tiny young ones ready to try tasting anything that looks interesting out of your sight very long. What's so weird about this, as one chap thought, since Steve McQueen ate bugs in the movie "Papillion" after all? Aha, you thought I strayed from the main topic. Well that child may be a bug exterminator some day. Big demand in the future, believe me!

In what follows, Salt Lake City, having no warning except heavy-duty thunderstorms approaching – I can't think of anything more lethal than heated cities shedding off a million BTUs every second. This fact may take time to register with the orthodox.

IT WILL TAKE TIME BUT IT WILL SINK IN ONE OF THESE DAYS.

Although I wrote Al Gore a five page letter in April 1998 (14 months ago) a follow-up from a previous missive sent to him two years before explaining how city heat during the day, thousands of autos, baked hardtop streets, concrete and brick absorbing solar heat and retaining it, factory stacks, etc…, therefore intensifying tornadoes should they cross over cities, now that the odds favor more destruction than ever in view of expanding metropolises, the following does not revolve over that fact at all. Enough is enough.

In an effort at some semblance to chronology, pertaining to more twisters striking at busy cities I must add Salt Lake City's experience August 11, 1999. It appears, this shocker, as bringing out from various articles nothing new in the way of enlightenment. But you, some day, may have the inner satisfaction of an "Aha, I know why!"

should you read what someone has to say, as they did this time: "This is very unusual." End of explanation. The expert goes back to his computer and that's that. I left out a very significant reality, other than heat, that upon which tornadoes feast, which, should the occasion arise for an opportune "Aha" where I'm certain your friends will be impressed at your knowledge (I'm just having a little fun) you must pick up where degreed experts leave off. Of course it is rare for a tornado in Salt Lake City!! Why now, that is the question. We have two lady researchers who spent a very long time and came up with my idea of material – no blurring contentions. We have a major increase (years, not one season) in humidity over one third of the USA from the west coast to, well, Utah at least. We have a major increase in humidity over one third of the USA (east side) from the coast to at least near the Mississippi River. Did the man with the braid, the title, some reporter scratching out a few notes, attempt to mention this fact and thus indirectly pay tribute to two hard working ladies who surely would have been pleased? Of course not. And so, a surprise visit by a tornado (the public will forget about it in about four days) (not that city's residents of course) while the public remains oblivious to the reasons why; why there is so much flash flooding in the southwest and tornadoes where they shouldn't be. We covered excess pollution particulates increasing the amount of moisture air can accumulate. The Pacific's air, blessed with filth from China and India, they comprising one third of the world's population, (if you visit Hong Kong, bring a gas mask) battered our northwest states for three months, dumping water in spectacular amounts during winter's more vigorous manifestations but now a more tranquil summer drives equatorial heat further northward than ever. Along came a weak cold front from Calgary and Saskatchewan meeting the ingredients already overly described by yours truly, heat and water vapor in profusion, thanks to a huge Pacific high and thanks to it's peer system east of Mexico pushing record breaking heated air.

I can't help emulating two greats who have written books that keep the snickering crowd at bay, critics quite naturally conceding their superiority in politics and economics, namely, Arthur Schlesinger

Jr., and John Gailbraight. Of course I selected choice passages that I happily am about to expose you to for the very simple reason I have experienced exactly what they are talking about. True, theirs has a mischievous bent and of course, like I, they have people they respect but this is reality, this is bureaucracy, this is classic male giants who rose to positions but destroyed thousands through unbearable stupidity. Arthur Schlesinger: "Don't be infatuated over Credentialism." Gailbraight, my favorite, a fixed smile while William Buckley Jr., talked to him on a TV show much to my amusement, had this to say in his book, "The Age of Uncertainty" and all I can say is, how true, Mr. Gailbraight. The title of one chapter THE STUPIDITY PROBLEM, "If inheritance qualifies one for office, intelligence cannot be a requirement. Nor is it's <u>absence</u> likely to be a disqualification. On the contrary, intelligence is a <u>threat</u> <u>to</u> <u>those</u> <u>who</u> <u>do</u> <u>not</u> <u>possess</u> <u>it</u>, and there is a strong case, therefore, for <u>excluding</u> those who do possess it." Always remember that. If you are young, frustrated by frauds, move off. You'll find better people. I didn't say everybody fits in to the above Stupidity Problem category. Remember that I was only 20 when the first decorated moron affected me and I am very proud of having sprayed smelly beer on him, very proud. Great move on my part even though I was under the influence.

In the past few days, this being August 31, 1999, the Arctic region has begun to say goodbye to the sun, or vice versa. Temperatures in Barrow, Alaska, about 1155 miles from the North Pole have dropped into the 30s and with this transition from warm to cold, high pressure systems forming there in the refrigerating zone, several masses (not very cold) have broken loose and moved down across Canada through our north east. Hurricane Dennis now recessed into a 60-mph wind storm and to me (you don't have to buy it) represents Nature's effort to balance out this typical late summer-early fall imbalance between too distant conflicting temperature zones that make life-supporting weather possible. Unless, as I wrote before, you are not a believer and Nature to you is casually viewed as randomness – not approving there is a force we call God that devised controls I want to pursue what to me is rational. This summer of 1999 our high pressure voluminous mass over the tropics, that being the other zone, still driving heated air northward <u>all</u> <u>around</u> <u>the</u> <u>world</u> has not been as expansive as last year and consequently represents less resistance to invasive cold thrusts from the north pole. If, and it occurred more frequently five decades ago and back in time, the Arctic circle as a home ground creating huge blocks of frigid air <u>had</u> <u>not</u> been subjected to a warming trend <u>as</u> <u>it</u> <u>has</u> during the decade of the 90s then the colder air masses launched southward by centrifugal forces would still be creating through a process of balance seeking, larger and more destructive land-falling hurricanes. Recently there are less. Statistically the 90s only produced four major hurricanes (winds over 110 mph) that struck land. In contrast, going back to the 50s in increments of decades, six in the 80s, four in the 70s, six for the 60s and during a cooler decade, the 50s had 9 hefty powerhouses. Why then less major (Force 3) property devastating hurricanes in a period from roughly 1990 to 1999 where due to increasing heat (you can't deny that) we have more tornadoes but less hurricanes of great force? It is my theory and while, as with everything meteorological, it will take a long time to prove its validity I'll stand by my guns. Already explained many pages back based on the principle of a Nature continually seeking balance, the colder the Arctic region crating major masses on the move, hell bent to go south the more it is necessary for some mechanism to remove what I know to be a huge barrier to remove out of the way, vertically and horizontally potently obstructive in a sense, resistant to displacement, that being summer's buildup of heated moist air covering over one third of our hemisphere from the equator north to Chicago, to Bermuda, to Greece (Paris remained nice – those lucky Frenchmen), to the Himalayas, to Central China, to Japan, to the Hawaiian Islands (temps there nudging 90 this summer and still about 88) – (remember I keep records and some folks do not!). I'm not an office man staring at computers; the world's thermometers are my focus. Now, no Great Swami am I nor am I decorated with the seductive badges of credentialism but an article I just read and probably one and a half million others has cranked up this Lafitte the Pirate to fire a few shots away. I am giving you the Big Picture – take it or leave it. It is as follows and it is a repeat. <u>No</u> <u>way</u> can Nature rely on the usual storm formative factor of cold air meets warm air, cold front bends, warm front partnering in to form a low pressure system setting off the means by which precipitation is shed upon us and also (remember?) remove warm damp air and ship it back to the poles. No way can such a system be created in semi-tropical zones!! Therefore, the only means to accomplish the same end, namely, vacuum up as much tropical air as quick as possible and make room for fresh batches of Arctic air in its programmed intent to provide a large portion of our mid latitudes with recycled fresher air, is to create an effective vacuum machine. Next, the same article, annoying as are some facets of it, their correctly including we have had so far in this 90s decade over 10,000 tornadoes but, and here is where I reacted negatively, the writers give the idea we just might not have accurately counted tornadoes some decades ago. Without saying so they could be leaving the reader wondering if there is anything unusual going on. I imagine the Carbon Club finds this gratifying. The olarchy behind Bush love it. It helps their cause. PC on my part but the writers did not intend that.

Alright, it's time for me to be assertive. They're not covering all bases these experts out to enlighten us. They instigate controversy over how we count twisters, how many tornadoes instead of why we have more of them and illogically overlook the increased heat factor prolonged abnormally seven or eight months instead of yesteryear's four to five months. First of all tornadoes as separate entities or spawned by thunderstorms are a product of heated air, somewhere above the 80 degree level, along with plenty of tropical moisture (you know all that of course) (cold fronts lifting air, etc., etc.,) consequently, ask yourself this question: Has the weather in this 90s decade been the warmest in a long time or not? Of course it has! <u>Therefore</u> <u>we</u> <u>have</u> <u>had</u> <u>more</u> <u>tornadoes</u> even in winter; incredible as that is. Inversely <u>we</u> <u>had</u> <u>to</u> <u>have</u> <u>less</u> <u>in</u> <u>the</u> <u>50s</u> for example, because the weather was cooler. Additionally as deceptively perceived is this preoccupation with world wide infra-red measurements convincing those anxious to deny there is warming, indicating the earth's average temperature has only increased one degree in the past 100 years. Fine, that's the average for an extremely voluminous air jacket wrapped around our planet but that does not mean all continents have been spared from prolonged seasonal bouts of record breaking high temperatures! We have to consider the ocean's stability also. We have the southern hemisphere involved in affecting averages.

I'm somewhat fed up reading attention-grabbing prose that intentionally or not hinder the environmental and ecological cause. Giving that the blessing of beautiful October (of course it is, I was born that month) and the refreshing arrival of early winter air, surely invigorating as it is and several months to forget what has happened I can imagine interest in what will drastically affect us with more intensity and great harm will fade until we get beat upon again.

I HAVE FOCUSED ON INCREASING FREQUENCY OF DESTRUCTIVE WEATHER

Knowing who I would like to reach I have put the case of earth warming and its deceptively slow advances in fuller more persuasive form. Having neither the patience nor time to innovatively structure subtleties in every paragraph depriving myself of acceptance by Shakespeare buffs and English Majors, some authors spending years (smart and younger; I guess I'm not) and able to utilize every word in a pocket-sized dictionary (I still have to leaf through one) the apparent enemies of environmentalists being a target here and there as I wrote, and not an excuse for a military adroitness and Jack Webb's "Just the facts, Ma'am." I tried to warn you the Money First, Mortuary Body Counts Last crowd has skillfully endeavored to make you believe: "All periods of destructive weather are cyclical." Consequently their effectiveness is obvious to me – despite a few cries from the wilderness most folks expect things to return to normalcy, therefore the psyche Que Sera, Sera prevails. Trusting my limited style of communicating doesn't strike you as petulantly tainted (I'm not demanding attention – just passing on what is sincerity) I must again tell you a recent article of August 28,1999, in all probability read by over a million, flashy in meteorological highlights but missing some very important details has me wound up again.

Essentially I offer the reader a topic on a conversational basis, or "This is it from my perspective", "Respond or don't respond." As for experience, weather work methodology, the basics of which have not changed. Weather maps, instrumentation, atmosphere probing balloon-carried data transmitters and all that remains the same. They now have pinpointing Doppler radar and satellite imaging but predicting remains a function of one's brain and not electronic machines.

It seems to be a habit writers of attractive articles have of producing eye-catching titles, unquestionably interesting opinions but invariably leaving the reader uncertain as to whether or not the weather is abnormal, unless of course, you are an old man in Texas looking at his cattle standing knee deep in water or chickens perched on roof tops knowing they're not ducks who said, "Haven't seen anything like this in 70 years." Or someone commenting, while applying for a federal loan, "This will be my third home to be built, the other two got disassembled by twisters twice in ten years." Such writers are prone to leaving out very important details – not thorough but yielding fancy pens. I refer specifically to the penchant of experts to confuse (some intentionally and some not) readers with references to cyclical bad weather 300, 500 or beyond 1000 years ago using ice core values from Antarctica or Greenland. I have described to you the association of pollution buildup and bad weather or how Nature turns on her air cleaning machinery but the subject becomes lost in a sense if you take at face value articles that suggest this 90s decade is just another cyclical bout of warmer wetter weather, too dry for some, violent for others that has been going on millennia – therefore today's disturbing and costly mess is temporary – be brave, keep smiling, it will soon be over.

There is one flaw in this supposition, tentatively advanced to the effect they do not differentiate between climate vagaries of ancient times with ours today, and that is Nature has always been a polluter; therefore it was necessary for a certain development of heat and moisture, then downpours to occur cyclically or this planet would have been one lung-caking hellhole to live in. That's what Acid Rain is about also but it is destroying old structures in Europe. They, these students of ancient weather, overlook the fact, besides African dust, bacteria and whatever else is killing coral reefs in the Caribbean and Gobi desert particulates over our western states, that we always for milleniums have had a flock of active volcanoes belching soot, mercury, sulphur dioxide (I have to leave whatever else to volcano experts). You see my point, I hope.

Please ignore the tripe promoted by fund seekers over weather's mysterious side…not so. Recognize climate is but heat and cold in conflict. I have laid out the preponderant realities of what thousands of thermometers have shown and what has been recorded. You take it from there. At least you will know who is lying and who is ignorant. Too bad there will not be the moment of consequences and accountability some day for leaders who choose to remain indifferent while privately approving institute pundits highly skilled at laying down smokescreens.

Should you be interested in an honest straightforward report on world affairs, try BBC*. 15 minutes – that's all – then repeated every half-hour. Unlike some of our entrenched progeny of pirates we are heirs to, lying as they do, acting out the role of sincere educated icons (you know, Yale, Harvard and so on) but ready to support wealthier plunderers, these English broadcasters are a breath of fresh air. Oh, we have fine people, of course – you know who they are I'm sure, but…I listened to one here just recently who made me wonder. Dan Quayle is protesting what he angrily calls micro-managing of the stock market by Alan Greenspan. I can't be sure of his motives or if he was born into wealth but it sure was a juicy bonanza in the Roaring Twenties for certain rapacious club members, untethered as it was and repeating it's usual unexpected dive into oblivion for millions as it did more than a half a dozen times before. Such a comment on my part still has to do with what we have to contend with whether it be climate problems or our nation's economic welfare.

Just to throw another add-on here, and by the way, how can I be cheerful after reading "Inside Congress" by Ronald Kessler; it was, well, an analogy is called for, like having an agent showing you a house but there are skunks in the cellar, such is the odor all prevailing. Let me point out some more evidence of how screwed up our planet's weather is. Now, everyone knows a major portion of Brazil is oxygen producing carbon-dioxide absorbing jungle, it's dense forest drenched as it usually is seasonally. It didn't rain this year during it's Winter months nearly ended and Spring about to come into play. Instead, they had a drought. Forest fires have spread and some are out of control simply because not very clear-headed Brazilians have been busy razing trees and undergrowth using scorched earth methods. Anxious for cultivation their population growing by leaps and bounds while many children roam the streets, abandoned by dim-witted poverty stricken fornicators where base gratification is foremost on minds a small army of desperate workers destroy what took millennia to build – and with it the disappearance of so many beautiful creatures. Then we have California – it burns too. Of course, unruffled and seemingly oblivious to what is strange, meteorologists refer to repeated bouts of flash flood causing rains as the Monsoons – as if the southwest was more or less another India or Bangladesh. Wyoming, of all states, just had an exhibition of tornadoes. And this time of the year??

There was a time when our nation's infrastructure improved by leaps and bounds even during one period when society was on the verge of revolution – labor challenged masters and poverty had reached a breaking point. But, despite flaws and what economic system does not have them, the populace looked up to our government. Dams were built, highways and bridges improved transportation, California found it's Gardens of Eden, arid lands having rested for milleniums responded with abundance, vast fields of agriculture attracted hordes of migrant workers year round following irrigation projects. Then, of course, thousands joined the rest of America when the Tennessee Valley Authority, constructed at great cost, provided electricity and other benefits. After the Dust Bowl of the 30s trees were planted for a thousand miles north and south.

I am not going to ramble on with philosophies of the past or nostalgic reminders of imagined visionaries, (a delusion anyway) there not being any in sight for now. Unlike historians whose biases are too frequently obvious, honoring party heroes or stripping unfavored politicians of all medals and stripes, I find it more expedient and simple of course to explain bursts of progress are really but invention forced by necessity. It still requires a lot of legislators encouraged by public support to do great things.

Here and there an appeal or a warning is heard over the eventual outcome of pollution. One time it is some official of the United Nations and another, perhaps a scientist – Al Gore took a shot at it. All it amounts to, no stirring reaction from anyone, is scattered attempts and brief statements – no effective messages, trying to reach our busy citizens, in a hurry chasing wealth and internationally, masses of retards anxious to kill each other. A little concentration, some logic needed; it is obvious necessity is a reality. But, there is a problem – no unanimous belief that there is a necessity. Additionally, no respect for government as it stands now. Symbolically speaking, how can there be invention?

One deterrent to recognizing the truth is the effectiveness of propagandists influencing several hundred lawyers (busy garnering up money, per "Inside Congress", a book that makes mine seem generous) and of course, many of our affluent who really do not care. There is insufficient education over the matter of our changing climate. Resolve and perseverance is lacking – Al Gore symbolizing environmentalists packed his tent and moved to less combative environs. As with the mentioned occasional squeak, President Clinton, while making a speech in Auckland, New Zealand last week managed to include an appeal to nations – "We must reduce the Greenhouse Effects of carbon dioxide." Unless we reduce the use of fossil fuels this is impossible. Furthermore, the powers that be and have a good grip on government affairs will see to it that no such action will be taken for a very long time. Proof? In 1992 we made a halfhearted commitment to 140 nations gathered in Rio, "We, the United States have intentions of reducing carbon dioxide, etc…etc…" From that point on with unprecedented indifference we fill the skies with ever increasing quantities of CO2. Again, rising ocean levels 50 or 100 years from now is the argument being used to persuade. It's convenient – all legislators will be retired. Tongue in cheek, I do believe Clinton and many in high places or just American citizens are oblivious to one of the many, many indicators of ecological changes due to earth warming such as the rapidly expanding worldwide habitats of disease – spreading mosquitoes multiplying in greater numbers, therefore expand northward everywhere.

No mention by anyone for months of our ozone layer's condition of course. Another hole has developed but the media just glossed over it a few days ago. It proves what I have suggested. Someone claimed the ozone layer would heal. There is something unrealistic about aerosols and freon being solely responsible. Something very powerful even when highly diffused in the troposphere such as acidic emissions have to be contributing. Experts can't even dither over the matter and if, very occasionally, write something interesting most folks can't follow it. And just what do scientists have to say that I've heard and read for decades? "The evidence suggests…but, further studies are needed." The boondoggle goes on as described in the August 1998 issue of Readers Digest. Hong Kong experienced a typhoon (9-16-99) and chaos reigned. Everybody is making money there but the populace ran amok as windowpanes in skyscrapers exploded. I regret to say this but the second major polluter on this planet, China, will have to learn – all violent weather phenomena are heat-caused. Cities, spreading wide have become more vulnerable. Why not. Nature is firing away. TV shows, looking for drama (everybody loves to look at destruction going on as long as it's happening to someone else) use such titles as Raging Planet, Nature's Fury or Wrath. And yet, despite increasing frequencies of destructive weather a small army of scientists keep busy looking for formulas and proof.

I explained a simple approach to better understand what is going on in the way of increasing solar radiation but it doesn't appeal to lofty high tech thinkers. One expert wrote, "We have no tool to predict droughts." All one has to do is sometime in late Spring or early Summer just look at temperatures in the Caribbean and temperatures in Alaska, part of it in the Arctic Circle. If they're both experiencing temperatures above the norm, then no precipitation creating storms can develop that amount to anything. Without brisk cold there are no robust cold fronts and what low-pressure systems develop usually poke along north along the boundary line between the USA and Canada. In short, if you want to know if solar radiation is up (definitely has been for several decades) then what better place than thermometer readings in the Caribbean. This does not appeal to several thousand scientists who periodically pepper away with interesting reports but manage to add a touch of mysteries yet to be solved. I'm sorry but somebody is finding ways and means to keep 2,400 scientists busy.

In need now of checking my annoyance, prone to convoluting as I am just as persevering lawyers do, I must ask what is so disturbing about futuristic concepts and great projects keeping in mind the private sector financed by the government? But, the rub – confidence in our government has waned terribly. Furthermore I don't believe the anticipated new leader or his backers are really concerned.

ADDENDUM

No one is putting the pieces together – let me try. The world has experienced quite a year in deranged weather so far. Instead of raining in Brazil's forest they have had a drought and wildfires, most started by imbeciles, have wreaked havoc. Officials in Bangladesh and India according to BBC are looking for some new locations for cultivating rice, crops repeatedly ruined by floods. As for the U.S.A. increasing wealth shrouding reality of destructive weather and it's effects we have had our share of climatic problems. Two weeks ago some optimistic person said the drought would not affect food prices. Today, the 13th of September our Providence Journal reports a sharp rise in the cost of putting food on the table.

Having finished my book two weeks ago I just had to add more pages since New York City's 7 million residents were told to shut off air conditioners, stay inside while helicopters pulsating about spraying insecticides, swarms of mosquitoes totally frightening everyone since St. Louis type encephalitis has killed three, six hospitalized and 55 under medical scrutiny showing some symptoms.

You would think someone in the hierarchy of experts or government might just, not boldly of course (the credentialled institute ultra-conservative spokesmen will pounce) would suggest a changing climate is to blame. This is an international* problem and not just New York. You would think costly and dangerous weather going back quite a few years, would be mentioned culminating in what any darn fool knows, in conditions all over the world mind you (already described in book) that allows extremes in insect reproduction and their spreading further northward in leaps and bounds. Tunnel vision – that's one explanation – the other, pertaining to political evasions is a study in deception, bamboozled and ignorance.

A New York expert in entomology announced, "It's just a case of bad luck" if that city has become a haven for mosquitoes. He took no risk with that one I'm sure. But, that's typical of most scientists. There's no such thing as the Big Picture with these people.

*Four months ago BBC reported 10 had died of encephalitis in Malaysia, one in Japan. Now, September 13, the toll has risen to 160, some survivors are mentally impaired. Some cannot recognize family members. Try to imagine that.

The subject of increasing earth warming (see graphs at beginning of book) has receded in the background – chastened environment-oriented politicians (there are some) don't want to make waves having already been laid siege to by astute English Majors, part and parcel of elite institute propaganda mills. Even though El Nino disappeared May, 1998 (that's over 14 months ago!) it served Carbon Club protectors well – they and Limbaugh types made environmentalists the butt of jokes. Typewriter artists really knocked out effective smokescreening material – earth warming was a myth. Despite extremes in weather and utilities not able to keep up with air conditioning demands the public remains somewhat effectively sedated. The effect of brilliant writing by disingenuous pundits is lasting.

Oratory skills – that's another "quickie" seduction. The might of the pen is known but so is polished bull excrement. I'm amazed at the effectiveness of the spoken word! A local radio talk show host was so elated, having succeeded into interviewing G. W. Bush for 15 minutes had callers phoning in, ecstatic over what "W" had in mind concerning revamping the public school system, parents to be paid with vouchers, private schools and all that. The governor is referred to as "W"! Fits in with zoom thinking. People phoned in accolades, "Gosh, he's the greatest" or "He can't miss!" One poor person was diplomatically blown away with his opinion asking the host how such a complex program would be managed and what would be the cost to taxpayers. The buoyancy, the optimism and a fast pitch by "W" was overwhelming! And, defense spending, that turned up the Encores! Let's play sheriff on a greater scale (he didn't say that). Lots of money to be made. That's one reason the Roman Empire collapsed – they were, these soldiers of ancient times, everywhere, from England to Egypt, including Jerusalem.

It appears Bush may become our next president. I do not fear this of course – besides who am I. We need to improve defense but not expand armed forces to budget-sapping extremes.

When Texas and maybe 35 other states get heavier doses of Nature's Fury (good TV title) will "W" and his backers be asked to explain the consequences since it appears that they do not regard environmental issues as worthy of public discussion.

VERMONT   10-4-99

State adds fish to list for mercury danger

BURLINGTON — The Vermont state Health Department is urging pregnant women to limit how much tuna they eat because the fish can contain potentially harmful levels of mercury. Vermont has long urged pregnant women to limit their intake of some freshwater fish. Also, pregnant or nursing women should not eat swordfish or shark, the department said in its new health guide; the same advice goes for children younger than 7. (AP)

In the meantime I'll record temperatures. Come late next Spring, I'll tell you as I did three months ago if we will get baked again. It's quite easy to figure out in advance. I choose the practical route. Let

2400 scientists ply the universe thinking somewhere in this churning atmosphere a Nobel Prize winning formula is bound to be discovered. Cost, about 200 million a year.

A late insertion dated October 4th. I report and add one item after another paying no mind to organized writing with designs on readers' emotions necessary in novels. This is not fiction. My first motive using this excerpt is to establish mercury's presence is far from new. The second, based on weather knowledge, there is a lot more area involved than just Vermont and New Hampshire! We just don't have enough expert investigators to go around. The rather large group we now have are very busy. Additionally, I know exactly where mercury is coming from.

The motivating force behind this literary undertaking of mine reached more passionate levels these past few years what with worldwide weather taken on a turn for the worse and yet no hint of serious concern that amounts to anything, as if what happens elsewhere is not a grave symptom of our atmosphere from North Pole to South Pole (not just here or another time over China, or Bangladesh or Mexico) undergoing an alarming yet insidiously slow transition from life supporting normalcy to what I see as a permanent destructive (records being set for the unusual) pattern with periods of welcomed respites not lasting as long as they use to. Trusting you will see me as a reporter rather than someone on a mission I have compiled everything there is to compile for your assimilation and your judgement. But, annoyance rather than some emotional urge started me off – I couldn't quite believe what I was reading.

Irritated at the idea, dreamed up by bureaucrats and enthusiastically endorsed by a state or states anxious to extract more money from tax payers considering employee numbers have practically quadrupled according to Arthur Schlesinger Jr., some rather fat pension plans and frequent raises increasing inclinations to dig up more plans for us to live with, the idea testing auto emissions would bring about a remarkable reduction (it will not) (we covered that) of pollution, particularly nitrogen-oxides (Nox) and it's summertime contribution to surface level ozone formation. The figures published for everyone to accept without question as to who and what and how much we auto owners were responsible for, astonished me. No mention was made of some 6,000 diesel powered trucks rumbling through Providence, for example, every day and some at night, except Sunday, specializing as they do with belching out black soot with it's high content of toxins I am being blamed for, every time the driver merely depresses the pedal one eighth of an inch (if you understand how diesel power works) Route 95 at times looking like the truck section of the New Jersey turnpike. Nope, no word of this. Then, they tried to get rid of me. Some guy called me up after I just had to tell them I knew something about weather, prevailing currents and all that, the so called science of meteorology not having changed at all in a half a century except warning systems are enhanced by radar pinpointing and satellite imaging. I had dared to tell an official over the phone (he called me, I didn't call him – surely I posed a threat to their plans) that we here in RI just as other northeastern parts or wherever, were heirs to big time polluters west and southwest of us. Well that didn't do. One year later, as I've already written, an article they produced pointed out the reality of west to east transport of filth eventually finding it's grimy abode over a widespread area of the Atlantic. "Would you like to work for us, travel and lecture to senior citizens so as to prepare them for auto inspection and what will be the requirements, etc…" That's right, dear reader; offering me a job so as to shut me up and side in with their cause. Then I received 40 pages of their emission standards to be imposed on us – inspectors, so help me, if they had to conform to these complex rules, would have to spend at least one year in college! But, don't get me wrong, testing autos will boost the economy and it might just put a nick in the pollution content of the air. In the meantime Bravo for those states that have not adopted auto testing. Other than make life miserable for poor people who have to get along with tired junkers, surely to fail testing, and who by the way do very little driving, their heaps parked behind some red-bricked sweat shop factory (they're the ones who should be inspected!) all day while soccer moms (forgive me, it's getting stale) phones in faces and neck-tied men in a hurry, rack up 15 to 20,000 miles a year with the biggest polluting machines ever conceived by man, the excuse plugged by manufacturers you can now tow boats or trailers with these gas hogs driving the cost of gasoline up. I drive on my job, stop and go, some of it feeling the slipstream of tankers barreling by me even though I'm motoring along at 65 m.p.h. on Routes 195 and 95 and in four years I've yet to see more than three SUVs or massive pickups (God they're huge!) towing anything. But, that's democracy, the affluent or careless with money, must be free to express themselves – two and one half to over three tons of vehicular mass is very important to them. When they step out of these monster boxes on four wheels in a supermarket parking lot you can see noses tilted slightly upward, proud as all get out, and I suppose, as I've heard a few say so on radio, willing to register disdain for we inferiors and our reasonably sized vehicles. Like me or not, I'm not alone in this sour view of what is preposterously a lousy trend, increasing carbon-dioxide emissions as it has to and consuming an excess of fuel.

Bellyaching aside, since 47 dollars every two years having someone stick a sniffer tube up my car's tail pipe will not kill me I was more inclined to be moved by the increasing onslaught on the human race by Nature the frequency and intensity climbing upward exactly in unison with climbing sales of 10 to 15 miles per gallon biggies (15 if there is a tail wind) and set out with an agenda hoping as moot as my expectations are to raise the awareness of a reader to understand and believe what I know is a deranged weather pattern, referred to as "Whacky" by a prominent magazine. But, a thorough explanation, other than a vague but not necessarily believed by all comment, such as "Blame earth warming" while President Clinton told the world from Auckland, New Guinea we must reduce the Greenhouse Effect and the stock market reacted upward because 51% of car sales, very expensive models, very profitable, no, no total explanation has every been published. My first and only rejection by a big time New York publisher following the usual submission plea and samples contained a brief message, "Topic does not fit publisher's mission." I wonder if I will find one who just might see the merits of setting aside prejudices and profit concerns (important of course) and think as I do that books are the foremost source of information and imperative a bible of sorts, always on shelves for another review, providing the emphasis is on truth instead of maintaining harmony. Mission is what? Is there a healthy balance between love and concern for a country that I'm positive will experience serious problems weather-wise and some reasonable bias for whatever political philosophies they harbor? Accept this appeal – I can change my name – I'll probably have to anyway. You can, publisher, delete, modify and spruce up composition as long as the general theme is not disturbed. We will have to in sensible incremental steps make changes but right now there is not much interest for reasons I've already belabored.

I do not regard ecological changes, however not moving anyone, in fact what follows as clues of changing conditions may draw snickers, as extraneous and boring. Reports I file away, some inserted in past pages, represent pieces, if you will or sometimes statistical analysis that contribute to the Big Picture. Here is a wee bit, trusting you will not deem it unnecessary: Suddenly fifty percent of Tucson's population began experiencing dripping noses and sneezing tormented by allergies never experienced on such a grand scale before. According to Paul Harvey, who I trust implicitly, even buying an electric shaver he advertised (I'm satisfied with it) this condition started in late September. Reason for folks crowding into allergists' offices? There was excess pollen in the air because humidity levels are now averaging 33% instead of the age old 30%! Delicate is the balance of ecological plant and life forms already overworked as a subject as I have, therefore something disappears and something unwelcome moves in. Not the case here. Aside from an upsetting incident, temporary as miniscule bits of reproduction driven pollen is, we have African bees, now quite at home in a nation (I call your attention to this) whose <u>southern</u> <u>tiers</u> are gradually becoming Africanized, meaning too much heat and too much rain, best expressed as a serious problem by a Texan who said, "How fast can you run." An elderly man died. He couldn't run.

Excessively elaborating as I am, the point of all this is, I hope you respect the now forgotten findings and the report made by two lady researchers who established excess heat and humidity affects two thirds of our country and this is certainly not temporary as the trend upward started several decades ago. To eliminate what has been dubiously glossed over and also, best of all, happy to expose smoke screening tactics by English Majors who, since their pet El Nino took it's usual powder, have been silent, waiting for another excuse to attack environmentalists…I mean even the reasonable ones, I fastidiously list and explain every clue, however little newspaper space is devoted to them. Humidity levels have risen here but also all over the world. Tucson, one dry hot place but it don't matter; 450 miles from the damper Pacific air and yet it's humidity reaches far inland and accounts also for monsoon activity in Arizona and thereabouts.

Here is something that struck me as rather strange, realizing of course we can learn something new for years to come. Of course medical personnel are probably familiar with odds and ends that I am ignorant of. It has to do with mosquitoes and the transmission of a disease that you would not expect in northern latitudes. However, this report by radio BBC who I fervently believe are a breath of fresh air, not at all interested in what politician sniffed cocaine, will not cause palpitation of altruistic hearts or protests by Animal Rights proponents. All penguin residents of a southern England zoo, 27 of them, died from malaria and shortly thereafter a few more in a London location. A traveling veterinarian involved with investigating diseases explained such an affliction, having killed close to a million human beings last year in South Africa, is quite common in penguins and other species he rattled off and that of course, increasing numbers of mosquitoes exacerbates the rate at which such and such a disease will expand it's contaminating potential. Here, we have dead birds all over the place and while the odds of contracting encephalitis are about the same as winning a national lottery (hell of a way to draw comparisons) it will be more than disconcerting when in coming years the presence of more mosquitoes will surely be.

I am mainly interested in convincing filling-in material determined to open up some closed minds. I have to be thorough – earth warming is not a myth as a loquacious radio talk show host who learned mind conditioning requires much repeating insists. Yet, I'm not sure if the far reaching effects of more heat and humidity all over the world is clear in the minds of many. I confess I have convoluted certain details – but my motives are quite different.

I don't know. Years ago, along with everyone else I had fate in our nation's populace who in various emergencies that I am more than familiar with showed their worth as sober contemplative people. I'm not sure anymore. Some time ago, Al Gore, in a fit of anger told members of Congress they had their heads in the sand. I wonder if he knew what many of their constituents had in mind.

AGAIN, ANOTHER CONNECTION SKIPS BY. DRASTIC DEADLY CHANGES FOR THE WORST OVERLOOKED – THE YOUNG HAVE NO COMPARISON TO JUDGE BY.

It appears from the public's point of view that the abnormal has become normal – a delusion. Becoming inured, distant disasters in a vast land is not disturbing – someone else's problem. Torrential rain, that's putting it mildly. How much is too much? One melted iceberg's worth perhaps? A slice of land almost 600 miles long and 30 miles wide inundated, a part of it after four days still not drained off – that's the equivalent of several icebergs the size of forty mile long Rhode Island. Melted ice that is. From the glorified halls of intellectual gigantism, unquestioned credentialism, even political analysts alert to find fault with environmentalists becoming overnight meteorologists. At least we in the Navy called newly arrived college graduated officers as "Ninety Day Wonders." But these people sporting double deck diplomas take the cake, regrettably keeping the public confused – their mainstay, El Nino, tossed into the chaotic arena of truth versus lies for as long as the Pacific phenomenon was out there kicking up a few storms well out to sea and horrors of horrors, tides* were just one foot above normal for a while during 1997-1998. But tacticians they were – and, very effective. Not the slightest hint of a casual warning about earth's warming.

We, meaning Washington D.C., (read Readers Digest August 1998 and see what I mean) state experts in "this" but not a related "that", researchers in no hurry, the media choosing what sells (can't blame them) do very little in the way of consistently keeping the public informed of a progressively worsening problem. No technical details are given. Piecemeal, all dole out bits of information but I don't believe "balance" in Nature is clear to them.

Tit for tat, abuse Nature and experience it's fury. The Goebbels did their harm. Back into the sand went heads in 1998. Al Gore quit the cause, savaged by callous elitists. Write something plausible, disentangle chaotic agglomerations of conflicting views on causes and effects, whose to blame and so on and you can count on research fund oriented specialists to repeat again and again the next-to-impossible to prove must be proven. Twenty years from now they still will be saying, "More studies are needed." While the subject of flooding

Report of tides from California acquaintance.

continues below let me stray into another area of ironies. Certain medical entities (I don't know who they are) declare this or that toxin is carcinogenic (animal rightists despise how they come to such conclusions) but hesitate to issue reasonable assumptions to explain why American cancer rates climb steadily upwards. Again, thousands of researchers at work but 30 years from now, and I quote authentically many, they still will be stating "More studies are needed."

Give it some thought. So much water which must come from the oceans, up into the sky it's vapors go, then as solid sheets of water, down it comes, cascading torrents dragging human beings, homes and animals leaving little to salvage. It occurred in so many parts of our world this year, last year and the year before – Central America, Texas, our east coast, Arizona feeling something strangely new, monsoons no less, Bangladesh, India, China; over 45,000 drowned, rice fields have vanished, Washington and Oregon drenched and wind blown 90 days in a row, typhoons creating mud slides in the Philippines and now Japan (9/24/99), flash floods in the mid west, seven human beings swept away in city streets, record snowfall out west and in the Alps, and so on. Rising ocean levels 50 to 100 years from now has been drummed up. It serves to delay rational appraisals of what is immediate. Rock-a-bye Baby – keep the pacifier tucked in constituents' mouths.

It's only my opinion, but leave ship captains worry about a piece of Antarctica drifting in the South Atlantic – it will melt and the beaches will stay beaches. Tahitian girls in the future will dance on the same golden sand that their mothers now enjoy. My reasoning explained below.

Technically speaking, how really informed is the public? North Carolina has experienced a deluge, the cost of lives including the unprecedented burden bearing the brunt as it's residents are of a five billion dollar estimate of damages and after the news in much detail (how many drowned livestock, etc...) were you told why? Other than routine statements, i.e., "Floyd, despite reduced in size and intensity skimmed along our East Coast after lingering for a while, slowly at first then scooted northward into New York state." Has any expert come forth with something that just might be a reasonable contention? Consider; that much rain is not normal, hurricane or no hurricane! On the next page is why and furthermore more of the unusual to come in the future.

We were blessed, if you are interested in when, where and how great is the potential for more deluges, by a hard earned discovery transposed into a described revelation, publicized as a one day shot of news but bounced off most heads. Humidity has greatly increased over two thirds of America in the past 50 years and so have droughts in frequency as a consequence of rising heat. Two lady researchers dug up data from 113 weather bureaus from east to west. Of course, such a simple project is beneath the dignity of doctorated experts who get paid to chase impossible and unavailable mathematical formulas in the heavens above. It took two years. They meticulously ground out figures from what I imagine are rooms full of records. I'm quite familiar with what our filing systems looked like. These are my kind of people – down to earth – fundamentalists and not "High tech do nothing but talk fancy" types. A simple deduction comes of it all: More heat means more humidity. More moisture means more precipitation from more storms. "More precipitation" is too innocuous an expression – it can be deadly in quantity and extremely quick to turn towns and countrysides into the unbelievable – miles of flowing water, remains of weakly structured homes floating by and cars carried along like so many toys. This month 60 drowned, 300,000 hogs along with over two million chickens perished and of course, the heartbreak of ruined farmers. And the cost to taxpayers, selfishly speaking? Clinton, addressing huddled masses in shelters said, "If we don't have enough money in our federal emergency fund then I'll ask Congress for more." Perhaps Congress folks and Senators should donate since they've seen to it the subject of earth warming stay on the back burner.

Convoluting I may be (detailed all this before) but if you do not connect the <u>physics</u> <u>fact</u> of airborne filth (the Pacific and the Atlantic have periodically been loaded up with it) <u>with</u> <u>building</u> <u>up</u> moisture density in the atmosphere then slumber along with the rest of the world. Arizona, arid as it is; who ever heard of monsoons there? Right smack at the end of a wide valley about 150 miles long, through which damp Pacific air funnels along then collects against that state's sandy hills and of course downpours occur more frequently than ever. Again, Done That, You Got That: China, India, big time polluters, fill Pacific skies with the products of coal burning including more vehicles. Therefore monsoons over our southwest in summer and damn near monsoons over two of our northwest states Washington and Oregon during the winter. Unless of course, you believe walking in ankle deep water for three months and several million umbrellas sold as being normal.

When flood insurance is refused (it's getting tougher now) and/or property insurance cost an arm and a leg, food and clothing prices are driven up (cotton fields have taken a beating) (I dislike tightly woven polyester), electric rates go up as they eventually must, 5000 repairmen working overtime untangling a skillion miles of wires after removing downed trees with which you could build 10,000 new homes – that is not cheap. Already mentioned before, the cost of other effects of deranged weather, then I suppose some stirring of realistic thinking may begin – quite overdue as it is. In the meantime the invisible government one can identify as the Carbon Club olarchy will see to it you remain dumbdowned and of course, not interfering with the immoral trend that I view increasing consumption of gasoline as. Anything for a buck. Don't matter if hundreds die and lives are wrecked.

I experienced four hurricanes, involved meteorologically with two in the 50s, one in RI and three in Louisiana where I lived for almost 10 years, beginning in 1938, no more distant from their furies than five miles to 40 miles in the 60s. No way, I repeat, no way, did it rain the incredible amounts of today's storms!

Don't be lulled into total complacency. Recognize false prophets from now on. I shan't bore you with many not at all original suggestions I hoped to influence political dignitaries with. They responded. Of course, they claimed to be on top of the problem. I'm sure, contrary to propaganda, particularly over the radio waves, we will not have to sell apples on street corners. Economies will not be wrecked. However, advise your children readying to be married and home ownership in mind it might be prudent to keep a boat behind the garage and depending on where you live (about 35 states) dig a big enough hole in the backyard then attach a neat lid. Am I being facetious? Really? Why is the government financing emergency shelters all over the country (also, remember 2500 economists mentioned before) with unprecedented vigor – no delaying, believe me. Generously they just donated 300 thousand to small town Pawtucket to fortify the existing shelter. Too bad a politically appointed administrator of Federal Emergency Funding, namely, James Lee Witt (I wish I had a name like that – it helps) cannot reveal to all of us how he feels since I couldn't help but notice his shocked expression as he viewed tornado destruction earlier this year and I'm quite sure we will see more of him on TV in coming months or years until he is replaced. He has been more on the road or in helicopters than in his office of late.

As a final bit in my menu of food for thought in the hope you are a believer but having Done That – You've Read That, knowing as I do from casual conversations with folks, it is human to allow periods of decent weather as portending it's continuation and bad times but history to be forgotten, I must fashion again a simpler explanation (we outsider dummies are limited that way) of a Nature, unfeeling as it must be, is programmed to react and adjust. Cyclically it follows repeated patterns of behavior but I cannot predict it's timetables nor be a dreamer believing there is some way I can accurately anticipate intensification and lulls. You must agree Nature has arranged a livable climate – there has to be the control factor – adjustments have always been necessary. With increasing pollution there is accumulated heat and airborne excesses of moisture. It must be dispensed with. Nature then goes into a stage of worldwide downpours. Precipitation drags down particulates, acids and what have you. Following that cleaning cycle there is a period of normalcy. How long it lasts is unpredictable but I know it will be short lived, trying destructive times lasting longer as they have. Centuries ago it seems from limited determinations arrived at from ice cap core samples Nature reacted rather tumultuously (heat and much rain, etc...) or poetically speaking, unkindly. Those apparent lengthy periods of decent weather that followed are things of the past.

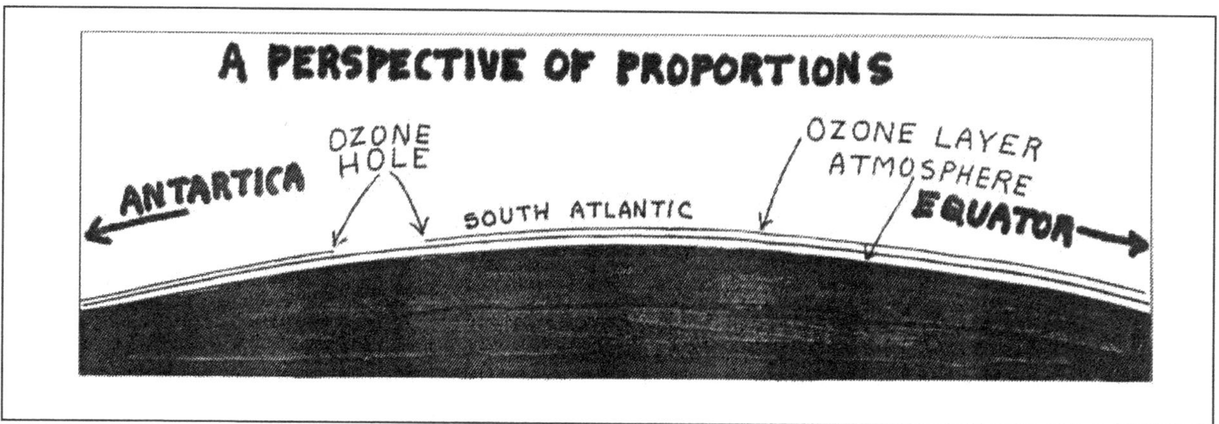

By the way, believers in rising ocean levels in the future (I'm not sold on it) should be reminded our planet's dimensions are always expanding. But, good question: Will it's water volume keep up with it? If we keep in mind Balance (my favorite theme) don't you believe climate would become too cold if polar caps were allowed to continuously build up and by necessity expand their perimeters, assuming I'm correct in believing an Entity has always been in charge? And so, one final poke at innate logic most of us are blessed with, is Antarctica breaking off it's edges truly a cause for concern? Specifically I'm referring to those who repeatedly fuss over expected changes in ocean levels – that's just a bit too speculative for me right now. I don't consider it proof of anything but after visiting a beach in Maine that I had frequented over 50 years ago I saw absolutely no indication the ocean had made any headway – the same area for sun bathers existed despite the fact I made available to you in graph form average temperatures have steadily increased in the past 30 years.

NEW HAMPSHIRE PROVIDENCE JOURNAL 11-28-99

Sea urchins dying mysteriously

PORTSMOUTH — Thanks to Japanese demand, the sea urchin became one of Maine's top seafood industries, but now the spiny little creatures are dying off mysteriously. Scientists and fishermen think it is more than overharvesting. The paramoeba suspected in lobster deaths also might be attacking the urchins or it could be global warming. Urchins had been thought to be near inedible, but the Japanese consider urchin roe a delicacy, establishing a new market virtually overnight in the early 1990s. (AP)

Based on a planetarium film of Earth, atmosphere and the stratosphere above it from a satellite camera the illustration above delineates our very thin atmosphere clinging to Earth's massiveness. Surely 6 billion people produce a great amount of pollution! Therefore, within this slender life-supporting layer of air there are adjustments being made by nature. One problem – they are becoming more frequent and definitely, increasingly destructive. I've undertaken a hard sell – reality is being avoided.

We have been, if I may include a reader as a partner, merely but analysists, putting together what Nature is telling us. When the time comes and believe me it will, let someone else sound the alarm bells. My main gripe; when will some prominent figurehead of world-wide respect put it all together and shake up this world? Waiting for a conglomerate of scientists, all having their own projects, proof necessary before any commitments are made, means resolve to stir up the human race has to come from some determined individual who is not shackled by political and business associations. Again, common sense, logic and fortissimo will have to do while dismissing the necessity of proof. Proof; that's quite a barrier in a way. Did not the Supreme Court order the EPA to tone down their punitive approach fining companies for polluting? The justices, falling back on the Constitution, claimed proof was lacking.

If you were to choose between land and sea to add up evidence of prevailing ecological damages I would suggest the ocean first of all. We have migrating finned denizens of the sea traveling well beyond customary limits. Along shore lines insignificant (except to biologists) little creatures are disappearing. Why list other symptoms – you know all about bizarre mass whale beachings, red tide and bacterial invasion eroding food chains in shallow waters from New England to Florida. The excerpt below from yesterday's newspaper (Nov. 28, 1999) as news, will interest but a few. I'm sure newspapers from inland states all the way to Nevada are not interested.

Some folks in Maine will lose their jobs, and so, off to another occupation. Downsizing busted millions. Eventually we will have to help coal miners re-adapt. So what, says the world. The main preoccupation is on job losses. It has to be. If lobsters, droves of them dying (details in later pages) the affluent don't care. Pay $10 a pound and to hell with it. By the way I didn't know sea urchins were a prized food. Spiny? That doesn't appeal to me.

Each ecological change on the down side is cause for concern. Oceans are our life support systems keeping air temperatures stable, furnishing the moisture for precipitation and of course, providing seafood. Something is very wrong when shore life is disappearing and crustaceans are diminishing in numbers along both coasts. Why the mystery if sea urchins are dying? Biologists, by the nature of their responsibilities must determine why. I wonder though, why the mystery part? Warmer water than usual, particularly with a stronger sun beating down on a few inches or a foot or two of it, lapping over heated sand and rocks, a pool left behind as low tide goes out, incubates more of what is proving very harmful. Why is it a mystery when we know bacteria proliferates with more heat? How monotonous my insistence. But, it's all bits of evidence and the pile is getting bigger. Raising anxiety levels is not at all my objective – just reviewing facts.

Then we have the characteristic hesitancy of the medical profession, except the American Lung Association who were loud and clear in the mid 90s badgering our state legislatures. They wanted something done about surface ozone formation in summer, insisting asthmatics were affected, some children suffocating to death. Yesterday, again an article from another medical establishment down south, also citing the same 160% increase in asthmatic cases except no mention made of nitrogen-oxides, ozone and pollution outdoors. This time households are not the healthiest places in the world. The article pointed out the probability of indoor dust, rug deterioration particulates, mites, even cockroaches included as contributing to the increase in asthma. I don't mean some doctor who made that statement doesn't realize the effects of outdoor pollution. But, it doesn't seem quite right not to also mention various oxides drifting around from industrial sites and vehicular emissions, particularly diesel. Unintentionally it blurs distinction of what toxins should concern the public. The article probably reached far and wide from east to west coast and just what effective impression does it make if no mention is made of acidic substances in the air drifting from one area to another?

The same overly cautious attitude by all scientists guarantees we will wait and wait for convincing conclusions.

A little boasting is due. I have to drop the subject of spiny little creatures and other clues of deteriorating ecological systems I collect. Well back in the book, pages written months ago, I claimed the effect of torrential rains in abnormal quantities cleans the air in many regions, therefore in it's final arrival at the north pole it has less pollutant type resistance to cooling. In short, some degree of normalcy can be expected in terms of a colder winter coming up. That's what I wrote – no superior intellect required. If you watch Fairbanks, Alaska's temperatures in November this year, it is substantially lower than last year. That's all you need to be a general forecaster. That's the source of cold weather and shipment of frigid air is express-shipped here, particularly the northeast. Amaze your friends. Tell them, next Tuesday you expect a cold snap or whenever. Just give the shipment about five days. Nothing to it. Or, come June and pay attention to Jamaica, Miami and Bermuda. Five degrees above normal? These facts are listed in decent newspapers. You can expect sizzling heat over much of America one month in advance. Nothing to it. Well, don't try to be specific about how much snow or will there be a northeaster over your area. The game was never a sure thing. It takes time to be good at it and even then, nature throws curves at meteorologists. But, we've accomplished something haven't we. Five days after Fairbanks's temp spikes down to a low point we get up and scrape the ice off car windows. Then, you should know why it is going to be colder this winter.

I also wrote we should have across America a more normal year (mid 1999 to 2000) and less hurricanes. I don't know about tornadoes, because the ozone layer's condition up there is not quite as intact as it should be, meaning resumption of air being heated near the equator during Spring, working it's way over our southland and combine that with what I expect will be more active cold fronts; that can spell serious trouble.

My most desired wish: A government funded program (to hell with the anti-government affluent) to convert coal fired plants to natural gas should have first priority. Not only are such obsolete plants filling the air with tons of acidic toxins and mercury posing a serious menace to health but in all probability keeping the ozone layer in a thinned out condition.

Now scientists fear the flood has also created an ecological time

TODAY'S TOPIC

PROVIDENCE JOURNAL 11-30-99

bomb that could bring disaster of a different sort, and they fear for the biologically rich waters that separate the famous Outer Banks from the Carolina mainland. This complex of sounds, bays and inlets makes up the second-largest estuary in the country, after the Chesapeake Bay, and is one of the nation's most important incubators of marine life.

Rarely have ecologists confronted such a striking example of what can happen when a first-order natural disturbance is combined with a first-order disturbance of the natural world by humans.

"I guarantee you there have been floods like this in the past," said Dr. Robert S. Young, a coastal marine geologist at Western Carolina University, "but I can also say with the same amount of assurance that there has never been a flood like this with the potential for this much ecological impact. Never."

Governor Bush is under siege by the Sierra Club along with EPA giving his state the worst rating possible. Texas is the number one polluter in America. It also enjoys the reputation for delaying cooperating with EPA. Houston is now reported as topping Los Angeles for dirty air, including that state releasing to the atmosphere one of the deadliest toxins, mercury. Allegedly, per a radio talk show host, petroleum companies deny the existence of global warming. You see what we can expect. You understand of course, what we are up against.

A world whose population grows at an unprecedented rate exacerbates conditions leading to territorial conflicts. Economics under duress, many regions never having progressed beyond third world poverty stricken status have had their natural resources such as oil provide international cartels with extremes in wealth and power. Oil is the gold everyone is after. Russia is suspected of having other designs besides ridding Chechnya of terrorists, allegedly having it's eye (they're not alone) on accessibility to more oil and shorter shipping and receiving routes for same. Pakistan, floundering economically is now trying to adjust following a coup d'etat but nervously are involved over the same gold rush going on. We also have to protect our interests and so, President Clinton has signed an agreement to build, (Iran and Turkey I believe, are partners) an oil pipe line across parched lands while ragged peasants there in the Middle East shuffle about just barely able to eke out an existence. Will such a venture help them? I don't know. My point of this is: On the surface, nations seem to be in accord over the eventual necessity of reducing the use of fossil fuels while the power structures, the Transnational entities behind the oil business, show no inclination to agree. Quite a dilemma.

I had to take advantage of the following excerpt as backing what I have insisted upon – never have we had over this world such spectacular quantities of water dumped on the human race. Additionally if I am super critical of scientists, particularly meteorologists, it is because I've yet to read or hear an explanation other than the vague and usual "Oh, earth warming might be to blame." That will not do. Redundant again but I must repeat, excess humidity provides the necessary water vapor and it has to come from the oceans. Or why, releasing much of their waters to the air, levels cannot rise dangerously. But, the melting iceberg titillators are still at it despite findings to the contrary.

Dr. Robert S. Young has my gratitude and respect. No tiptoeing through the tulips. Hit em, rock em, sock em, that's what the American psyche reacts to! Reports of a 5 billion dollar federal assistance granted to North Carolina was a shortsighted estimate. That's a lot of money but the cost has mounted much higher. It doesn't take into account thousands who will not ask for loans nor the psychological damage a multitude of North Carolineans will live with. And consider the effect on prices of everything.

Certainly I'm justified criticizing. Central to my concern are tornadoes. What have residents of Tornado Alley been told, for example? In the past decade their frequencies have hit the doubling index. Yet, only last year, 1998, I heard an expert on TV say we average 600 a year when in reality, the figure has gone as high as 1,300 in 1992. Have residents of cities spreading out, therefore more vulnerable, been told what part an anemic ozone layer plays in allowing more heat over southern regions? To chafe my nerve ends more (I don't get livid) some credential bedecked expert had the audacity to say the cause of tornadoes was unknown.

If I weary anyone it's because they don't quite understand the importance of looking upon Nature's ways differently – a combination of religious appreciation of God's creation and also humility and respect for a power that can pick up or slam your worldly possessions with forces of water helter-skelter. Numb with misery survivors poke around for a few valuables. Not too many of us relate I'm afraid.

You did not know the atmosphere has been programmed in it's genesis millions of years ago, to take action when it is overburdened with pollution. That is my exclusivity that I lay claim to. If the human species goes on and on for another decade or two postponing what has to be done then Nature will do it for us, except with a very, very heavy hand.

My sights, and mere requests by the way, are not short-termed ones. I envision grave expensive problems in the next 10 and 20 years. You have in this book something permanent and useful. You have all the facts you need to know about climate peculiarities. No mysteries as it is professed.

It didn't make the newspapers but my favorite radio station AM 1290 reported the efforts of several biologists and volunteers determined to rescue over 100 sea turtles stranded on a Cape Cod beach the week of November 15, 1999. When they were asked about funding for such a project requiring hauling them off to an aquarium, slowly thawing out these creatures who evidently were no longer able to fend for themselves, feeding water to ward off dehydration and since they're vulnerable to pneumonia in the condition they were in also requiring anti-biotics, the head biologist replied he really didn't care about the money part. "I love all creatures – it doesn't matter how much time it will take – we'll go on until the job is done." Since whales, even porpoises have driven themselves up on beaches before, not warranting a media vigil I would not bother adding to this book such a commonplace story except there is a connection to be made here with climate change.

Straying from Cape Cod's bit of drama, involving genteel people and not often seen creatures for a moment, lobsters are dying in droves (already written about) (but) in Long Island Sound but it may not be for the same reason turtles drag themselves on to sandy beaches appearing as if they had given up on living. Scientists are perplexed as to what may be going on deep down at the bottom of coastal waters. Worried lobstermen, their livelihood at stake, some already have quit, are pulling up trapped active live lobsters but intermixed with some quite dead and others acting very sluggish. If this advice I heard over radio AM 1290 (the most advanced I have ever experienced!) is not in circulation there may be reasons I've more than made brief mention of – anything that may arouse concern over detrimental climate changes (we've got super brains in charge) doesn't sit well with certain folks. Got to keep Americans floating along in Never-Never Land undisturbed. This is what was recommended should you find yourself looking down into a glass-encased aquarium, live lobsters moving about for you to select: "Don't pick one just because it's size meets your approval. Poke it. If it doesn't move as much as it should, leave it there. Go for the healthy ones." That's right – that was the latest advice and the end of it. So far scientists are quite perplexed as to what the problem is but cannot find anything bacterial or any bottom deposits of toxins. According to the latest report they plan to test blood samples.

Getting back to turtles – they overstayed their summertime visit simply because wind-driven warmer waters stayed around longer. No need to go on and on repeating what has happened to what was once normal weather systems. Unfortunately there was a sudden drop in temperatures, winds switched around, cold water currents upwelled and a degree of hypothermia set in. Instinctively these turtles headed for what I have to suppose was the last resort.

Remember, certain people in high places think we're stupid. I didn't make that up. Also, it's not necessarily the White House but as a prominent Washington, DC, journalist (named some pages back) wrote for all in the USA to read, there are thousands of unimportant office managers who just love to wield rubber stamps. Information that should be passed on to the media for you and I to be informed of ends up stored away after "Secret" in red ink is stamped across the top of each sheet. Also the titillation-seeking factor afflicts some eager purveyors of information on the Internet. Icebergs in the oceans are not just menaces to shipping (I leave that to the Navy and NOOA) but someone is spreading the idea (again!) ocean levels will rise drastically as these icebergs melt. That's what this book is about – separating nonsense from reality.

I'm treating the subject of Global Warming in an honest manner, no sensationalism in mind, believing there are a lot of intelligent people out there. As a president of long ago, Andrew Jackson, once insisted upon, it is wrong not to let citizens in on secrets of a non-military nature. How can they vote intelligently otherwise?

I have not written anything endearing, no charm offensive, compelled to criticize so many in positions to show their worth as sources of information who seem very inclined to remain silent, since there is a lot going on weather wise of an unusual destructive nature. In proper perspective, all that occurs in our northern hemisphere, not just America, is sufficient evidence that the abnormal is to stay. Today, November 15, 1999, a meteorologist

had this to say about hurricane Lenny: "We have never seen a hurricane of this magnitude unprecedented in the direction it is moving." To this I would like to add, and so late in the season. Another comment:

> Lenny's lateness in the season and easterly path left even experienced observers agape. "It's unheard of," said veteran meteorologist John Toohey on San Juan's WOSO-AM radio. "This is a once-in-a-lifetime event."

t was not with any intent of reaching for lofty levels of expression if I stressed the need to keep thinking in terms of the Big Picture. We err otherwise if we do not view an October catastrophe drowning over 60 human beings in North Carolina as also being part of the same worldwide condition contributing to a huge cyclone of immense power snuffing out the lives of over 7000 in India and to my dismay, no mention made of 450 Vietnamese meeting the same fate two weeks later. Along with most adult Americans who read newspapers or watch TV news, I would have known nothing at all of this last display of Nature at it's worse had it not been for BBC radio. The connection I know to exist between the above three incidents occurring over a range, latitudinal, some 12 thousand miles escapes the scrutiny and analysis of experts who would do well to volunteer information. Unprecedented, that's a word that has been used more than usual – we should be told why so much water being dumped, killing so many and destroying vast stretches of farm lands, villages; millions homeless. Vietnam's plight, BBC newscasters praising the remarkable show of that country's inhabitants turning out in droves to help each other, deserved some newspaper space. Is someone here afraid of the truth?

In matters of global warming, particularly on an international level, considering rumblings of an accusatory nature, details explained below, 80 percent of polled Americans are supposedly concerned about the environment but few want to hear too much about the details however.

I set out writing with clarification in mind but knowing very well without clear definitive details periodically disseminated by the power structure, the carousel of controversy will keep going round and round. Internationally we have protestors becoming angry at major polluting industrialized nations but they are the little guys on the sidelines of world affairs – some idea of what this growing problem is about, explained below, thanks again to BBC. No thanks to our media who know what the majority is not that interested in. It is not enough to blame Earth Warming for weather going haywire. That's getting old and is being kicked around. All it takes to dissuade acceptance and approval of planned action, although that doesn't necessarily mean the end of prosperity, is a few conservative think-tankers to blat out over the radio for millions to hear "There isn't a shred of evidence to support global warming. World temperature averages have only rise 1 to 2 degrees in the past 100 years." I thought this pitch was over since El Nino disappeared but I heard it again one night late October over a radio station that reaches 400 other stations spread across the United States. If I could pull all those heads back out of the sand again, talk to them, they would understand we have two hemispheres with different temperature distributions. Two atmospheric entities more or less independent of each other. We are concerned about our half of the planet which has (emphasis, please!) 70 percent of the world's land masses and much more of the world's population. The southern half has over 70 percent of the world's sea surface area. It has a greater potential to absorb more carbon-dioxide than ours. It has the cooler or stabilizing effect of oceans. A stated world average of temperature trends is deceiving. I can't be writing page after page about Nimbus 7 zooming silently around this planet averaging out temperatures measured from pole to pole and how Antarctica is quite different from the Arctic. The two are not quite the same in their buildup of winter's cold and it's duration. Why do you suppose a lady doctor afflicted with cancer could not be rescued from Antarctica for four months? Had she been stuck in the Arctic regions such a problem would not have existed – not for four months!

I must repeat again since it is November 15, 1999, and several months have passed since I covered the same problem that follows. A fact: (You can't be accustomed to facts – nobody makes statements as I do) The world's encircling equatorial zone is warmer than usual. We can blame that first of all on a stronger sun because it's rays are subjected to less filtering up there in the lower layers of the stratosphere as they once were. Jamaica's temperatures, even halfway through November are 5 degrees above normal or hovering around 90. A late season hurricane (Lenny) developed just south of that island for that reason. Vietnam and southern India are also under the same latitudinal solar influence plus tons of airborne pollution over lands and previously reported over the Bengal and Malaysian sea areas. Then of course, not publicized at all but already explained by yours truly we have more humidity all over the world contributing to incredible quantities of water being dumped here and there killing so many. November 14, 1999: 27 drown in southern France, same story, too much rain and a river overflowed. As with our apathetic attitude (not all) that did not merit any newspaper space just as Vietnam did not. Here,

I was treated to the latest exciting news from a fast talking frentic type personality stating jubilantly, Bradley was the American man's choice because he played basketball with the Knicks and Al Gore might be the ladies' choice. Isn't that an intelligent way to choose presidential candidates! I like sports and pay extra for all TV cable channels but good grief, must a president be an athlete?

Eventually the issue of international agreements on curbing pollution will be a moral one. Right now we are not quite up to it. A prevailing indifference can be blamed by the influences of certain factions who don't give a rat's rear end about anything else but building up power requiring amassing more wealth. I don't mean an uninformed public is totally to blame but those storming forward with gold in mind, the affluent and wealthy probably would be happy to hear 160 nations meeting in Bonn, Germany, the week of October 25, 1999 to hassle out who, how to and when would commit themselves to at least have a meeting of minds concerning the reduction carbon dioxide, seemingly ended up with no newsworthy agreement. It started in 1992 and they're still at it. But, did our newspapers describe this Bonn conference? Of course not. I choose to push the morality theme because that will be a choice we will have to make eventually.

According to BBC, certainly not from our good looking newscasters (It must be tough not to be pretty – all the weather girls are model types) this is one of the highlights of the 160 nations meeting in Germany: Small island countries and some existing as coastal nations in warmer latitudes accused major powerhouse nations such as ours, China, and India as being responsible for more hurricanes, typhoons, flooding, destruction and deaths. I know they're right but we're not about to admit that. Most certainly, nobody in America, as far as I can tell, has been told of this accusation. Then, the same protestors turned their ire on Saudi Arabia who they insisted were employing delaying tactics (that's my interpretation) accusing this oil rich country's representatives of trying to shuffle scheduled presentations, or the Arabs' turn on the firing line. So much for another conference. I thought our newspapers would devote a few inches of space to such an important affair, but no, nothing. I wouldn't be writing such a lengthy adjunct if I didn't find this attitude intolerable along with insane weather elsewhere not deemed worthy of more detailed public notification. That this world pollution problem is not to go away and therefore a head of steam is building up internationally, protesting nations making that clear, should at least arouse more interest than it has. But, harmony over truth, comforting news preferred, many fearing they'll be selling apples if we reduce the use of fossil fuel ever so slightly, therefore no politician wants his ass shot full of arrows pushing for a consensus on what has to be eventually. Oh, Bradley and others speak a few words (very few) about their concern but I know nothing positive and concrete is in the works. Did you read anything to indicate disapproval of the Chamber of Commerce glorying in what they happily announced (boy, that got to the press!) as a defeat for the EPA and a victory for businesses when, for the second time, a court decision denied that government's branch request to have some of it's powers reinstated after being chopped down five months before by Supreme Court Justices? Of course the environmental issue is a big headache and there is no easy solution but perhaps we should first consider rising cancer rates and what have you, asthmatics choking to death and so on. Benzene, mercury and acid rain in drinking water doesn't seem to be worrying too many of our citizens.

It sounded as a reasonable idea. I had great hopes when President Clinton proposed last January, applauding representatives standing up, except for the usual Scrooges, he would push for federal financing enabling filth-emitting coal burning power plants to convert to natural gas. Problem is, everybody who makes a few more bucks than most, their egos bloated, imagining they're able to run this country by themselves and downing government, loquacious radio talk show hosts drumming up more of the same, guarantees support from those who do most of the voting will be lacking for some time to come. In all, gross national cynicism runs rampant at this time.

I'm sorry to be so negative. Yet, who knows, can we start something worthwhile? I gave you the Big Picture, sticking with rules-based explanations, some elementary physics and shared with you many years of experience and contemplation, trial and error, searching for truth although making no pretext this was a lofty high level struggle. No mystery at all – meteorology is not astro or geophysics. There is order and control to be discovered with a little effort if we toss out randomness. I don't lose sleep over the problems I have reviewed with you. I trust you will not also – the end is not near. But standby for books predicting population exterminating weather calamities – tabloids are already at it.

Although I'm not the only one who was aware of spurious campaigns against environmentalists I feel vindicated. I could not resist adding another excerpt from an article by Kevin J. Sweeney, chairman of Ozone Action, an environmental group focused on global warming published today (12-11-99) by the Providence Journal. Throughout the book I have hammered away at deceptive themes promulgated in the finest manner by spin- doctors excelling in journalistic skill, backed by special interests. I wondered if the reader would trust in my sincerity. Let the wild radio hosts chew on their cuds over what this article implies! It's a kick in the pants for the Carbon Club.

Such a change of heart by the big man of Ford Motors puts other power moguls in an exposed awkward position. I imagine G. W. Bush's campaign bunch are uneasy. If G. W. becomes president he may find himself heir to a hot seat eventually since internationally this global warming problem promises a day of reckoning and diplomatic pressure from a lot of nations. When, I don't know but the Senate and Congress will have to face reality and approve a step by step program without upsetting the economy. The hue and cry by protestors in Seattle and others spread out nationally emphasizes various creatures in trouble, trees and other important matters but little if any mention of disastrous weather beyond the norm. Need I push the subject anymore? Despite misinformation and the resulting scornful attitudes this article encourages me – it is the beginning. Truth will out eventually.

WILLIAM CLAY FORD JR., chairman of the automaker bearing his name, announced that he is pulling his company out of the Global Climate Coalition. Though the announcement came quietly in a letter sent on Monday, it may well be the kind of strong and clear statement that can end the debate over global warming.

The GCC is a group of fossil fuel producers, energy providers, car companies and trade associations whose singular goal is to spread confusion about global warming. When 2,500 scientists agreed on a statement that global warming is a real and urgent phenomenon caused by humans, the GCC drummed up a small handful of dissenters.

Spending millions on advertising, lobbying and other efforts to discredit the best science, the group's small clique has had a huge presence. They have allowed political leaders to hide behind the false notion that there is dispute on global warming in the scientific community. They have provided ready sources for journalists who, under the guise of objectivity, lazily assume that "another side" to the science should be presented in each story on the topic.

The Ford Motor Co.'s withdrawal exposes this deception. Ford is tacitly admitting that it is folly to continue clouding the debate and that discussions instead should focus on policies and products that can stop or slow the warming trend. In walking away from the GCC, the auto giant has pulled back the curtain to show that science has never clouded this debate. It was the sophisticated advocacy and self-interest of panicked industrialists that did so.

I now feel better. No one to my knowledge has dared to accuse purveyors of misinformation as I have and dozens of loaded pages at that. Finally someone with heart and feeling has written the truth that this excerpt represents.

Here we are just a few days before Christmas 1999 and greeting us in the morning newspapers are stark images of Venezuela's unprecedented ordeal truly of nightmarish proportions, thousands drowned, relatives not expected to be found buried forever under yards of mud, death tolls impossible to estimate – that country's government requesting we send 10 thousand body bags, hardly enough from mounting counts while children wander the streets crying for water. Unlike meteorological experts whom I believe cautiously wait for proof of this or that (impossible anyway) and furthermore most in lower levels of the profession hampered from expressing views by certain rules, Mike Wallace covering a peculiar tornado incident in Oklahoma wherein (I believe) a private sector weather man was convinced a twister was readying to strike a certain area but was forbidden by some mentioned restriction to broadcast a warning, I do not believe I am audacious if I tell it like it is. The outcome of a tornado wreaking havoc called for an interview by one of Mr. Wallace's partners who, appearing as the very polite prosecuting attorney, questioned some rather young weatherman, looking uptight, standing and not relaxed in a chair who was asked why he, belonging to what I think was the weather bureau (couldn't follow rapid fire details) had failed to warn of this tornado when someone else supposedly had phoned him convinced a twister was due. No need to describe explanations given. The point is: Everyone from politicians scared to push action needed to curb global warming, hustling votes and of course, Mr. And Mrs. Success, clutching their stocks, adding to their 401s, and who do most of the voting not wanting to hear it anyway, to scientists dilly-dallying insisting we must wait for more studies to be done (God, decades will go by!) the point is, high level scientists are also scared to climb up on the proverbial platform and do some shouting as I am doing through my typewriter.

Throughout the book I point to causes and effects with confidence. Can I prove it? I care less. If a simple equation I gave you, e.g., Pollution equals heat and more humidity equals more flooding doesn't explain so many disasters, then we'll have to wait for a clairvoyant from outer space to finalize what should have been concluded years ago, my impression having been formed from reading wishful and dramatically expressed high level pursuits of complex formulations that are not at all available. But it surer than hell justifies a small army of researchers, already tabbed as I have in need of some serious culling.

Let us be realistic, one more shot at it, by first understanding masses of cooler air picking up tons of our nation's pollution, moves eastward over the Atlantic, merge into and adds to what is conveniently called the Bermuda High, etc... Having already worn out that explanation and get to my point; all weather maps reveal air from America winds in an expansive clockwise circulation and quite heavily loaded with particulate accelerated moisture pickup from the Atlantic, flows therefore eastward after a four to five day voyage, across the Caribbean Islands and also, of course, moving over northern South America. Venezuela, mountains to the west, Caracas not too far to the east of these Andes and rather prominent hilly regions south of that city acts as a wedge, warm damp air rolls in over rising land masses, more or less trapped in this wedge and is lifted up. For two weeks it rained. I also noted in my daily scanning of weather station reports, Alaska and the Caribbean of special interest, in fact Jamaica four degrees above normal, that San Juan, Puerto Rico had quite a lot of rain day in and day out for several weeks but apparently not dangerously so.

Some of the above is a repeat of previously described atmospheric vagaries on the abnormal side but if 20 thousand human beings may have died then I feel justified in reviewing again a conditions existing throughout the world, namely, extra humidity levels, not to the point of discomfort this time of the year, but a major factor over warmer latitudes in both hemispheres. If we and the world's leading nations do not arrive at some consensus recognizing Nature is on a rampage and has no timetables in the next ten years (not 20!) then there is a very serious morality problem. But rational I must remain – we can at least consider alleviating the problem of global warming since it might be near impossible to curtail it back to what it was once. Here we are consuming gasoline with more leviathan vehicles and Asia reported to be increasing their demands along with a tremendous number of belching diesel powered trucks – my outlook is somewhat bleak. In fact I'm inclined to believe attitudes of transnationals go beyond immorality. I would say it borders on insanity. There is a limit to greed particularly when thousands will die.

All I had hoped for was to change perceptions. From there the possibility of more open minds

About the Author

Forecaster in naval aviation eight years, tutored for one year by MIT meteorologist for a ten year stay in the U.S. Navy. Weather researcher motivated by falsehoods rearing up in the media in 1993, disagreeing overplayed merits of auto emission testing, amateurish reasons given by non-meteorologist types, but skilled with the pen, letters from state and Washington dignitaries included in my literary effort to substantiate my valid experience, no fancy degrees but one very determined tenacious truth-seeker for the sensible reader to evaluate. Eight years travelling in New York state teaching electronic analysis of combustion engines, deploring from that experience the voracious test site inspectors – all this describes in technical details from which the reader can decide who is doing what to whom in this controversy-ridden topic of pollution. Exclusively relies on recorded established facts and figures, not theorems, nor dilly-dallying along with what has become show-boating mystery-oriented endless fact chasing paid by taxes.

www.ingramcontent.com/pod-product-compliance
Lightning Source LLC
Chambersburg PA
CBHW080345030726
47598CB00009B/2632